Usability of Complex Information Systems

Evaluation of User Interaction

Usability of Complex Information Systems

Evaluation of User Interaction

Edited by
Michael J. Albers and Brian Still

CRC Press
Taylor & Francis Group
Boca Raton   London   New York

CRC Press is an imprint of the
Taylor & Francis Group, an **informa** business

CRC Press
Taylor & Francis Group
6000 Broken Sound Parkway NW, Suite 300
Boca Raton, FL 33487-2742

© 2011 by Taylor and Francis Group, LLC
CRC Press is an imprint of Taylor & Francis Group, an Informa business

No claim to original U.S. Government works

International Standard Book Number: 978-1-4398-2894-6 (Hardback)

Library of Congress Cataloging-in-Publication Data

Usability of complex information systems : evaluation of user interaction / editors,
    Michael J. Albers, Brian Still.
        p. cm.
    "A CRC title."
    Includes bibliographical references and index.
    ISBN 978-1-4398-2894-6 (hardcover : alk. paper)
    1. Human-computer interaction. 2. User-centered system design. 3. Computer
software--Evaluation. 4. Information storage and retrieval systems. 5. Web site
development. I. Albers, Michael J. II. Still, Brian, 1968-

    QA76.9.H85U68 2011
    004.2'1--dc22   2010027447

Visit the Taylor & Francis Web site at
http://www.taylorandfrancis.com

and the CRC Press Web site at
http://www.crcpress.com

Contents

List of Figures vii
List of Tables xi
Foreword xiii
About the Editors xvii
Contributors xix
About the Contributors xxi

Section I   Comprehending Complexity: Solutions for Understanding the Usability of Information

1. Usability of Complex Information Systems 3
   Michael J. Albers

2. Combining Rhetorical Theory with Usability Theory to Evaluate Quality of Writing in Web-Based Texts 17
   David Hailey

3. Language Complexity and Usability 47
   Nathan Jahnke

4. Innovation and Collaboration in Product Development: Creating a New Role for Usability Studies in Educational Publishing 67
   Tharon W. Howard and Michael Greer

Section II   Theorizing Complexity: Ideas for Conceptualizing Usability and Complex Systems

5. Mapping Usability: An Ecological Framework for Analyzing User Experience 89
   Brian Still

6. Usability and Information Relationships: Considering Content Relationships and Contextual Awareness When Testing Complex Information 109
   Michael J. Albers

7. Continuous Usability Evaluation of Increasingly Complex
   Systems 133
   Vladimir Stantchev

8. Design Considerations for Usability Testing Complex Electronic
   Commerce Websites: A Perspective from the Literature.................... 157
   Julie Fisher

Section III Designing for Complexity: Methods of Conceptualizing Design Needs of Complex Systems

9. An Activity-Theoretical Approach to the Usability Testing of
   Information Products Meant to Support Complex Use 181
   Heather Shearer

10. Designing Usable and Useful Solutions for Complex Systems:
    A Case Study for Genomics Research... 207
    R. Stanley Dicks

11. Incorporating Usability into the API Design Process 223
    Robert Watson

Section IV Practical Approaches: Methods for Evaluating Complexity

12. Tapping into Desirability in User Experience...................................... 253
    Carol M. Barnum and Laura A. Palmer

13. Novel Interaction Styles, Complex Working Contexts, and the
    Role of Usability 281
    *David Golightly, Mirabelle D'Cruz, Harshada Patel, Michael Pettitt,
    Sarah Sharples, Alex W. Stedmon, and John R. Wilson*

14. Information Usability Testing as Audience and Context
    Analysis for Risk Communication 305
    Donna J. Kain, Menno de Jong, and Catherine F. Smith

15. Usability Testing, User Goals, and Engagement in Educational
    Games.. 333
    Jason Cootey

Index.. 363

List of Figures

Figure 1.1   Information required by different groups. 4

Figure 1.2   Basic 5Es diagram. In practice, the size of each circle varies depending on the situational context. 8

Figure 1.3   5E diagram for a museum. 9

Figure 2.1   Schema describing how an author might approach a writing task. 27

Figure 2.2   An evaluator can use a similar schema for examining a text. 28

Figure 3.1   Failure rate by task language complexity (lower is better). 53

Figure 3.2   Correct first clicks by task language complexity (higher is better). 55

Figure 3.3   Little linguistic knowledge can be successfully taught. 59

Figure 4.1   A visual approach to grammar instruction 71

Figure 4.2   Block's "balance of responsibility" 80

Figure 4.3   The accommodationist model of product development 83

Figure 4.4   A redesigned constructionist model of product development 84

Figure 5.1   Deer–wolf food chain. 94

Figure 5.2   Simple system diagram. 95

Figure 5.3   Ecosystem map of software. 96

Figure 6.1   Relationships and understanding a situation 115

Figure 6.2   Moving from a mental model to contextual awareness 116

Figure 6.3   Global and local views of a situation. Reality of the relationships for the information space 119

Figure 7.1   Overview of the usability evaluation methodology. 135

Figure 7.2   Process model of system analysis. 136

Figure 7.3   A health care scenario where increasingly
                  complex systems can be considered.................... 139

Figure 7.4   A sample event-driven process chain within the
                  application scenario. 141

Figure 7.5   Overview of the ASUR components in the
                  health care scenario. 142

Figure 7.6   Project landscapes and project portfolio
                  management (PPM). 144

Figure 7.7   Activities within project portfolio management
                  (PPM). 146

Figure 7.8   Acceptance of results of the project portfolio
                  management (PPM) process in the portfolio
                  decision... 147

Figure 7.9   Hierarchy of project portfolio management (PPM). 147

Figure 7.10   Decision criteria for project investments......................... 148

Figure 7.11   Major usability problems of increasingly
                  complex project portfolio management (PPM)
                  systems in the view of our respondents........................ 148

Figure 7.12   Maturity levels of the respondents: Bars in
                  every dimension denote the occurrence of a
                  certain maturity level 149

Figure 7.13   Result of the portfolio process in the Microsoft
                  Office Project Portfolio Server (MOPPS) as an
                  example for an increasingly complex project
                  portfolio management (PPM) system. 152

Figure 8.1   Usability-testing design considerations and
                  actions for complex systems............................... 173

Figure 9.1   1992 Food Guide Pyramid. 182

Figure 9.2   Unofficial (labeled) MyPyramid graphic........................ 183

Figure 9.3   The activity system. 189

Figure 9.4   Levels of activity in the activity system. 191

Figure 9.5   Cultural-historical activity theory (CHAT)
                  triangle.. 192

Figure 9.6   Subject–object example. 197

Figure 9.7   Activity system example....................................... 198

Figure 11.1   A simple application written in C that calls
                  a library function. 229

Figure 11.2   The definition of the cputs( ) function. 230

Figure 11.3   Usability-testing design considerations and actions for complex systems. 230

Figure 11.4   Definition of the socket. Select( ) method for C#. 236

Figure 11.5   Code example that illustrates how the socket. Select( ) method is used in a C# application. 237

Figure 12.1   Example item (one of six faces) from the faces questionnaire. 258

Figure 12.2   Card choices across all documentation designs tested. 266

Figure 12.3   Comparison of card choices related to ease-of-use category for current site, P1 and P2. 267

Figure 12.4   Tag cloud of negative language for network monitoring system. 269

Figure 12.5   Frequency of cards chosen for client's hotel reservation site compared with competitor's site. 271

Figure 12.6   Positive language for the current site versus the prototype site. 272

Figure 12.7   Negative language for the current site versus the prototype site. 273

Figure 12.8   Frequency selections for prototype, current, and two competitor sites. 275

Figure 12.9   Number of participants selecting the same cards to describe their experience with the new call center interface. 276

Figure 13.1   Example of a multi-user, projected display. 290

Figure 13.2   Example of an early single-user, head-mounted display (HMD). 290

Figure 13.3   An abstraction hierarchy example for rail signaling. 297

Figure 15.1   Greek polis: Location for local politics and discourse in ancient greece. 335

Figure 15.2   Neverwinter nights. The game can be 'modded' by the gaming community. 337

Figure 15.3   Complex plot: Too many clues and too dense. 344

Figure 15.4   The polis: Otis speaks with a citizen on the polis in the Otis in Stageira test. 347

Figure 15.5   Otis the miner: Dialogue interface to persuade
                   Otis in Otis in Stageira. 348

Figure 15.6   Polis massacre: Paladin leaves bodies on the
                   polis in Otis in Stageira. 351

Figure 15.7   Fairy forest: The user's avatar accesses the
                   reward system in Otis in Stageira. 351

Figure 15.8   Survey numbers: Students who enjoy gaming
                   1–4 on a scale of 10. 352

Figure 15.9   Survey numbers: Students who enjoy gaming
                   5–7 on a scale of 10. 353

Figure 15.10   Survey numbers: Students who enjoy gaming
                   8–10 on a scale of 10. 353

List of Tables

Table 3.1   Albers's Criteria to Human Language 50

Table 3.2   Tasks Sorted by Language Complexity 51

Table 3.3   Failure Rate by Task Language Complexity
             (Lower Is Better) 53

Table 3.4   Correct First Clicks by Task Language Complexity
             (Higher Is Better) 55

Table 3.5   Unconscious Performance Is Superior 61

Table 5.1   Comparison of Usability Approaches 104

Table 7.1   Group Profile 142

Table 7.2   Summary of Usability Evaluation Results 143

Table 7.3   Group Profile of Enterprises for the Continuous
             Evaluation of Increasingly Complex PPM Systems 147

Table 7.4   Data Matrix of the Experimental Setup of an
             Increasingly Complex PPM System as Compared
             to a Real Project Landscape 150

Table 7.5   Coverage of the Domain Services as Usability
             Aspects of the PPM System (Excerpt) 153

Table 8.1   Websites and Tasks Used in Usability Testing 163

Table 8.2   Participant Details 167

Table 8.3   Usability Test Instruments 170

Table 9.1   Leont'ev's Basic Activity Theory Vocabulary 190

Table 9.2   Cultural-Historical Activity Theory Vocabulary 192

Table 9.3   Research and Design Stages of MyPyramid 194

Table 9.4   Methods of Conducting Activity-Theoretical
             Research for the 1992 Food Guide Pyramid Revision ... 200

Table 11.1   Examples of APIs Provided by Software Vendors 228

Table 11.2   Software Development Kits (SDKs) Offered by
             Software Vendors 229

Table 11.3   Usability Definitions 234

Table 11.4 A Summary of the Twelve Cognitive Dimensions Used to Evaluate API Usability.. 238

Table 12.1 Complete Set of 118 Product Reaction Cards................. 260

Table 12.2 Card Choices for Teachers' Professional Development Website... 269

Table 13.1 Chapter Summary: A Summary of Technical and Contextual Complexity, plus Methodological Challenges for Each Case Domain.................................... 285

Table 14.1 Summary of Comments... 317

Table 15.1 Overview: Organizes Test Methods, Project Names, and Example References... 342

Table 15.2 Bug Test Examples; Three of the 131 Bugs Found in Aristotle's Assassins.. 346

Table 15.3 QA Walkthrough Examples: Three Examples from Avalanche Software.. 346

Table 15.4 Learning Principles: Gee's (2003) Descriptions for the Six Selected Learning Principles.......................... 349

Table 15.5 Survey Results: Students Answered Questions about Their Test Experience... 355

Foreword

Whitney Quesenbery

Usability has come a long way. We can easily find signs of progress in the practice of usability evaluation and user-centered design, and the actual usability of many of the applications and websites we use.

Signs of progress: In a February 2010 column on his long-running Alertbox, Jakob Nielsen reports that usability of the Web is improving by 6 percent a year, based on his ongoing evaluations of 262 websites. Despite the persistence of sites that simply don't get the importance of user experience, the leaders in almost every industry pay attention to how easily people can use their sites. They have to. People—regular people—have gotten used to shopping, banking, booking travel, and staying on top of news stories from the Web. It's simply part of daily life.

Signs of progress: As both a cause and a result of the growth of the Web, the growing ranks of people in user experience (or UX, as it's abbreviated)—information architects, interaction designers, and Web designers—talk about design informed by data about the people who will use their products. User research up front, usability evaluations during design, and analytics to monitor behavior are all a normal part of the UX toolkit.

Signs of progress: The old attitude that "it's okay for enterprise systems to be difficult because we train them" is dying. People have started to wonder why the Web applications we use at home are easy but the tools we use at work are still confusing. Why have a complicated search page when Google's single search box works better than your enterprise search? Or struggle with an expense reimbursement system that is not as easy as your home accounting software? As information and communications technologies reach into every industry, even those that may not have used computers in the past, this demand for easy work tools will only grow.

Complex systems are the next frontier, as information and communications technologies become more and more embedded in our physical, social, and work environments.

This book continues the conversation about the evolution of usability, asking how we can design and evaluate these complex systems and the complex work they support. One of its valuable contributions is the wide range of views on complexity in these chapters. The case studies look at many different types of complexity:

- A complex work environment, requiring collaboration among different people or a goal sustained over time, and often in the face of distractions, interruptions, and planned pauses
- A complex information context, one with no single answer, where the data change dynamically or where the *best* answer may rely on other aspects of a fluid environment
- A complex technology, in which people use many different applications in their work and collaboration
- A complex topic, requiring advanced technical or domain knowledge

Some of the most compelling case studies blend all of these complexities. Barbara Mirel used a medical example in her 2004 book *Interaction Design for Complex Problem Solving*: Groups of health care professionals must collaborate on medical care, using several different clinical systems to help them keep track of the changing status of each patient.

A timely example from news headlines is the struggle by U.S. agencies to answer a seemingly simple question: Should this person be allowed to board an airplane? Unfortunately with this question, as with the subject of many complex systems, failures seem obvious in hindsight. But the work of sifting through the data to answer this question and then acting on that answer requires not only powerful data-mining capabilities but also an implicit collaboration across many types of organizations.

Even systems that might seem simple are, in fact, complex. The shopping interface for an e-commerce system may not be complex, but the databases, business processes, and logistics behind it certainly are. Medical decisions may come down to a simple answer, but cancer patients and their doctors have to sift through a large body of constantly changing information about treatment options and clinical trials, and evaluate that information against the patient's specific situation.

Even after facing the challenge of designing such a complex system, we are then faced with the challenge of finding out if we have been successful. Both Mirel (2003) and Ginny Redish (2007) have raised the question of whether usability studies constructed around observing one person working on a single task in the constrained environment of a laboratory can be effective in evaluating the usability of a complex system.

Many of the chapters in this book describe research into new methods and measurements. Others look at touch points in the design and development process, asking whether we are creating the right conditions for success. Usability professionals have long maintained that final summative evaluation is the least effective way to make a system usable, while early involvement can make a real difference. Chapter 11, by Bob Watson, goes all the way back to the beginning and asks whether the development tools and application programming interfaces (APIs) we use contribute to poor usability.

Complexity is not an entirely new problem. Challengingly complex applications have existed for a long time, and design and usability teams have worked to make them both useful and usable. What is new is the recognition that the usability and usefulness of a complex system are more than just the sum of its parts. In fact, Chapter 10, by Stan Dicks, suggests that the solution is not new at all: A good user-centered design process of iterative design and evaluation can ensure that not only are all the parts usable, but also they fit together into a useful system.

The authors of this book look different aspects of designing and evaluating complexity. Taken together, they are a step in developing a deeper understanding of what it takes to make complex systems work. Usability has come a long way. But there is farther to go, and this book helps point the way.

References

Mirel, B. (2003). Dynamic usability: Designing usefulness into systems for complex tasks. In M. J. Albers & B. Mazur (Eds.), *Content and complexity in information design in technical communication* (pp. 232–262). Mahwah, NJ: Lawrence Erlbaum.

Mirel, B. (2004). *Interaction design for complex problem solving: Developing useful and usable software.* San Francisco: Elsevier/Morgan Kaufmann.

Redish, J. (2007). Expanding usability testing to evaluate complex systems. *Journal of Usability Studies, 2*(3), 102–111.

About the Editors

Michael J. Albers is an associate professor at East Carolina University (ECU), where he teaches in the professional writing program. In 1999, he completed his PhD in technical communication and rhetoric from Texas Tech University. Before coming to ECU, he taught for eight years at the University of Memphis. Before earning his PhD, he worked for ten years as a technical communicator, writing software documentation and performing interface design. His research interests include designing information focused on answering real-world questions and online presentation of complex information.

Brian Still, PhD, is an associate professor of technical communication and rhetoric at Texas Tech University. He has directed the Usability Research Lab there since 2006. Brian's research interests focus on user experience, technology and memory, online communities, and open source software development and adoption. In addition to publishing in technical communication journals such as *IEEE Transactions on Professional Communication* and the *Journal of Business and Technical Communication*, Brian has coedited a collection, *Handbook of Research on Open Source Software* (IGI Press), and authored a book, *Online Intersex Communities: Virtual Neighborhoods of Support and Activism* (Cambria Press).

Contributors

Michael J. Albers
Department of English
East Carolina University
Greenville, NC 27858
albersm@ecu.edu

Carol M. Barnum
English, Technical Communication,
and Media Arts Department
Southern Polytechnic State
University
Marietta, GA 30060
cbarnum@spsu.edu

Jason Cootey
Department of English
Utah State University
Logan, UT 84322
jason.cootey@usu.edu

Mirabelle D'Cruz
Human Factors Research Group
University of Nottingham
Nottingham NG7 2RD
United Kingdom
Mirabelle.Dcruz@nottingham.ac.uk

Menno de Jong
Vakgroep Technische and
Professionele Communicatie
University of Twente
7522 NB Enschede
The Netherlands
m.d.t.dejong@utwente.nl

R. Stanley Dicks
Department of English
North Carolina State University
Raleigh, NC 27695
sdicks@unity.ncsu.edu

Julie Fisher
Caulfield School of Information
Technology
Monash University
Victoria 3800
Australia
Julie.fisher@infotech.monash.
edu.au

David Golightly
Human Factors Research Group
University of Nottingham
Nottingham NG7 2RD
United Kingdom
David.Golightly@nottingham.ac.uk

Michael Greer
Pearson Higher Education
Lafayette, CO 80026
michaelsgreer@comcast.net

David Hailey
Department of English
Utah State University
Logan, UT 84322
david.hailey@usu.edu

Tharon W. Howard
Clemson University Usability
Testing Facility
Clemson University
Clemson, SC 29634
Tharon@clemson.edu

Nathan Jahnke
Department of English
Texas Tech University
Lubbock, TX 79409
njahnke@gmail.com

Donna J. Kain
Department of English
East Carolina University
Greenville, NC 27858
kaind@ecu.edu

Laura A. Palmer
English, Technical Communication,
    and Media Arts Department
Southern Polytechnic State
    University
Marietta, GA 30060
lpalmer2@spsu.edu

Harshada Patel
Human Factors Research Group
University of Nottingham
New Lenton, Nottingham NG7 2RD
United Kingdom
Harshada.Patel@nottingham.ac.uk

Michael Pettitt
Orange
Bristol, Avon BS12 4QJ
United Kingdom
Michael.Pettitt@orange-ftgroup.com

Whitney Quesenbery
Whitney Interactive Design
High Bridge, NJ 08829
whitneyq@wqusability.com

Sarah Sharples
Human Factors Research Group
University of Nottingham
Nottingham NG7 2RD
United Kingdom
sarah.sharples@nottingham.ac.uk

Heather Shearer
Department of Professional and
    Technical Communication
Montana Tech of the University of
    Montana
Butte, MT 59701
hshearer@mtech.edu

Catherine F. Smith
Department of English
East Carolina University
Greenville, NC 27858
smithcath@ecu.edu

Vladimir Stantchev
Public Services and SOA Research
    Group
Berlin Institute of Technology
D-10623 Berlin
Germany
Fachhochschule für Oekonomie und
    Management
Bismarckstraße 107, Berlin
Germany
vladimir@stantchev.com

Alex W. Stedmon
Faculty of Engineering
University of Nottingham
New Lenton, Nottingham NG7 2RD
United Kingdom
alex.stedmon@nottingham.ac.uk

Brian Still
Department of English
Texas Tech University
Lubbock, TX 79409
brian.still@ttu.edu

Robert Watson
Microsoft Corporation
Redmond, WA 98052
bob.watson@microsoft.com

John R. Wilson
Human Factors Research Group
University of Nottingham
Nottingham NG7 2RD
United Kingdom
Network Rail
London N1 9AG
United Kingdom
john.wilson@nottingham.ac.uk

About the Contributors

Carol M. Barnum is professor of information design, director of graduate programs in Information Design and Communication, and director and cofounder of the Usability Center at Southern Polytechnic State University. She is the author of six books, with the newest one being *Usability Testing Essentials: Ready, Set...Test!* (Morgan Kaufmann/Elsevier, 2011). She teaches a course in usability testing and directs projects with clients in the Usability Center. The studies in her chapter (Chapter 12) come from some of these client projects. She is a Society for Technical Communication (STC) fellow, a 2009 recipient of STC's Kenneth Rainey Award for Excellence in Research, and a 2000 recipient of STC's Gould Award for Excellence in Teaching Technical Communication.

Jason Cootey lives in Logan, Utah. He is a doctoral student in the Theory and Practice of Professional Communication Program at Utah State University's Department of English. He researches North American genre theory and internal software documentation, with a particular emphasis on game development practices. Much of his research comes from consultation with local software developers. In addition to research on game development documentation practices, he has also researched community narratives in massively multiplayer online role-playing games as a new genre of online text (text interpreted with new media research). He is currently researching agile documentation practices in the software development workplace. Specifically, his research indicates that traditional internal documentation is not compatible with agile practices; his research demonstrates agile documentation in practice. He balances his gaming with endurance cycling, Taekwondo, and wrestling with his children.

Mirabelle D'Cruz is European research manager for the Human Factors Research Group, Nottingham University. She has a BSc (Hons) in production and operations management and a PhD on the structured evaluation of training in virtual environments. She has been a member of the Virtual Reality Applications Research Team (VIRART) since 1993 and is an associate consultant for the Institute for Occupational Ergonomics (IOE). During this time she has been involved in several projects with national and international industries. She has been active in fifteen European projects for over ten years, in most cases leading the work on behalf of Nottingham. She was the integration manager for the network of excellence INTUITION (virtual reality and virtual environment applications for future workspaces; IST-507248-2) with fifty-eight partners around Europe and is a founder member of the European Association for Virtual Reality and Augmented Reality (EuroVR).

Menno de Jong is an associate professor of communication studies at the University of Twente, Enschede, the Netherlands. His main research interest concerns the methodology of applied communication research. In the context of document and website design, he has published on formative evaluation and usability testing. In the context of organizational communication, he focuses on methods for communication audits and image and reputation research. He is the program director of the undergraduate and graduate programs of communication studies at the University of Twente, and editor-in-chief of *Technical Communication*.

R. Stanley Dicks, after an initial stint in academia where he earned his doctorate at Ohio University and taught at Wheeling Jesuit University, enjoyed a seventeen-year career as a technical communicator and a manager of technical communication groups at United Technologies, Burns & Roe, Bell Labs, and Bellcore (now Telecordia Technologies). He has now entered into a second tour of academia at North Carolina State University, where he has taught for thirteen years, and where he directs the MS in Technical Communication Program and coordinates the undergraduate minor in technical and scientific communication. His book, *Management Principles and Practices for Technical Communicators* (2004), is part of the Allyn & Bacon series in technical communication.

Julie Fisher is an associate professor and associate dean of research in the Faculty of Information Technology at Monash University, Melbourne, Australia. Among her many research interests is the area of usability, particularly usability and website design. She has published in leading information systems journals such as the *European Journal of Information Systems* and *Journal of the American Society for Information Science and Technology*.

David Golightly holds a doctorate in psychology, with particular examination of problem solving and the role of experience. Subsequently he was manager of human factors and systems usability activities for mobile telecoms and a blue-chip financial services provider, responsible for user-centered design for interactive systems, capturing user requirements in the design process, and evaluating prototype designs. His work in the Human Factors Research Group at Nottingham University utilizes his expertise in knowledge elicitation, task analysis, focus groups, interaction modeling, and knowledge management. He is currently working on a range of projects looking at human factors for transport and travel technology. He was the founding chair of the Ergonomics Society's HCI Special Interest Group, and is a chartered psychologist.

Michael Greer is a senior development editor at Pearson Higher Education. Since 2000, he has worked to develop, design, and publish more effective and usable textbooks for college English courses. Prior to joining the editorial team at Pearson, he was senior editor and director of acquisitions and development for the National Council of Teachers of English. During an academic career that preceded his move into publishing, he taught courses in literary and cultural studies at Georgia Tech, Illinois State University, and the University of Illinois. He has published and presented papers on literacy, publishing, and usability at several national conferences, and his research interests include video game design and culture. He has also taught online graduate courses in professional and technical writing through the Department of Rhetoric and Writing, University of Arkansas at Little Rock. He lives with his wife and two cats in Colorado.

David Hailey teaches and researches evaluation practices for digital media. His research includes testing procedures and test planning for user preference and cognition, plus content, rhetoric, metadata, and code qualities in digital media. In addition to his research, he teaches hardcopy and digital publication practices in the Professional and Technical Communication Program at Utah State University (USU). He directs the Interactive Media Research Laboratory (http://imrl.usu.edu) at USU. His projects there include a variety of independent learning modules and interactive simulations for the College of Engineering, approaches to preserving critical skills in danger of being lost for the Department of Energy, and instructional workshops for the U.S. Environmental Protection Agency.

Tharon W. Howard has directed Clemson University's Usability Testing Facility since 1993 and is professor of English at Clemson, where he teaches graduate seminars in user experience design, visual communication, and digital rhetoric(s). He is the community manager for UTEST, an online community for usability-testing professionals; and, in recognition of his contributions to the field of usability studies, he has received awards from the Usability Professionals' Association and the Society for Technical Communication. Howard has published in the *Journal of Usability Testing,* the *Journal of Technical Communication, Computers and Composition,* and many other journals and edited collections. He is also the author of *Design to Thrive: Creating Social Networks and Online Communities That Last* (Morgan Kaufmann, 2010); *Visual Communication: A Writer's Guide,* 2nd ed. (Longman, 2002); and *A Rhetoric of Electronic Communities* (Ablex, 1997).

Nathan Jahnke is pursuing a PhD in technical communication and rhetoric at Texas Tech University, where he currently serves as assistant director of the Usability Research Laboratory. He is also CEO of Taiga Software LLC, maker of Taiga Forum. His research interests include the user-centered design of Web applications and human language as a complex system in the context of usability engineering.

Donna J. Kain is an associate professor in the Department of English at East Carolina University (ECU) and specializes in technical and professional discourse. Her current areas of research include risk and emergency communication and conceptions of genre in public policy contexts. She is co-investigator with Catherine F. Smith (principal investigator) and Kenneth Wilson of a project funded by a National Oceanic and Atmospheric Administration (NOAA)–NC Sea Grant to study hurricane risk and hazard communication in eastern North Carolina. She also serves as the director of communications and outreach for the North Carolina Renaissance Computing Institute ECU Engagement Center (Renci at ECU). She is currently working with Renci at ECU collaborators on a usability study of storm-tracking maps.

Laura A. Palmer is an assistant professor in the Information Design and Communication Program at Southern Polytechnic State University. As a senior associate at the Usability Center at Southern Polytechnic, her focus is on data collection and analysis. In addition to usability research, her interests include media theory and technologies, information literacy and its implications for pedagogy, and the use of quantitative methodologies in the humanities.

Harshada Patel is a senior research fellow in the Human Factors Research Group within the Faculty of Engineering, University of Nottingham. She is a chartered research psychologist and member of the British Psychological Society, graduating from the University of Sheffield with a BSc (Hons) in psychology and a PhD in cognitive/developmental psychology. She has been active in European projects for nine years and is currently a researcher on the EU-funded integrated project (IP) CoSpaces, which aims to develop a fundamental understanding of collaboration and design new technologies to support distributed, collaborative work in the aerospace, automotive, and construction industries. She has previously worked on the European Commission–funded projects INTUITION (IST-507248-2), VIEW of the Future (IST-2000-26089), and Future_Workspaces (IST-2001-38346), and the *Engineering and Physical Sciences Research Council* (EPSRC)–funded project FACE (Flightdeck and Air Traffic Control Collaboration Evaluation). Her main research interests are in user requirements generation, the usability and evaluation of new technologies, and understanding, modeling, and supporting collaborative work.

Michael Pettitt is a user experience professional currently working within the telecommunications industry in the United Kingdom. His work involves the application of a range of usability techniques and research methods within product development, and he is currently involved in embedding end-to-end service design, prototyping, and evaluation approaches into the project management life cycle. Prior to this he was a research fellow with the Human Factors Research Group at the University of Nottingham, where his work covered complex usability issues in the development of collaborative working environments and in novel interaction devices. He holds a PhD in human–computer interaction from the University of Nottingham, and an MSc in information technology and a BA in English from Loughborough University.

Whitney Quesenbery is a user researcher and usability expert with a passion for clear communication. She works with companies from the Open University, IEEE, and Sage Software to the National Cancer Institute. She has served on two U.S. government advisory committees: for the U.S. Access Board (TEITAC) updating the Section 508 accessibility regulations, and for the Elections Assistance Commission (TGDC) creating requirements for voting systems for U.S. elections. She and Kevin Brooks recently published *Storytelling for User Experience: Crafting Stories for Better Design* (Rosenfeld Media). She has chapters in *The Personas Lifecycle* and *Usability Standards: Connecting Practice around the World*, and is proud that her chapter, "Dimensions of Usability," in *Content and Complexity* turns up on so many course reading lists. She is a fellow of the Society for Technical Communication (STC), and was honored with a Usability Professionals' Association (UPA) President's Award.

Sarah Sharples is an associate professor in human factors and head of the Human Factors Research Group at the University of Nottingham. She has been at Nottingham since completing an MSc in human factors in 1995, and completed her PhD in human factors of virtual reality in 1999. She has worked on a number of projects focusing on the design and implementation of novel technologies in industrial and research contexts, with particular focus on 3D visualization technologies, transport technologies, and mobile and location-based computing.

Heather Shearer is an assistant professor of technical communication and the director of writing at Montana Tech, where she teaches courses in rhetoric, research methodology, and technical writing. She has presented at Modern Language Association (MLA), the Conference on College Composition and Communication (CCCC), and Association of Teachers of Technical Writing (ATTW). Her research focuses on activity theory, especially as it applies to the usability testing of health care information products. Heather also explores the connections between activity theory and rhetorical history. To this end, she is currently working on a study that examines the organizational communication practices of utopian communities in the United States.

Catherine F. Smith is professor of English/technical and professional communication, East Carolina University, and adjunct professor of public policy, University of North Carolina at Chapel Hill. She is author of *Writing*

Public Policy: A Practical Guide to Communicating in the Policy Making Process, 2nd edition (Oxford University Press, 2009). Her current research focuses on public safety communication, specifically extreme weather risk and emergency communication. She is principal investigator with co-investigators Donna J. Kain and Kenneth Wilson of a project funded by a National Oceanic and Atmospheric Administration (NOAA)–NC Sea Grant to study hurricane risk and hazard communication in eastern North Carolina.

Vladimir Stantchev is a senior researcher and head of the Public Services and SOA Research Group at the Technische Universität Berlin. He is also a senior lecturer at the Fachhochschule für Oekonomie und Management in Berlin and is affiliated with the International Computer Science Institute and with the Information Service Design Program of the University of California, Berkeley. His current research activities focus on SOA-based integration concepts and assurance of usability and non-functional properties (NFPs) in complex application domains. Stantchev holds a master's degree from the Humboldt University at Berlin and a PhD from the Technische Universität Berlin, both in computer science. He is a member of the IEEE Computer and Education Societies, the Association for Computing Machinery, and the German Computer Society.

Alex W. Stedmon is a lecturer and course director of the MSc Human Factors in the Faculty of Engineering at the University of Nottingham. He is a registered ergonomist, chartered psychologist, and fellow of the Royal Society for the Encouragement of Arts, Manufactures and Commerce. With a background in sociology and psychology, he completed a PhD in human factors investigating speech input for real-world and virtual reality applications. He has published extensively in areas such as aviation, military training systems, transport applications, and augmented and virtual reality. He has worked on key EU projects developing novel interaction metaphors for virtual reality applications such as VIEW of the Future (VIEW: IST 2000-26089), Project IRMA (FP5-G1RD-CT-2000-00236), and, more recently, Sound and Tangible Interfaces for Novel Product Design (SATIN: FP6-IST-5-034525).

Robert Watson developed software professionally for seventeen years. During his tenure as a software engineer, he developed software for many products and designed many application programming interfaces (APIs) before becoming a programming writer. As a programming writer, he has been writing about software for software developers for over six years. He holds a master's of science degree in human-centered design and engineering from the University of Washington with a focus on user-centered design, and he has been studying API usability for the past two years.

John R. Wilson is professor of human factors at the University of Nottingham, and also manager of ergonomics delivery at Network Rail. He holds a degree in engineering and MSc, PhD, and DSc in human factors, and is a chartered psychologist and a chartered engineer as well as a fellow of the Ergonomics Society. He has been manager, principal investigator, or grant holder on over seventy-five major grants from research councils, the UK government, the European Union, and public bodies, and has carried out research or consultancy for over one hundred companies. He has authored over 500 publications with more than 300 in refereed books, journals, or conferences and is editor-in-chief of *Applied Ergonomics*. He was awarded the Ergonomics Society Sir Frederic Bartlett Medal in 1995 and the Ergonomics Society President's Medal in 2007, and was the 2008 recipient of the Human Factors and Ergonomics Society Distinguished International Colleague Award.

Section I

Comprehending Complexity: Solutions for Understanding the Usability of Information

A fundamental problem inherent in traditional approaches to usability evaluation is that evaluators overlook seemingly hidden complex problems. Whether it be in the discovery stage leading to the construction of testing, the actual evaluative stage of testing, or post-test analysis, problems related to unforeseen complexity go unnoticed, only to make themselves known when it is often too late, typically when the product in question has been released for use.

The authors in this section of the book deal directly with hidden complexity in unique ways. David Hailey's study of more than a hundred experienced professional writers determined that none could find serious problems in simple Web pages because the users in question lacked the ability to identify the different genres in digital media. Nathan Jahnke's testing of writing reference software led him to conclude that the inordinately high task error rate found among novice users stems from the fact that they cannot comprehend and thus use the formal language of composition instruction that is typically made available for explaining writing process, terms, and other related information. Both Hailey and Jahnke offer possible solutions for dealing with such problems of complexity, as do Tharon Howard and Michael Greer. Building on a pilot test of a college writing handbook that revealed a great deal of unexpected complexity, Howard and Greer document in detail the process of user experience testing that they argue should be adopted for dealing successfully with the evaluation of complex user-centered systems.

1

Usability of Complex Information Systems

Michael J. Albers

East Carolina University

CONTENTS

Examining the Definition of Usability 8
Syntactic, Lexiconal, and Pragmatic (How-To Layers)................................. 11
Semantic (Functionality Layer) 12
Conceptual Layer 13
Conclusions.. 14
References.. 15

Many traditional usability tests focus on function (essentially the button pushing) and not on the process of how the information is used. Recent work in usability calls this model into question, arguing that it is too divorced from reality to provide useful information (Mirel, 2003). In its place a new approach has been called for (Redish, 2007), one that recognizes that most users operate, or carry out their tasks, within complex systems that present multidimensional challenges—layers of changing depth that, unfortunately, traditional usability methods often cannot adequately measure.

Yes, it might be easy to record, for instance, time on task, but if the assumption underlying this measurement is that the user performs in isolation single actions that can be timed, the data collected from such a process hide turmoil beneath. The fact is that deeper issues, ones that cannot be measured with typical approaches, are increasingly responsible, as users engage with ever more complex systems, for whether a website, software, or some other product, application, or interface is usable.

Multiple audiences also complicate usability test issues (Figure 1.1). With complex information, the information needs and presentation formats can radically vary between audience groups. The system is expected to effectively communicate high-quality information to each group, although the information needs and background knowledge vary with each group. The too common mantra "Write for the novice and the expert will understand it too" is not true; providing information designed for the lowest knowledge audience group impairs comprehension in the other groups (McNamara, 2001). In Figure 1.1,

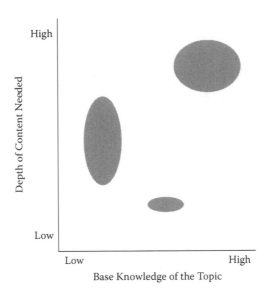

FIGURE 1.1
Information required by different groups. The shaded areas correspond to the knowledge an audience group has on a topic and the amount of content they need on a topic. (Adapted from Albers, M. (2003). *Journal of Technical Writing and Communication*, 33(3), 263–279.)

for instance, the novice would appear close to the origin of the X axis, an area that doesn't even correspond to one of the three defined groups in this particular graph. Obviously, in this case, testing with novices would not yield valid results. Instead, the tests would have to address all three audience groups with somewhat different tests to fit their respective information needs.

Let's consider an example from the groupings in Figure 1.1. In a business proposal report, there may be two main audiences. The executives who have to approve the proposal may be in the left-hand oval. They have a low base knowledge and only need a medium depth of information. The people assigned to implement the proposal fall into the upper right hand oval. They understand the subject and need a great depth of understanding to properly implement it. Of course, for the report, separate sections are easy to write. Health care information often is parsed this way to allow for different documents offering different levels of understanding for different audiences.

But what if cost limitations mean that only one manual can be produced? Some people want to make full use of the product; others only need certain features or only make use of it at limited times. Given that there will be different audiences, different needs, and different levels of knowledge involved in the use of this manual, usability testing of it would be essential to ensure that its multiple audiences are properly served. But such testing, not reflective of a complex systems methodological approach, would fall short of doing this. As Dicks (this volume) asserts,

> As the product gets larger and more complex, usability-testing methods prove increasingly inadequate for testing usefulness. This has recently been pointed out (Albers, 2004; Dicks, 2002; Mirel, 2003; Mirel & Spilka, 2002; Redish, 2007), but most of us have experienced it on our own when performing tests of larger systems. Our traditional, one- or two-hour usability tests seem to indicate that everything about a system is working fine; yet end users are not able to use the system to complete tasks successfully.

Focusing on usability of individual components makes it relatively simple to construct quantitative measures. The time to complete a component task or total number of clicks makes straightforward measures that provide some level of usability of that component. However, quantitative evaluation of a complex design is suspect since the quantitative evaluation often privileges one aspect over others without having a solid theoretical foundation on which to base the privileging (Sutcliffe, Karat, Bødker, & Gaver, 2006). Instead, those creating product designs for complex systems embrace flexibility and usability, and work to ensure that flexibility meets user needs (Shearer, this volume).

Mirel (2003) comments on how although user-centered design methods do work in context, they all carry the assumption that a bigger task is a sum of its individual components, leading to a design focus on supporting discrete tasks; she strongly questions the validity of this assumption with complex problem solving. These discrete tasks map directly onto the one- or two-hour usability tests that Dicks talks about. Unfortunately, although the design and usability of the individual components might be very efficient, the overall process may not be efficiently supported. As the system increases in complexity, the risk of usability failure increases. Although we lack any empirical evidence, we're willing to bet that the risk increases at least exponentially with system complexity.

A fundamental problem we must address is whether Mirel is correct in claiming that the usability of complex design problems is more than the sum of the individual components. Actually, we are sure most researchers would consider it a strawman argument to claim otherwise. Usability testing of a complex system must be regarded as fundamentally different from simple usability testing. Accordingly, we must rethink how we test, entertain new ideas, and also employ new methods (Redish, 2007; Scholtz, 2006). As Mirel says (2003, p. 233), "[C]omplex tasks and problem solving are different in kind not just degree from well-structured tasks." Complex information is multidimensional. There are no simple answers, there is no single answer, and there is a dynamic set of relationships that change with time and in response to situational changes (Albers, 2004). (See Jahnke, this volume, for an expanded description.) Discussing methods of communicating complex information, Mirel (1998) follows the same line as Conklin's (2003) wicked problems when she points out that analyzing complex tasks requires seeing more than a single path:

> This broader view is necessary to capture the following traits of complex tasks: paths of action that are unpredictable, paths that are never completely visible from any one vantage point, and nuance judgments and interpretations that involve multiple factors and that yield many solutions. (p. 14)

But fundamentally how can this be done? How can we test usability if the sum is greater than the parts when traditional usability methods tend to focus on the parts? We need to avoid—and change managements' and programmers' views of—a belief that usability can be defined as a sum of the parts. It is the concrete aim of the subsequent works found in this book to address these questions, doing so from a unique disciplinary perspective, one arguably ideally equipped for examining complex information design and the corresponding usability evaluation of it.

Complicating matters further is the fact that the usability team may be forced to work within certain limiting parameters, such as time and budget constraints, company politics, or other external factors. If they were to argue for a more complex approach, one that might seem more convoluted and time-consuming to a client, they would require solid empirical evidence to make the case for complex systems testing. They need data to push decision makers to accept more than a discounted, quick, and simple solution to addressing user needs.

In a simple system, where the whole is the sum of the parts, the individual elements tend to exist independent of each other. However, within a complex system, the individual elements have high levels of interaction and multiple feedback paths between each other, the user, and the environment. Any meaningful usability measurement needs to capture, or at least account for, those complex interactions. For example, Speier (2006) points out that most research into presentation formats has concentrated on simple information tasks, rather than the complex information tasks of normal real-world decision making. Howard's (2008) research with composition handbooks shows that even what is normally perceived as a simple task, writing a properly formatted article citation, is actually a complex information task. Mirel (2003) reported a study of a system for recording administering drugs to hospitalized patients. If the patient took the entire dose on time, system interaction was trivial (highly usable), but if any factor resulted in the patient receiving only a partial dose or experiencing a delay in taking a dose, the system interaction became extremely frustrating for the nurses (very unusable).

In any situation, a usability team could try to create a full interaction tree, but for anything beyond a trivial system, it is impossible to test that entire tree. For a complex system, creating a complete interaction tree itself becomes impossible. As a result, the sheer size of the interaction space itself precludes thorough testing, much less trying to reproduce the dynamic external factors in a usability test. Even relatively straightforward systems have enough features so that testing all of them would require too much time for the return on investment (ROI). The standard guidance is to pick and choose the features to test in order to focus on the areas that *must* have good usability.

Now a simple test task might require a nurse to enter just one dose at a given time. But chances are that if this were tested, nothing problematic would be found. In fact, because of this the task might not even be included in a test because the usability evaluator might, as Howard (2008) suggests, determine that nothing of interest would be found. But if we consider the usability of drug dose recording along with Speier's (2006) findings of an overfocus on component-level tasks, they should serve as a warning that we need to consider how our current usability methodologies scale with complexity. Do the methods capture the usability within the complexity of actual system use, or have they overly simplified (for a trivial example, consider simply minimizing clicks) or ignored aspects of the situation (for example, dealing with partial doses given two hours apart) that render any usability results suspect?

Postmortem studies of major system failures can identify where the usability tests failed to account for these scaling issues. The problems identified with the San Jose (California) Police Department's computer-aided dispatch system exhibit these scaling issues (Santa Clara Grand Jury, 2005). A desktop test of the system would be trivial. However, when the user is expected to deal with maintaining situation awareness, driving, and manipulating a keyboard, multiple factors compound and the complexity greatly increases. But such postmortem identification of problems is better suited for any resulting lawsuits than for improving usability. Usability is not retrospective work done after a "crash and burn." It is prospective work performed within a highly charged dynamic development environment. Methodology scaling needs to be considered and understood so that development teams can have faith in the results. Unfortunately, we seem to lack much research or published practical experience with the scaling issues of both determining the test cases and handling the complexity of a real-world test case. Golightly et al. (this volume) provide three case studies looking at issues of how to handle this level of testing.

Some health care researchers claim we may be suffering from a forest and trees problem when viewing issues of usability of complex systems. The complexity of health care information in areas such as intensive care is generally accepted as a simple fact. Munira and Kay (2005) suggest that this simple acceptance be reconsidered. In their study of intensive care unit (ICU) work processes, they found it was not the work itself that caused the complexity, but the way in which information was used and managed. From the electronic medical record (EMR) standpoint, they found the underlying process to be very similar at different sites, but the information flow gave each site a very different external appearance. A usability team faces trying to differentiate between the two, but this is a problem since the usability team typically works for a vendor and may or may not have any contact with hospital staff, and the hospital can highly customize the system. There is a chance that two hospitals with very different external appearances are actually using the same software. And, to really complicate the test design, the usability team may be working early in the development process, long before it can be exposed to the realities of an ICU environment.

Examining the Definition of Usability

The International Organization for Standardization (ISO; 1998) defines *usability* as the "effectiveness, efficiency and satisfaction with which a specified set of users can achieve a specified set of tasks in a particular environment." Unfortunately, it has left the definitions of *effectiveness*, *efficiency*, and *satisfaction* undefined and open to interpretation as to what achieving them means.

Quesenbery (2003) presents one attempt at trying to operationalize the issues of the preceding paragraphs. She defines the usability of a system as how well a person can use the system to accomplish his or her task, and describes the 5Es of usability: effective, efficient, engaging, error tolerant, and easy to learn (see Figure 1.2).

More importantly for effectively modeling the usability situation, she also acknowledges that although Figure 1.2 shows balance, in reality each element shifts in relative priority for different applications. For a museum site, she transforms the drawing as shown in Figure 1.3. Each individual project requires independent consideration of the relative size of the five factors. For example, if a person uses an application constantly, whether it is easy to learn may be less important in comparison to the need for it to be highly effective and efficient.

For a software system such as one used by customer service representatives, *efficient* will have a dominatingly large circle; *engaging* will be quite small; and, depending on training, *easy to learn* will vary between systems. Unfortunately, this model once more provides us with a good model but with little guidance on how to actually decide on the relative size of the circles. Assigning the wrong relative values will compromise the usability test plan since it may direct focus to areas of lower priority.

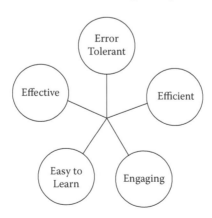

FIGURE 1.2
Basic 5Es diagram. In practice, the size of each circle varies depending on the situational context. (Adapted from Quesenbery, W. (2003). In M. Albers & B. Mazur (Eds.), *Content and complexity: Information design in software development and documentation* (p. 84). Mahwah, NJ: Lawrence Erlbaum.)

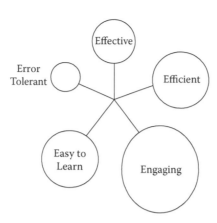

FIGURE 1.3

5E diagram for a museum. (Adapted from Quesenbery, W.(2003). In M. Albers & B. Mazur (Eds.), *Content and complexity: Information design in software development and documentation* (p. 93). Mahwah, NJ: Lawrence Erlbaum.)

For poorly performing systems, the exercise of determining what the circles should look like and testing to determine what they do look like can help to identify problems with the design. If the system as implemented fails to have a 5E diagram, which looks relatively close to an ideal one for the software's situational context, then part of the usability problem has been identified. And the usability engineer and developers can focus their efforts on making changes that will also make the required changes to the diagram.

Operationalizing the use of Quesenbery's five dimensions, especially for complex information as viewed in the light of Mirel's, Speier's, and Dicks's research, leads to many questions. What level of task? How about task sequences? How about sequences with highly branching tasks? What about handling real-world interactions, such as interruptions or pauses while waiting for more information (i.e., the lab results are not yet returned, or the person who has the information you need is out today)? In addition, as Fisher (this volume) points out, we also need to consider the reliability and validity of the tests. This point can prove problematical since the number of people tested typically is much too small to reveal problems in either reliability or validity.

Such operationalization gets complicated since the criteria for each E can be different and often conflict. Shearer (this volume) considers how activity theory provides one method of operationalizing the issues when she looks at the complex usability issues of the new version of the food pyramid. For another example, consider how for a medical patient record system *efficient* often means technology-based and quicker data entry and faster access to all information, while *effective* means better decisions and improved patient outcomes. Of course, while not always exclusive, it is possible that a slower system with better information design will lead to better patient outcomes. But we need more empirical research into how complex information is communicated in real-world situations and more fundamental research to both generalize the empirical finding and work to develop effective information presentations based on those generalizations (Sutcliffe et al., 2006).

We doubt if anyone involved in usability strongly disputes that the 5Es provide a good coverage of the goals of usability, even for complex information. On the other hand, the relative importance that different usability people would assign to them in any specific situation may vary widely. Hailey's (this volume) findings call into question how well even usability people can evaluate the contextual level of the information, primarily because they lack a clear model of how to do so. Moving forward as a discipline and a profession requires defining methods of narrowing those differences. Quite simply, given the same starting scenarios and audience groups, the usability tests, objectives, and methodologies for addressing that specific situation should be relatively consistent.

Defining the relative size of the 5Es requires consideration of both the situation and the audience. Unfortunately, both situation and audience are complex entities themselves, and the relative size changes within them. To take a step back, we've been tossing around the terms *situation* and *audience* for several paragraphs. Now let's take time to clearly define them, then consider how they relate to developing usability tests and how we can move forward.

One of us (Albers, 2004) has previously defined a *situation* as "the current world state which the user needs to understand. An underlying assumption is that the user needs to interact with a system to gain the necessary information in order to understand the situation" (p. 11). However, the current world state of a situation is not static. Situations have a beginning, middle, and end. They can have undesirable or error conditions. In addition, the information can be context sensitive when the meaning or interpretation of any change (or even the absence of change) is quite sensitive to some but not all the details of the current situation. And memories of similar situations in the past raise another complication since both complex information and people's memories and reactions to a situation have a history; past events and choices can, in other words, influence current choices.

Audiences include people with a range of knowledge about the topic, the system, and the frequency of use. Even for a trivial grouping with just high, middle, and low for each of these three dimensions, there are nine component subgroups. And with high, middle, and low sections of each subgroup, there are twenty-seven components or sections for each audience. Of course, this assumes that these twenty-seven components constitute a single group of people who have a consistent set of tasks (e.g., these are just the floor nurses for an EMR system). With more different audience groups (following the more standard definition of grouping by primary tasks), each one may have its own twenty-seven different audience components. The fact is that if we were to expand the EMR system to include physicians, multiple types of therapists, administrative staff, and so on, the twenty-seven multiplies rapidly. Luckily, most audiences don't have a full twenty-seven component groups and there is significant overlap between audience groups. However, every potential component needs to be determined for every potential group in order to design meaningful tests.

Finally, there are issues of what level of testing the usability test performs. A traditional usability test that was attacked in the first paragraph of the chapter focuses on button pushing. But for complex information, the button pushing is secondary to information use. HCI research has defined a layered model with *conceptual, semantic, syntactic, lexical,* and *pragmatic* layers:

> Each layer has its own characteristics in respect of intuitive interaction. The *conceptual* layer describes the main concepts (mental model of the users) of the interactive system, e.g. spreadsheet applications, text editors, graphical tools. The *semantic* layer defines the functionality of the system, sequences of user actions and system responses. The *syntactic* level defines interaction tokens (words) and how to use them to create semantics. The *lexical* layer describes the structure of these tokens and the *pragmatic* layer describes how to generate them physically by means of user actions and I/O elements, e.g. displays and input devices. (Naumann et al., 2007, p. 130)

Each level provides a different take on both the human–information interaction and the usability. As such, the most efficient and effective testing methods change. We'll now look at how the various layers can affect usability testing of complex information.

Syntactic, Lexiconal, and Pragmatic (How-To Layers)

These three levels—syntactic, lexiconal, and pragmatic—make up the lowest levels and the direct interface manipulation. Here is where issues of button layout and selection and information entry sequences are relevant. However, these are not the main elements of concern regarding the usability of complex information systems.

Traditional usability testing involves exploration almost exclusively of these three layers. If a person has trouble with the data manipulation, usability tests still need to be performed that reflect these three layers. However, focusing on these layers within a complex information system will not ensure the production of a usable system.

On the other hand, these layers cannot be ignored since they substantially contribute to the look and feel of the system and the resulting user satisfaction and perceived user-friendliness. In addition, and this is very important for complex information interactions, high user satisfaction and perceived user-friendliness tend to result in better decisions (Moreau, 2006). Barnum and Palmer (this volume) discuss the use of product cards to help capture the user satisfaction level with a product.

Semantic (Functionality Layer)

The semantic layer, representing sequences of user actions and system functionality, is where usability for complex information really begins. Because most complex information interaction is not procedure driven, there is no "correct" path. Consequently, there is no easy way of checking off perceived usability as good/bad or passed/failed. Instead, the usability test needs to analyze the interaction as a whole to determine the whys that drove a person along the chosen path. Both Dicks (this volume) and Howard and Greer (this volume) encountered variations of this problem in their studies.

In addition, actual use of a system will produce surprises. Expecting and handling these surprises should be seen as part of the design process (Sutcliffe et al., 2006). Especially with complex information, where people are going to get creative at interacting with the system as they work toward their goals, usability tests need to consider just how creative people will get. Kain, de Jong, and Smith (this volume) found multiple issues of how people interpret hurricane evacuation information that conflicted with the intended purpose of the documents.

Traditional usability can end when the person finds the information, since it is assumed to be the end goal. With complex information, the end goal is *using* the information, not simply possessing it. The sequences of user actions that lead users to the information and the actions and decisions made with the information all must be considered. Unfortunately, complicating this analysis is how people respond to large amounts of information—they tend to ignore it (Ganzach & Schul, 1995). Woods, Patterson, and Roth (2002) considered this as a "data availability paradox":

> Data availability is paradoxical because of the simultaneous juxtaposition of our success and our vulnerability. Technological change grows our ability to make data readily and more directly accessible—the success[. A]nd, at the same time and for the same reasons, the change increasingly and dramatically challenges our ability to make sense of the data available—the vulnerability. (p. 23)

Reflecting on Woods et al.'s (2002) work brings up issues about the usability of complex information and how the analysis fits into improving the designs of the semantics layer. Is complex information usability fundamentally about data overload and ways to avoid it? With complex information, there is a substantial amount of information that needs to be communicated and that can very quickly overload a person—a problem most pronounced when people don't have the knowledge or prior experience to efficiently process the information. In-depth consideration of these issues and their resolution requires more research.

Conceptual Layer

The conceptual layer connects the mental model of the users to the main concepts of the system. A mental model (the literature also uses terms such as *cognitive model, cognitive schema, mental schema,* and *scripts*) corresponds to the cognitive layout that a person uses to organize information in memory (Johnson-Laird, 1983). The mental model helps to make connections among disparate bits of information (Redish, 1994). In overly simplistic terms, a mental model is a template in the mind, built on previous experience, that contains a collection of known information and relationships for a particular class of situation. All new information gets fitted into this template.

With complex information, mapping user goals onto mental models and ensuring the information system reflects that mapping are much more difficult than with simple information. Simple information has a 1:1 mapping with easy-to-define paths to a defined end point; a correct path can be defined, and a usability test can ensure this path is easy to follow. With complex information, there is not a 1:1 mapping and there are multiple paths with no clearly defined end points (Albers, 2004). With these much looser constraints, the nature of usability testing needs to change. A special issue of the *Journal of Usability Studies* (vol. 3, no. 4; 2008) and Redish (2007), among others, have started to address this issue; a major goal of this book is to continue this exploration of needed changes and practices.

For a simple system, the first step is often defining the user's data needs. With complex information, the first step is to define the communication situation in terms of the necessary information processes and mental models of the users. Rather than providing data, the users need information that they can use to gain a full understanding of the situation, a concept called *contextual awareness* (Albers, 2009). With contextual awareness, the person is able to both understand how the information relates to the current situation state and predict future developments. Based on this understanding, he or she can make informed decisions about how to interact with the situation, know how it will change based on that interaction, and know how the information within the situation relates to each other (Albers, this volume).

Mental models and prior knowledge are relatively easy to activate through the use of titles, headings, subheadings, and lists of key words or concepts (Duin, 1989). The users will be interacting with the information, and populating their mental model, to comprehend the situation. This understanding comes when they interpret the information with the proper mental model, which means the system must present integrated information that matches the goals and information needs of the user as closely as possible. Based on their mental model, people quickly evaluate a situation by placing available information into waiting slots, form their goals, and define their information needs. Obviously, if the mental model is wrong, the goals and information needs are wrong for the situation, although they may be right for the incorrect mental model.

A significant factor in usability testing of complex information is to ensure that the proper mental model gets activated and that the presentation supports easy population of that model. The currently active mental model strongly influences how the information gets perceived and interpreted. Two major problems are how, once activated, people do not switch mental models, and how they interpret information until the disconnect with reality is so profound it cannot be ignored (Einhorn & Hogarth, 1981). Many accident postmortem reports find people ignoring highly salient information because it was not considered relevant for what they thought was happening—as opposed to what was really happening.

The errors occur when the proper information has been correctly perceived, but the relative importance of those information elements has been misinterpreted (Jones & Endsley, 2000), often because the information did not receive proper salience for the person's current goals (Albers, 2007). An interesting design problem here is that people tend to ignore information they consider irrelevant; they see the information, but since it doesn't fit their current understanding of the situation, it is disregarded. The people suffer from tunnel vision and a desire to privilege confirming, rather than disconfirming, information. The design of a usability test must consider these factors and attempt to tease out and identify places where they may affect a person's interaction with the information.

Conclusions

The usability of complex information is itself a complex problem and not one that lends itself to easy answers. Yet, many people do try to address the issues with a simple-information and easy-answer mentality. As a result, they try to reduce the scope and avoid confronting the complex information usability issues head-on. For usability test results to reflect the real-world operation of a system, the complexities of the design, the information, and the human–information interaction must all be considered and made part of a usability test's evaluation, data analysis, and design change recommendations. Thus, it comes as no surprise that addressing the usability of complex systems requires an ongoing and iterative approach through the development cycle (Stantchev, this volume).

We hope this book will help to clarify some of these issues and make it easier (although by no means easy) to design and perform usability tests of complex information.

References

Albers, M. (2003). Multidimensional audience analysis for dynamic information. *Journal of Technical Writing and Communication, 33*(3), 263–279.

Albers, M. (2004). *Communication of complex information: User goals and information needs for dynamic web information*. Mahwah, NJ: Lawrence Erlbaum.

Albers, M. (2007, October 22–24). Information salience and interpreting information. Paper presented at the 27th Annual International Conference on Computer Documentation, El Paso, TX.

Albers, M. (2009). Design aspects that inhibit effective development of user intentions in complex informational interactions. *Journal of Technical Writing and Communication, 39*(2), 177–194.

Conklin, J. (2003). *Wicked problems and fragmentation*. Retrieved from http://www.cognexus.org/id26.htm

Dicks, R. S. (2002). Cultural impediments to understanding: Are they surmountable? In B. Mirel & R. Spilka (Eds.), *Reshaping technical communication* (pp. 13–25). Mahwah, NJ: Lawrence Erlbaum.

Duin, A. (1989). Factors that influence how readers learn from test: Guidelines for structuring technical documents. *Technical Communication, 36*(2), 97–101.

Einhorn, H. J., & Hogarth, R. M. (1981). Behavioral decision theory: Processes of judgment and choice. *Annual Review of Psychology, 32*, 53–88.

Ganzach, Y., & Schul, Y. (1995). The influence of quantity of information and goal framing on decisions. *Acta Psychologia, 89*, 23–36.

Howard, T. (2008). Unexpected complexity in a traditional usability study. *Journal of Usability Studies, 3*(4), 189–205.

International Organization for Standardization (ISO). (1998). Part 11: Guidance on usability. In *Ergonomic requirements for office work with visual display terminals (VDTs)* (ISO 9241-11:1998). Geneva: Author.

Johnson-Laird, P. (1983). *Mental models*. Cambridge: Cambridge University Press.

Jones, D., & Endsley, M. (2000). Overcoming representational errors in complex environments. *Human Factors, 42*(3), 367–378.

Journal of Usability Studies. (2008). [Special issue]. *3*(4).

McNamara, D. S. (2001). Reading both high and low coherence texts: Effects of text sequence and prior knowledge. *Canadian Journal of Experimental Psychology, 55*, 51–62.

Mirel, B. (2003). *Interaction design for complex problem solving: Developing useful and usable software*. San Francisco: Morgan Kaufmann.

Mirel, B., & Spilka, R. (2002). *Reshaping technical communication*. Mahwah, NJ: Lawrence Erlbaum.

Moreau, E. (2006). The impact of intelligent decision support systems on intellectual task success: An empirical investigation. *Decision Support Systems, 42*(2), 593–607.

Munira, S., & Kay, S. (2005). Simplifying the complexity surrounding ICU work processes: Identifying the scope for information management in ICU settings. *International Journal of Medical Informatics, 74,* 643–656.

Naumann, A., Hurtienne, J., Israel, J. H., Mohs, C., Kindsmüller, M. C., Meyer, H. A., et al. (2007). Intuitive use of user interfaces: Defining a vague concept. *HCI International 2007, 13,* 128–136.

Quesenbery, W. (2003). Dimensions of usability. In M. Albers & B. Mazur (Eds.), *Content and complexity: Information design in software development and documentation* (pp. 81–102). Mahwah, NJ: Lawrence Erlbaum.

Redish, J. (1994). Understanding readers. In C. Barnum & S. Carliner (Eds.), *Techniques for technical communicators* (pp. 15–41). New York: Macmillan.

Redish, J. (2007). Expanding usability testing to evaluate complex systems. *Journal of Usability Studies, 2*(3), 102–111.

Santa Clara Grand Jury. (2005). Problems implementing the San Jose police computer aided dispatch system. Retrieved from http://www.sccsuperiorcourt.org/jury/GJreports/2005/SJPoliceComputerAidedDispatch.pdf

Scholtz, J. (2006). Metrics for evaluating human information interaction systems. *Interacting with Computers, 18,* 507–527.

Speier, C. (2006). The influence of information presentation formats on complex task decision-making performance. *International Journal of Human-Computer Studies, 64*(11), 1115–1131.

Sutcliffe, A., Karat, J., Bødker, S., & Gaver, B. (2006). Can we measure quality in design and do we need to? In *Proceedings of DIS06: Designing interactive systems: Processes, practices, methods, & techniques 2006* (pp. 119–121). Piscataway, NJ: IEEE.

Woods, D., Patterson, E., & Roth, E. (2002). Can we ever escape from data overload? A cognitive systems diagnosis. *Cognition, Technology, & Work, 4,* 22–36.

2

Combining Rhetorical Theory with Usability Theory to Evaluate Quality of Writing in Web-Based Texts

David Hailey
Utah State University

CONTENTS

Introduction 18
    Even the Software We Use Has Evolved 19
    A Need for More Effective Metrics 19
    The Problem as I See It 21
Relationship between Usability and Writing 22
    But … 22
    A Test of Writers' Ability to Evaluate Texts 23
    The Subjects Began by Discussing Design 23
    The Subjects Moved to Discussing Navigation 23
    Eventually … They Discussed Writing Quality 23
        Effectively Discussing Writing Quality 24
    How It Could Happen 24
        Incomplete Rhetorical Filters 24
    Genres and Filters in Websites 25
Applying Genre Theory to Content Evaluation 25
    A More Extended Theory of Genres 26
A Genre-Based Heuristic 26
Combining the Components into a Schema 27
    Exigency and Purpose 28
    Knowing the Exigency and Purpose of a Text Is Critical for
        Describing Its Genre 29
    Exigency, Purpose, and Audience in Complex Information Systems .. 30
        Individual Pages with Independent Exigencies, Purposes,
            and Audiences 31
    Audience Need and Expectation Follow Naturally from Purpose 32
        Rhetorical Expectations 32
        Even More Subtle: Rhetorical Need 32

Complexity of Audience Analysis 33
Redish's Solution 34
Conceptual Structure of the Text 35
Conceptual Structures in Web Design 36
Conceptual Structures in Complex Information Systems 37
Physical Structure, including Medium 38
Expectations and Demands on the Author 38
Social Demands 38
Applying the Heuristic 39
Cognitive Walkthrough 40
Post-Test Interviews 40
Practical Application: At Page Level 41
NASA Does It Right 42
Final Point 43
Reader Take-Aways 43
References 44

Abstract

I suggest that while usability studies are becoming increasingly sophisticated in evaluating complex information systems, they are in danger of leaving content evaluation behind. In recent studies, I found that out of more than a hundred experienced professional writers, none could find egregious problems in simple Web pages. The problem seems to arise from a common inability among professional writers to identify the different genres they see in digital media. This chapter examines that problem and suggests solutions (tests) that can be introduced into usability studies or run in coordination with them.

Introduction

One might reasonably ask, "Why does a chapter on rhetorical analysis belong in a book about usability studies in complex information systems?" The purpose of this chapter is to answer that question and the additional questions *that* answer implies. As bodies of information and their sources have become more complicated, our ability to evaluate them has been forced to evolve, and although usability studies experts have systematically developed processes that evaluate the structure, navigation, and design of these documents, I suggest that analysis of the quality of content has not kept up.

Contemporary, digital documents may be constructed from multiple resources archived around the world. These documents often interact in real time with individual readers using dossiers that a variety of computers may have compiled. These dossiers are often stored in databases alongside millions of similar dossiers describing other readers. For example, at Amazon. com, different customers see entirely different landing pages; CBSnews.com automatically includes my local news and weather on its homepage (the weather is lifted from a sensor at the local airport, and the news is extracted from my local newspaper's database), and CNNFN.com automatically loads my stock interests and continuously updates them for as long as I am on the site.

A database scattered around the world may contain the entire legacy of a company and can deliver its content based on the specific needs of the individual reader. A reader accessing the document through a marketing landing page will have access to completely different information and in a different format from those available to a reader accessing a business-to-business (B2B) landing page, although both pages may mine the same content. Moreover, the formatting will be different when readers use computers, cell phones, personal digital assistants (PDAs), or iPods.

Even the Software We Use Has Evolved

As information developers, even if we do not want to move into the chaotic world of complex information systems, our software is pushing us along. The latest version of Dreamweaver (CS 4) assumes we will be developing pages using a separate cascading style sheet (CSS), JavaScript (JS), and XML pages, so that if we open an HTML page for editing, the software automatically opens the related CSS, JS, and XML pages. Moreover, Dreamweaver will combine the code from all of these pages into a single page that permits developers to study them in concert. When I export a video as a Flash file from Camtasia, I get MPEG4, JS, SWF (two of them), and HTML pages in a single folder that I can map from anywhere to anywhere. It is no longer possible to open an old-fashioned Help file in Microsoft Office or Adobe Creative Suite. Now we get massive databases that include articles, frequently asked questions (FAQs), step-by-step instructions, tutorials, and even videos, all imported from resources scattered around the world.

A Need for More Effective Metrics

In *Communication of Complex Information*, Albers (2004) introduces a hypothetical environment that "looks at the complex situation of providing astronomical information to the general reader, loosely defined as a nonprofessional" (p. 23). In his discussion, he introduces many of the ideas I will examine in this chapter. Briefly, he describes a complex document of this kind as having no "single answer," responding to "open-ended questions" with a "multidimensional approach," and providing "dynamic information" that responds to the reader's "history" in a "non-linear" format. As an example, he uses a hypothetical astronomical website. The

National Aeronautics and Space Administration's (NASA) collective of sites provides an excellent example of a comparable astronomical environment that fits perfectly within his description.

In the past, websites have often been described as having a hierarchical structure beginning with a landing page and radiating down through individual tiers. That model reasonably effectively describes the NASA home site, but breaks down when describing the entire information system. Although the home site (www.nasa.gov; NASA, n.d.) looks traditional enough, analysis of its code reveals links to XML databases, XSL style sheets, JavaScript, CSS, and Advanced Stream Redirector (ASX) pages, in addition to the scores of traditional XHTML pages all providing rich content (e.g., video, games, animations, quizzes, blogs, RSS feeds, and opportunities to participate in experiments). As complex and impressive as the NASA site may be, it is merely the center of dozens, perhaps scores, of satellite URLs comprising a comprehensive collection of individual homepages in diverse styles and design philosophies, all interconnected so it is possible to get to any of them from any of them—though not necessarily through a single, direct link. The sites are scattered like clouds around the world, but with the intensive interconnectivity that makes them a single, complex information system. It is very difficult to find all of these sites and may be impossible to evaluate them all in context.

In the face of such complexities, IT professionals and technical communicators have developed increasingly sophisticated metrics for evaluating quality in navigation, structure, and content. Even so, I believe that current initiatives are incomplete. A few years ago, I did a study involving more than a hundred professional writers examining simple Web pages. Out of more than a hundred writers, *none* was able to identify *any* textual failures in any of four Web pages, although all of the pages had significant and easily identifiable problems (I discuss this at length later in this chapter). While IT professionals and technical communicators continue to develop increasingly sophisticated tools for measuring quality of structure and navigation (including design and textual structures), there seems to be no good system for simultaneously evaluating the relevance and rhetorical quality of the content. Yet, virtually any page you turn to is designed (or should be designed) to get you to do something—maybe buy something, learn something, change (or confirm) your opinion, identify a solution to a problem, make a connection, book a connection, or just go to a next page. *Rhetoric* is the part designed to get you to do that thing, whatever that thing might be. I see this inability to evaluate relevance and rhetoric in digital content as an important problem we should address as we move toward increasingly complex information systems. As we develop more sophisticated metrics for tracking the quality of structure and navigation (usability), we should also be integrating more sophisticated metrics for measuring the relevance and quality of the rhetoric within the content.

The Problem as I See It

I suggest that the problem of technical communicators having difficulty evaluating texts in digital environments stems from two conditions.

- The Internet is a pastiche of cut-and-pasted texts pulled from places where they might have been meaningful and pasted into environments where they might or might not still be meaningful. Such texts are almost always on topic and so superficially may seem correct, but they are often written for the wrong purpose or audience or contain incorrect information for the rhetorical situation of their new environments. This condition can only be exacerbated as sites become more complex and as computers become increasingly involved (i.e., humans become less involved) in the production of content.

- As a community, we seem to have incomplete rhetorical filters for evaluating the content of writing in a digital environment. When looking at an Internet text, we often have no idea of its genre. A perfect example of this problem can be found in e-mails. Our graduate advisor will often get letters of application from students with "Dear Professor Grant-Davie:" in their salutations, but e-mails of exactly the same genre will begin with "Hi Keith."

In a recent conversation with an information systems manager at Lockheed Martin, I was told that among their greatest problems is "texts moved from document to document with no sense of context or regard for audience" (Meersman, personal communication, 2008). So far the only solution to that problem I have heard (and I have heard it several times) is "write for a generalized audience." But persuasive arguments directed at everybody are not directed at anybody. On the other hand, once a heuristic is available, it is a simple enough task to include text evaluation in usability studies while examining a complex digital document. It becomes a relatively simple task to examine for relevance and rhetorical quality, in addition to quality of structure and navigation.

So this is my claim in this chapter: A usability study without text evaluation is incomplete. Future studies must necessarily examine both structural and content quality. My intent in this chapter is to propose why such studies are important and how they might be done. Much of the time, I will be describing problems in ordinary websites, but the descriptions map directly to complex information systems (which I will also discuss). My justification for discussing problems in ordinary websites follows from Bloom's taxonomy. In his taxonomy, Bloom (1956) argues that learning is hierarchical. Learners who cannot grasp knowledge at the first level have no hope of moving to any of the higher levels of learning. Much the same thing applies with evaluating digital content. Technical communicators who do not know how to evaluate the simplest Web pages have no hope of evaluating the more complicated problems in complex information systems.

Relationship between Usability and Writing

The best-known usability theorists have long advocated adapting Web-based writing to meet needs implied by usability theory, and a number, including Janice Redish with *Letting Go of the Words* (2007), have written books specifically on that topic. In general, their advice on writing for a Web-based environment is straightforward: "[K]eep your texts short" (Nielsen, 2000, p. 100), "don't make [them] think" (Krug, 2006, p. 11), and "give people only what they need" (Redish, p. 94). These suggestions are in keeping with the results of more than two decades of research that clearly shows an impatient community of readers. According to them, if a text is to be read, it must necessarily be short and easily read.

But ...

Impatient readers may not be the whole problem. Some scholars suggest they are symptomatic of a larger problem. They argue that online readers read as they do, not because there is something endemic in digital texts that forces people to read a special way, but because they never learned how to read digital material. In "Online Literacy Is a Lesser Kind," Mark Bauerlein (2008) makes that claim and agrees with the usability expert suggestion that when faced with long readings, readers "don't" (p. B11). As unfortunate as it might sound, Bauerlein goes on to suggest that because they have been developing bad habits for so long, students may never learn to read well in a digital environment. The implication of Bauerlein's argument is that not only are texts currently carelessly read, but this level of Internet illiteracy may never be overcome.

Although I disagree with Bauerlein's (2008) argument that careless reading is an irreparable habit, I agree that in the current Internet environment Nielsen (2000), Krug (2006), Barnum (2002), Bauerlein (2008), and the others are correct: Web users do not read carefully. I go a step farther and make the claim that many technical writers have similar problems reading Web-based copy and even greater difficulties evaluating it, and for the same reasons. They have simply never learned how to do it, and they evaluate texts much as Internet readers read them.

One would expect a person who spends day in and day out writing in a digital environment to be unaffected by the problems afflicting casual readers. As it turns out, this seems not to be the case. Our graduate student body is made up of professional writers—technical writers, journalists, editors, communications directors, managing editors, Web editors, Web managers, and Web developers (to name a few)—with years of workplace experience and from around the world. A major requirement for matriculation into our program is to be a working, professional writer. So when I ask a graduate student in our program a question, it is generally answered by a professional writer.

A Test of Writers' Ability to Evaluate Texts

Every year from 2002 through 2007, in an ongoing (albeit informal) study of methods technical writers use for evaluating digital texts, I had my subjects evaluate four pages on a website: (1) the landing page, which was used to recruit sophomores but was originally written to impress Northwest Accreditation evaluators; (2) "Career Opportunities," which was used to recruit sophomores into a technical writing program but was originally written to recruit students into the literature program; (3) "Course Descriptions," which failed to describe any courses; and (4) "Philosophy," which repeated the landing page word for word.

The Subjects Began by Discussing Design

Typically, the students began by discussing design. They invariably presented conflicting opinions: "I don't like the color of the text," "I love the color of the text," "I hate the use of the paper metaphor," and "I like the way the shadow falls under the edge of the page." Being writers and not artists, they had no vocabulary or theoretical foundation for evaluating design. Still, they always began by evaluating the visual design of these pages.

The Subjects Moved to Discussing Navigation

Having exhausted the discussion of design, they began discussing navigation. The navigation was very straightforward (each page had three links and the texts were short and properly chunked, though there were no breadcrumbs), so there were few usability issues to discuss, and they quickly moved on.

Eventually ... They Discussed Writing Quality

Finally, the subjects began doing what they should have been able to do well: examining writing quality. Usually, they pointed out an embedded mechanical problem or two, but, to a person, they were consistently satisfied with the writing quality. Just as consistently, I assured them that every page had significant problems that as professionals they should be able to find. One page, for example, contains the following two sentences:

> English majors have found job opportunities in financial institutions, insurance companies, federal and state government agencies, the hospitality industry, universities, museums, and service organizations. They are employed as personnel and planning directors, administrative associates, marketing directors, technical librarians, wage and salary representatives, service correspondents, claims adjustors, and insurance agents. (*Career Opportunities*, 1998, para. 2)

When I told the subjects all the pages had serious problems, they looked more carefully at the text, but ultimately replied that they could find nothing wrong.

Effectively Discussing Writing Quality

Only when I finally made them identify the purpose of the page did they recognize the audience. The purpose of the page was to recruit technical writing students from a pool of sophomores, but the text was clearly meant to recruit literature majors. Worse than not providing correct information for the appropriate audience, the text provided information that might very well have repelled any potential technical writing student. An informed rhetorician analyzing the text in the context of its genre should instantly recognize it as directed at the wrong audience, but first the rhetorician needs to know the genre.

Out of more than one hundred subjects, only one has ever found any of the problems (and he had no vocabulary for explaining what he found). He felt that the sentences were too long, but he was at a loss to explain why that was a problem other than he felt it was a bad idea to have long sentences on the Internet. The text, however, was designed for Northwest Accreditation reviewers. For their original audience the sentences were not inappropriately long, but they were copied and pasted into a page designed to recruit sophomores. The sentences were inappropriate because they were being applied for the wrong purpose and the wrong audience.

How It Could Happen

As I pointed out in the introduction, I believe I have found two problems. That the Internet is a cut-and-paste pastiche of these genres needs little or no explanation. I am currently producing seven websites. In most cases little of the content is new in these projects. This cut-and-paste mentality is a habit of dubious quality but nonetheless common in the profession, and is perhaps as common in complex information systems, where the same biography stored in an XML database might simultaneously be used in a brochure, on a Web page, in a proposal, and in an annual report.

Incomplete Rhetorical Filters

That "there are incomplete rhetorical filters" is a more complicated issue. A variety of studies indicate that when we read, we read through different sets of filters. For example, Judy Gregory (2004) argues in "Writing for the Web versus Writing for Print" that "when readers encounter a text and identify it as belonging to a recognizable genre, they know how to deal with that text and what to expect from it" (p. 281). In some cases the mechanics of a text are important (a résumé), and in some cases less so (a few words of love on a Post-it® note). Sometimes, quality of type is critical (contemporary art book), and at other times less so (Shakespeare's 1623 *Folio*). Depending on the genre, vernacular prose or academic prose or even poetry may be appropriate. We read everything through these filters. But we assign the filters intuitively and based on experience and education (e.g., we have to learn to read documents such as scholarship, modernist novels, and postmodernist poetry, just as we have to learn to appreciate contemporary art and classical music).

Part of learning to read different genres is learning what to expect from them, what they should be doing—these are the filters. In a traditional world, when we begin to read hard copy, we load the appropriate filters, and we naturally evaluate as we read. We only know what filters we need because we know the genre we are reading—love letter, humorous greeting card, sympathetic greeting card, curriculum vita, mystery novel, fantasy, proposal, brochure, annual report, and so on. The thing I find with my professional-writer students is that when they read online, they seldom recognize the genres they are reading, and so they have no rhetorical filters for evaluating how the text should work and no vocabulary for discussing it. They cannot evaluate these texts because, like everybody else, they were never taught to read them.

Genres and Filters in Websites

Web pages are collections of many easily identifiable genres, but they are also collections of not-so-easy-to-identify new genres. In my graduate classes, I teach working, technical writers how to recognize genres they might have no problem recognizing as hard copy, but because many genres are new (e.g., *You Suck at Photoshop*; Bledsoe & Hitch, 2007), we often have to invent new filters for evaluating them.

Once the writers understand this process, they no longer have trouble evaluating the texts. In a new study, I now have undergraduate students evaluate the same, egregious four Web pages after teaching them processes I use for identifying the genre of the text. They consistently, immediately identify the problems, *and* they have a vocabulary for discussing them.

Applying Genre Theory to Content Evaluation

Contemporary genre theorists have long abandoned the notion that a genre is some kind of category that can be described in terms of similarities in textual structures. In "Genre as a Social Action," Caroline Miller (1994) describes "genre" as "a 'cultural artifact' that is interpretable as a recurrent, significant action" (p. 37). Bazerman et al. (2003) describes it as

> language in use. As language in use, texts mediate human activity, take on its meanings from its roles in the activities, and influence thought as participants make meaning of the texts within activities. (p. 456)

In "Anyone for Tennis?" (1994), Anne Freadman additionally suggests that genres require action, but she argues they also require a reaction from an audience.

In short, genre theorists do not see genres as static. Rather, genres are identifiable moments in ongoing conversations, and they evolve through time. There are events leading up to a proposal and events following, and these events help define the nature of the proposal. Somewhere, there is usually a request for proposal (RFP), and following the submission of the proposal there will usually be a response. The RFP will have begun with exigencies that may extend from an earlier proposal or from a change in political agenda, and the response to the proposal will often be the exigency for subsequent actions.

A More Extended Theory of Genres

One might argue that technical writing is meant to accomplish something—even if it is just to inform. Web analysts call this accomplishment *conversion*. The word describes a narrative, of sorts, where a person reads a body of text and changes. One way of measuring a genre is to identify what it is supposed to accomplish and measure the extent to which it accomplishes that thing. With traditional genres (e.g., grant proposals), determining what it is, what it is supposed to do, and whether it is successful is not difficult. With many of the new, interactive genres, identifying the genre, recognizing what it is supposed to do, and determining whether it does that thing can be harder. With the newer complex information systems (e.g., that of NASA), it can approach impossible, although it is no less important.

A Genre-Based Heuristic

Extending theories the genre theorists have proposed, I offer a heuristic for identifying and describing the genres of Web-based texts and a vocabulary that enhances their discussion. I suggest that if you can accurately describe the genre's various characteristics, you can effectively evaluate and parse the text, but you also have the language for discussing your results. There is no need to apply some kind of name to the text (e.g., mystery, grant proposal, or sales brochure) because many of the new, digital genres are as yet unnamed. A good description is better in any case. The key is to begin by recognizing the characteristics of the genres you are evaluating. Although there are many characteristics, I believe the most important ones are as follows:

- Exigency or need for the text
- Purpose of the text
- Audience needs and expectations
- Conceptual structure of the text

- Physical structure of the text, including its medium
- Expectations and demands on the author

If you can describe these six characteristics, you can effectively describe the genre, identify what it is supposed to accomplish, and determine whether the artifact accomplishes that.

Combining the Components into a Schema

It is also possible to demonstrate how everything comes together with a pair of charts (see Figure 2.1). The first chart depicts a schema describing how an author might approach a writing task. By knowing the purpose of the text, the author can identify and describe the audience. Knowing purpose and audience permits the author to identify an optimal structure for the text.

Purpose, audience, and structure each place a variety of different demands on the author. Once the text is complete, the author steps out, and the art becomes an artifact. To evaluate the text once the author is no longer present, the evaluater replicates many of the steps used by the author. Identify the

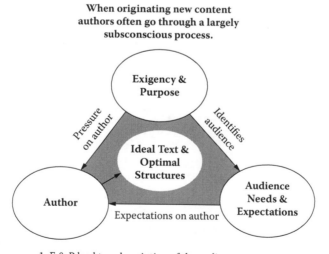

When originating new content authors often go through a largely subsconscious process.

1. E & P lead to a description of the audience
2. In combination, E & P and audience apply pressure on the author to produce ideal text and optimal structures.

FIGURE 2.1

Schema describing how an author might approach a writing task. By knowing the purpose of the text, the author can identify and describe the audience. Knowing purpose and audience permits the author to identify an optimal structure for the text.

exigencies and purposes of the text: That identifies the audience. Actually, this may be all you have to do. In the case of the "Career Opportunities" page example I use above, the instant we know the purpose of the page and the audience, it becomes clear that the page was never designed for that audience. We immediately know the page is problematic. For more complicated evaluation, you must drill deeper into exigencies, purposes, audience needs and expectations, and best structures—including rhetorical stance—to know why a page works or does not work.

Figure 2.2 depicts how an evaluator can use a similar schema for examining a text. The author will likely be gone, but the evaluator may find the original purpose of a text by asking stakeholders. Knowing the purpose of the text permits the evaluator to identify and describe the audience and best structure. The evaluator can then determine the extent to which the text meets its goals.

Exigency and Purpose

Exigency and purpose might seem similar but are quite different. Arguably, exigency is the force that precipitates the act, while purpose is the intended

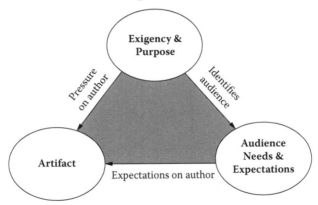

1. Author is absent.
2. Evaluator identifies E & P by asking principals or stakeholders.
3. E & P leads to audience and expectations on artifact.
4. Evaluator tests text based on combination of E & P, audience, and expectations on artifact.

FIGURE 2.2
An evaluator can use a similar schema for examining a text. The author will likely be gone, but the evaluator may find the original purpose of a text by asking stakeholders. Knowing the purpose of the text permits the evaluator to identify and describe the audience, then the structure. The evaluator can then determine the extent to which the text meets its goals.

result of the act. Understanding both is necessary for understanding the genre of the text. The exigency and purpose will be defined by stakeholders and may be intuitively understood with the initial writing, but are often not understood by later evaluators. Writers who are tasked with evaluating copy should know the exigency and purpose of a page before they can even guess at the audience.

With the rebuild of a marketing page on a website, the exigency might be the company's decline in sales, while the purpose of the text might variously be to unload old stock or hype a new product or develop a mailing list. Of course, because this is about communication, "hype a new product" is incomplete. The purpose of a communication always leads to a description of the audience—"hype a new product to …," "unload our old stock to …," or "create a mailing list of … ." Arguably, knowing the exigencies of a text always leads to an understanding of its purpose, and the purpose leads to a description of audiences.

Knowing the Exigency and Purpose of a Text Is Critical for Describing Its Genre

Clearly, recognizing the audiences of a document is important, but it is relatively simple to demonstrate that exigency and purpose are even more critical for describing the genre of a text. There is a very interesting spoof of Photoshop tutorials called *You Suck at Photoshop* (Bledsoe & Hitch, 2007). It presents itself as a series of twenty tutorial videos, but after a very few moments of watching the first segment, the viewer realizes it is actually a story about a neurotic (possibly psychotic) Photoshop guru who is losing his wife, hates his job and boss, and is in an advanced state of emotional devolution. The initial episode begins with the following:

> [M]aybe you have a photo of a Vanagon that your wife and her friend from high school spend Friday nights in.… Oh, hey look, I've got that one myself, right here, so let's go ahead and open that right up. That's what we'll start with. (Bledsoe & Hitch)

In a total of twenty episodes, he shows his viewers a variety of bleak events from his personal life: how to prepare an image of a wedding band for posting on eBay, how to remove a hypothetical someone (himself) from a security video taken outside a store where thousands of thumb drives have been stolen, how to regain a girlfriend he has flamed in a previous episode, and more. All in all, it is an exceptionally creative piece that splits the difference between an audiobook with its narrative stripped away and a satire of a typical Photoshop tutorial.

Within only a few months, these tutorials developed a significant following that speculated for more than a year on who was producing them (8.5 million people, according to National Public Radio [NPR]). The consensus was that this was a new and creative approach to telling a story. The episodes seemed to represent a completely new genre of fictional, first-person stories hidden in apparently traditional tutorials. In the end, completely different exigency and purpose were revealed in an NPR interview. The introduction to the interview spells it out: "Troy Hitch talks with Scott Simon about *You Suck at Photoshop*, a hit series of Web videos created to explore viral marketing concepts. Hitch is the creative director for the agency Big Fat Institute" (NPR, 2008). The text's relationship with the audience completely changes once it becomes clear this is marketing research. It is still very creative, and it is still some kind of hybrid that is in many respects like an audiobook and in other respects a tutorial, but it is also the staging platform for a study in viral advertising—and 8.5 million people had the genre and audience wrong. They thought they were the audience, but they were not; they were the subjects of an experiment. Its principal audience was the advertising community, and its purpose was to examine new marketing techniques.

In a different case, PowellsBooks.com advertises a two-volume set for $350,000. When I suggested their marketing effort for the book could be improved, the rare book manager replied,

> Because this geographical area was so important to the Lewis & Clark adventure, the 1814 in original boards makes sense for us to have as a showpiece or attraction.... Not that we would turn down a reasonable offer, but it is the only part of our rare book inventory that we did not mark down 30% this spring. (Berg, personal communication, 2009)

In other words, the primary purpose of the page is not to sell the book, but to enhance the ethos of the company. Without that information, an evaluator (me) would automatically be looking at the document as a sales (as opposed to marketing) tool. Without knowing the exigency and purpose, we can never know about the additional, complicated layers of complex documents, nor can we know the totality of their audiences. In effect, we can never be confident of any evaluation we do without knowing why the document was produced.

Exigency, Purpose, and Audience in Complex Information Systems

Albers introduces the idea of "purpose" in *Communication of Complex Information* (2004): "Assume the site's task is to provide information about various astronomical events and concepts" (p. 24). He goes on to explain that this purpose is too broad, and ends with the following: "[T]he scope needs to focus on the individual goals." In doing so, he shows how describing the purpose of a document automatically introduces its audience.

NASA's Mars Rover (2009b) site is designed to provide information about the Mars Rovers *Spirit* and *Opportunity*. It is possible, with some research, to identify some of its exigencies and purposes. NASA must annually report on how it has used the funding it received for each year. In this report, NASA speaks to the different exigencies for its many websites. In its 2008 *Performance and Accountability Report* (NASA, 2009c), it presents three specific exigencies: (1) a requirement that they keep the public informed; (2) an ongoing mandate to enhance student knowledge of science, technology, engineering, and math (STEM); and (3) a need to build a body of potential scientists who can become future employees. Toward the first end, NASA says,

> Every agency has increasing responsibility to demonstrate that it is a good steward of the assets, capabilities, and workforce entrusted to it by the American taxpayer. Part of this responsibility is good communication, not only to the incoming Administration that will guide the direction that federal agencies will take in the coming years, but also to the public. (2009c, p. 29)

In addressing the second point, NASA says,

> NASA anticipates that the exploration initiatives will spark the public's imagination and inspire the Nation's youth to pursue careers in science, technology, engineering, and mathematics as a result of their renewed interest in space. (2009c, p. 73)

And, in addressing the final point, it says,

> To meet long-term workforce needs, NASA's Education programs support internships and fellowships at NASA Centers, help inspire students at all levels to pursue STEM-related careers, provide professional development opportunities to STEM teachers, and develop interesting STEM content for the classroom, the Web, and informal learning environments like museums and community-based organizations. (2009c, p. 85)

Finally, it is reasonable to suggest that NASA needs to build and maintain an informed community of supporters, because if interest wanes or turns antagonistic, it impacts NASA's funding—$18.7 billion annually (2009a, p. ix). So although it lists three exigencies in its report, a fourth hovers quietly overhead.

Individual Pages with Independent Exigencies, Purposes, and Audiences

Arguably, most NASA–Jet Propulsion Laboratory websites will exhibit combinations of the exigencies listed above, but not every page will exhibit all of them. Within the exigencies of "inform the public" and "build and maintain an informed community," some pages are designed to publicize successes to a general audience not particularly interested in specifics, so every site has a section of daily press releases. On slow news days, a science reporter can always count on finding a story on one of the NASA websites.

Within the exigencies of "inspire students at all levels to pursue STEM-related careers," NASA has pages of activities that teachers may use and other pages (for more advanced students who want to become involved in

NASA research) offering internships and research opportunities. In addition, many pages contain simple, interactive games that preschoolers might enjoy. It becomes clear that even when the exigencies remain the same (e.g., encouraging interest in STEM topics), the purpose of pages might well change. The purposes of the teachers' page, the students' page, and the children's page are all different. To evaluate the quality of writing in any complex document, the evaluator must know all of the different genres on each page and so must know the exigencies and purposes applicable to those pages.

Audience Need and Expectation Follow Naturally from Purpose

I have already mentioned that many professionals in Web analytics (e.g., Tim Ash in *Landing Page Optimization*, 2008) call the process of persuading someone to do something *conversion*. Most if not all Web pages (even in the most complex documents) have the need to persuade their audience to do something. This is why an understanding of rhetoric is fundamental to analyzing writing for the Web. Teachers of writing universally recognize that if writers hope to persuade anybody to do anything, they must know whom they are trying to persuade. At the very least, audiences will have expectations as concerns style, information, and rhetoric, *and* they will have needs they often do not recognize.

Depending on the genre, the audience might expect more or less technical jargon (a computer tutorial) or poetic language (romance literature) or glib prose (dummies and idiots guides) or dense scholarship (*PMLA*) or accuracy in mechanics (a résumé); they might also expect unique and creative punctuation (creative nonfiction, poetry), or truth (a scientific journal) or dishonesty (a first-person novel with unreliable narrator).

Rhetorical Expectations

Expectations of structure and style in a text are easily identified and as simply understood; rhetorical expectations are more subtle and more complicated. For example, the visitor who comes to our "Career Opportunities" page (the above example) is not hoping to be talked out of becoming a technical communicator. That people are visiting the page implies they are encouraging us to persuade them to join us.

Even More Subtle: Rhetorical Need

Rhetorical need is different from rhetorical expectations. For example, students often arrive at classes with opinions that drive their expectations. They may expect their learning to be an extension of misconceptions they already hold, and they can be petulant when they discover the teacher is going to teach them something contradictory. But they often need to be disabused of what they already "know" because it is so often based on ill-formed opinions. The National Research Council (2000) describes it this way:

> Before students can really learn new scientific concepts, they often need to re-conceptualize deeply rooted misconceptions that interfere with the learning ... people spend considerable time and effort constructing a view of the physical world through experiences and observations, and they may cling tenaciously to those views—however much they conflict with scientific concepts—because they help them explain phenomena and make predictions about the world (e.g., why a rock falls faster than a leaf). (p. 176)

Not only is rhetorical need different from rhetorical expectations, but also the author often has to tread on unstable ground when filling the need.

Complexity of Audience Analysis

There are numerous descriptions of how to evaluate the audience. In a cursory review of a dozen technical communication textbooks, I find that all discuss audience analysis to some degree. Most devote whole chapters to it. Yet, I find surprisingly few technical writers who can effectively describe an audience for a page without significant coaxing. Whenever I ask my graduate students about the audience for the two-sentence example I introduce above, they invariably respond, "Students." When I explain that "Students" includes pretty much every literate person at some time or the other, and I ask for a more specific answer, they will universally reply with something like "Technical writing students." Actually, this offers no more information. There may be fewer technical writing students in the world, but the statement still says nothing to describe who they are. The people we attract to our program include creative writers interested in supporting their creativity by writing professionally, artists who write well and are interested in Web design, engineering students with no aptitude for math, science majors looking for a major that will permit them to have a career without having to get a master's degree, computer wizards who love both technology and writing, and a few who have no idea what they really want to do when they graduate but know they want a job. In short, naming the audience is not the same thing as understanding it. Each of the people on this list will have a different set of expectations. Some are looking for opportunities to be creative, while others are uninterested in being creative—they just want to study technology and write about it. In short, a single document designed to attract these different interests would take each of them (as opposed to all of them) into account. A list of possible careers would include jobs that each of them could see themselves happily doing. A well-understood purpose might sound something like "I need to create a recruiting page designed to attract (the entire above list) into our technical communications program." However, to reiterate, the improved page is not designed to recruit all of the audiences; it is designed to recruit *each* of the audiences.

It is critical to be able to describe each audience in great detail. If writers cannot do that, they have no hope of describing audience needs and expectations; and how, then, could they possibly know if those hopes and expectations are met? Unfortunately, I was never able to develop a heuristic that made it easy to teach students to consider their audiences more comprehensively.

Redish's Solution

More recently, however, I have modeled my instruction on Janice Redish's excellent heuristic for identifying and defining an audience. In *Letting Go of the Words* (2007), Redish describes this approach. She suggests six steps for identifying and understanding your audiences:

1. List your major audiences.
2. Gather information about your audiences.
3. Gather your audiences' questions, tasks, and stories.
4. Use your information to create personas.
5. Include the personas' goals and tasks.
6. Use your information to write scenarios for your site. (p. 12)

I think this list is one of the most important contributions in her book, although number 3 on the list ("Gather your audiences' questions, tasks, and stories") lends itself more to usability of the text than to its rhetoric. For my purposes, I have added a "number 3.5" to the list: "Identify your audiences' needs and expectations." My change is within the spirit of her suggestions, but I think it recasts her ideas in a more rhetorical direction. It seems to me that "questions" and "tasks" address what the audience wants to do more than what they need or expect rhetorically. Within the context of her discussion of audience analysis, the easy answer, "Student," is not such a bad place to begin, because if you continue following her prescriptions, you should eventually have an excellent understanding of your audiences.

Albers provides an example in *Communication of Complex Information* (2004). Beginning in Chapter 1 and reiterating in subsequent chapters, he describes possible audiences for a complex site designed for providing information about astronomy. He suggests that such a site might cater to three audiences: "knowledgeable amateur," "student researcher," and "general public." As his book progresses, he goes on to describe his audiences in much greater detail, and suggests that each is a different audience needing different content.

Earlier I alluded to different purposes for the various NASA pages. I suggest that among their purposes was to present "pages of activities that teachers may use and other pages (for more advanced students who want to become involved in NASA research) offering internships and research opportunities." NASA meticulously identifies its different audiences and presents content meant to effectively persuade them. For example, on their student's landing page (NASA, 2010b), the audience is broken into K–3, 5–8, 9–12, and higher education. The content, designed for teachers, includes a section where they can encourage students to consider careers in engineering, astronomy, or even astronautics. Very young students are encouraged to play games and puzzles. More advanced students are encouraged to design experiments they can send up in a space shuttle flight and apply for internships and astronautics camps. For example, one Web page states, "Student teams in grades nine through 12 must submit a research or flight demonstration proposal [to be conducted aboard the space station] to NASA's Glenn Research Center in Cleveland by Friday, Feb. 19, 2010" (NASA, 2010a). In a sense, students are being encouraged to move from Albers's "general public" to his "student researcher," and the teachers are encouraged to help them. I believe NASA makes an excellent model of effective rhetoric being used in an exceptionally complex information system.

Conceptual Structure of the Text

Conceptual structure, as I use it here, is that structure commonly and naïvely identified as "the genre." It is the structure that makes mysteries "mysteries," science fiction "science fiction," and proposals "proposals." The plot of a novel represents a typical conceptual structure. A traditional mystery can be depicted using Freytag's elements of drama. A pyramid rises through a series of rising actions until the story reaches a climax, followed by a denouement. With a traditional mystery, the rising action continuously points to the elements of a previous story. The point of the mystery is to discover that previous story.

Some novels have very complicated structures in which the reader is carried back and forth through time. In such structures, readers are expected to hold the various parts in their heads, putting the stories together much like puzzles.

Technical communication contains similarly identifiable structures. If the purpose of a proposal is to solicit funding from the National Science Foundation (NSF) and the audience includes NSF reviewers, and it is submitted via the Internet, the structure will be different from a proposal submitted to a local funding agency where the details of the proposal and agreements might actually be worked out at the local beef-'n'-brew.

Online help contains a whole collection of structures that can help define different genres within the greater genre (super-genre?) of online help. For example, users of Word in the past received online help via HTML Help, and WinHelp before that. Currently, WebHelp on the computer provides help for the simpler problems, but much of the new help now comes directly from the company database, and that largely comes in the form of articles originally designed for tutoring and troubleshooting. With the advent of the new CS4 series of applications, much of Adobe's online help is presented as a series of very-high-resolution tutorial videos loaded from http://tv.adobe.com

(Adobe, 2010). The structures are different, though the purposes are largely the same. One of the exigencies growing from more than a decade ago is the need to save money by having texts perform multiple functions (single sourcing). The current process involves many different texts being used in a variety of different places in a variety of different documents for a variety of different purposes—complex information systems. These divergent texts may be inserted at document, section, paragraph, sentence, or even term levels into Web pages, printed materials, cell phones, or PDAs—reformatted for each application.

Conceptual Structures in Web Design

In all cases, conceptual structures are nested and layered, but in the case of digital documents, the layering and nesting have become more chaotic and misleading. The chaos becomes particularly difficult to overcome when the mixed metaphors being used to describe the communications inhibit under-standing. The conceptual structure of a website is typically "remote place" (a remote site). The metaphor is so strong that even writers who produce websites are often resistant to the suggestion that websites are documents and not places at all. Redish (2007) argued that she prefers not to use "reader" as a description of the Web page's audience "because these people 'use' websites; they 'use' web content" (p. 24). On the other hand, whereas "users" still book airline tickets and look up dates and trivia, "readers" may spend several hours a day drifting from news story to news story on MSNBC.com, CNBC.com, and ESPN.com. They might read every review on the new BMW diesel engine. They might spend a whole day reading academic articles, or an evening interacting with twenty-first-century literature on their computers or cell phones. This fact supports my disagreement with Bauerlein's (2008) argument that people will never learn to read digital media. I think we are already learning to read these new texts.

Time and again, Barnum (2002), Nielsen (2000), and Krug (2006) refer to the site as "the product," and consistently describe it using a physical-struc-tural metaphor. From the point of view of the reader, these are excellent metaphors. They represent a profound strength in interactive media, where people can stroll through their documents. We do seem to visit remote places as we browse (mixing the metaphor) the Internet, and websites do resemble Nielsen's "house with many doors." From the point of view of the website author, however, these mixed metaphors lead away from being able to evaluate the critical genres that occur on every page. The struc-tures imputed to places (studs, floors, lighting fixtures, walls, wallboard, windows, and doors) are incompatible with the structures of genres (para-graph, argument, rhetorical stance, demands on the author, introduction, persuasion, and reader), and the need to write or evaluate within the place metaphor creates a chaotic collection of contradictions the author has to sort through. This collection of contradictions becomes particularly opaque if authors have thoroughly embraced the place metaphor—they become blind to the nature of the texts that make up structures. I posit that places are not typically described in terms of genres, and that the filters we apply to evaluate places (space, atmosphere, and navigation) do not apply in the same context to documents.

The structure of "place," however, is only a metaphor; to misquote Gertrude Stein, "There is no place" there. The structure we see on the monitor is a popup book of sorts, based on a code designed to describe how the page should look—a popup book we import into our computer along with all of its varied components.

Earlier I claimed that people do not currently know how to bring useful filters to digital media—that they are functionally illiterate on the Internet. This is where I believe it happens and why. The conceptual metaphor of place overwhelms all other metaphors, and since a place is not a genre, the entire concept of the document is skewed. Moreover, even if authors want readers to imagine themselves in a conceptual place, they should keep in mind they are creating documents. Just as importantly, evaluators should also recognize they are evaluating documents.

Conceptual Structures in Complex Information Systems

Although the image we see on the monitor is real enough, it is produced by an underlying set of documents that call up a collection of disparate elements from a variety of different places, and in the end it rests on our computers. As the developer well knows, the real site is a collection of directories or folders containing images, SWF files, video files, and written content often extracted from matrices (e.g., ASP/SQL) or other documents (e.g., XML) or physical devices (e.g., weather stations), along with HTML, CSS, ActionScript, JavaScript, and pages of code describing how the files should be displayed. The site is not a place but a collection of folders and files, perhaps stored in a central location, but just as possibly spread across the whole Earth.

Actually, the developer who looks more carefully recognizes there are no folders; they are simply conceptual constructs used to make the way the computer works meaningful and organization of the site more possible.

In the past, a Web page contained a few photographs, perhaps an animated gif, some links, and a body or two of text. With simple, WYSIWYG software, children in elementary school could produce Web pages. All of the resources could sit in a single folder, and the link to an image could be a single file name with no path necessary. Now, when I load a page and a cookie tells the server who I am, it configures the page specifically for me by downloading different clumps of text and other data from sources around the world. The idea of folders no longer applies. In many cases, the server extracts a single number from some remote XML document (temperature perhaps) and forwards it to me. Moreover, the server might configure it differently for my computer, PDA, Kindle, cell phone, or iPod.

The structure of the Web has evolved toward increasing complexity and automation, and, in the effort for more effective management of communication, it will continue to do so into the foreseeable future. In this new environment, the metaphorical structure of the Web becomes an amazing and somewhat frightening image of something produced by a black widow rather than an orb spider. In this metaphor, the idea of quality of writing can become lost. In the end, being able to set aside the place and Web metaphors and seeing websites as a collection of related texts configured in identifiable genres make it easier for the writer to identify purposes and audiences for each page and evaluate them for rhetorical stance, voice, and argument— quality of writing as opposed to quality of structural metaphor.

Physical Structure, including Medium

Slide shows, whether on computer or on a projector, share the same conceptual structure. In either case, they are slide shows. In studies done in the late 1990s, scholars demonstrated that changes in media typically made no difference in cognition. During this time, a research team I was on was able to demonstrate statistically that changes in media make a difference in cognition if the new media force a change in genres (Hailey & Hailey, 2003). If changes in media force changes in genre, cognition can be severely hampered. On the other hand, changes in media often open up new genres and new opportunities for learning, alongside new exciting literary opportunities (new kinds of tutorials, simulations, and other independent-learning modules, such as interactive novels on the computer).

One of the reasons why physicists say that communication can never go faster than light speed is because, as far as we know, all communication has a physical structure. We "read" texts based on patterns in their physical structures. This is where the idea of "genre" seems to overlap the idea of "medium."

Expectations and Demands on the Author

Expectations and demands on the author come from all quarters. The stakeholders in the company expect their money's worth; the audience generally expects some level of honesty, and often enough these are not compatible. In addition to ethical and other sociological demands, the author faces a plethora of technological demands. To the extent that evaluators can identify these demands, they can determine the extent to which the document meets them.

Social Demands

Certainly, the law and professional standards can demand the writer meet ethical benchmarks, and the audience might have certain expectations of honesty, but the audience might also expect a level of obscurity (poetry) or dishonesty (a first-person mystery novel). Depending on the genre, failing to meet the ethical expectations of the audience can have dire consequences. If an author pens a memoir and the audience expects truth, the author is cruelly treated when the audience discovers the memoir is filled with fictions. On the other hand, if the audience expects the authors to be unreliable and the text to be a spoof (e.g., *You Suck at Photoshop*; Bledsoe & Hitch, 2007), the authors receive a great deal of praise from audience and critics alike for the quality of their dishonesty.

Ethical expectations can be very subtle. Authors of online help might feel they have met their ethical needs if they merely present the information accurately, but this is not necessarily true. Authors might also recognize that anybody mining information from online help will be stressed to some extent, and under such circumstances, authors might contribute a rhetorical component designed to ameliorate the reader's stress. Something I have noticed with Adobe CS4 online help is how much more difficult it has become to use. Typing "import PSD" into Dreamweaver CS4's help-file search engine retrieves 24,700 different listings from Adobe's corporate site and from around the world, mostly in the form of questions asked in discussion forums. This is Adobe's "Community Help," and it offers more obstacles and confusion than help. Dreamweaver offers a second approach to finding help on the Subject Index. After opening directories and subdirectories within the index until arriving at "Importing Images," the user will still find nothing about importing PSDs (although Dreamweaver CS4 *does* import PSDs). A user of Dreamweaver trying to import a PSD into a document will more likely find the solution by trial and error. Imagine the impact of this documentation on an already frustrated user. The designers of this documentation (an example of a complex information system breaking down, by the way) are serving neither the company nor the user.

Applying the Heuristic

If I were to explain how to do a usability study, I could only explain in general terms. Since conditions change from document to document, there is no way of predicting the needs of the examiner. I might explain how a cognitive walkthrough works, but I could never explain what specific questions to ask or tasks to request. Similarly, I could suggest a heuristic evaluation (expert analysis, for example) that includes evaluation of structure, navigation, text relevance, code, and metadata (perhaps using a Delphi methodology), but the specifics would vary depending on the specific conditions. In the end, the people doing the study are forced to think their ways through their projects. Similarly, I cannot offer a step-by-step recipe for evaluating content, nor can I offer more than suggestions for how such an evaluation can be integrated into a specific study. Instead, I will describe two approaches I use and explain how I integrate text evaluation into them.

Cognitive Walkthrough

A cognitive walkthrough typically begins on a landing page, while the text evaluation I recommend begins in the innermost recesses of the document, so a comprehensive evaluation of textual and rhetorical relevance is more difficult to integrate. Still, some text quality–related questions can be asked during the walkthrough and in a subsequent interview. For example, in reference to the paragraph below, I asked subjects to find a group's mission statement. Once they found the mission statement, I asked, "What does it mean?" Here is an example of one such statement:

> Our mission is to build capacity in technology education and to improve the understanding of the learning and teaching of high school students and teachers as they apply engineering design processes to technological problems.

At first, the mission statement seems to make sense (it is certainly on topic), but on closer examination it quickly dissolves into techno-babble. By asking, "What does it mean?" as subjects find segments of copy, I can see whether they are getting the message I intend.

I will also allow subjects a period to explore without direction. As they explore, they describe their feelings, which gives hints about how they are impacted by the rhetoric ("Hmmm … that looks interesting …" or " I wonder what this does?"). But if we keep in mind that rhetoric is designed to get people to do things, within a behaviorist model, we can test to see if it does that. It is a simple task to simply follow the subjects' click streams to see whether their paths meet expectations (do they click the expected button? Do they stop and read the page we want?).

Post-Test Interviews

Each subject can be interviewed. I also hold post-test interviews. In these, I try to ask questions that identify how a section of a page impacts the subjects, or fails to impact them. To avoid "poisoning the well," I usually ask subjects questions in a manner that parses attitude without necessarily making the purpose clear. This can be done by asking the question obliquely, allowing the subject to wander into the answer, or hiding the important question in a number of unimportant questions.

Ad agencies often use both techniques. For example, an agency will sometimes permit subjects to view a TV pilot they suggest is being considered for future distribution. Subjects might view the show and answer questions about its quality, theme, and so on. Mixed in with the questions might be questions about the advertisers. *These* are the important questions. Testers want to know the viewers' attitudes about the ads, whether subjects remember the names of the important products, whether they remember what the ads promise, and so forth. It is not difficult to adjust this tactic for evaluating the quality of rhetoric on a Web page.

Practical Application: At Page Level

In addition to integrating content evaluation into usability, user cognition, and similar tests, it is useful to run independent content evaluations within the document. This begins with recognizing that websites are seldom specific genres for singular audiences. Rather, they are made of scores of different genres that may apply to as many different audiences, and, if done right, they apply to each of the audiences as opposed to all of them. Any page will contain more than one. Perhaps a page has a drop-down menu, search engine, splashy photograph, welcome message, plus current events blurb or calendar—these are all different genres. A different page might be something right out of a traditional catalog. The heuristic I am now recommending for evaluating quality of content applies to the individual genres within the page and not the whole site (or necessarily the whole page).

Various usability experts suggest that sites are layered and that the layers nearest the homepage tend to be structured for navigation ("landing pages"), whereas those farther from the homepage are more textual. Redish (2007) describes this structure as home leading to pathway pages, which lead to information pages (p. 29). Arguably, landing pages need less in the way of text analysis and more in the way of usability analysis. Such pages might contain little more than links and images (e.g., www.CBSNEWS.com; CBS, 2010). All landing pages have important rhetorical goals, however, and although they might seem to have little text, *that* text still needs analysis.

It probably goes without saying that the information pages require the most comprehensive content evaluation. These are longer and more comprehensive, and yet are often composed entirely of cut-and-paste snippets from a variety of sources.

Earlier, I mentioned two different pages on the Powell's Books site, both designed to market very expensive and very old books. In neither case did the page effectively do what it was supposed to. In contrast, the pages in the used book section work perfectly. Understanding why helps explain the problem with the pages that fail.

> Staring unflinchingly into the abyss of slavery, this spellbinding novel transforms history into a story as powerful as Exodus and as intimate as a lullaby. Sethe, its protagonist, was born a slave and escaped to Ohio, but eighteen years later she is still not free. She has too many memories of Sweet Home, the beautiful farm where so many hideous things happened. And Sethe's new home is haunted by the ghost of her baby, who died nameless and whose tombstone is engraved with a single word: Beloved. Filled with bitter poetry and suspense as taut as a rope, *Beloved* is a towering achievement. (Powell's Books, 2009)

A dozen glowing reviews from Powell's readers follow this quote. Although some writers might argue with the quality of writing in the above quote, nobody can argue that it is addressed to the wrong audience or being used for the wrong purpose. It takes only a glance to recognize this text is meant to sell the book.

To market the used books on their site, Powell's Books uses the original release notes from the publishers. When the books were published, the publishers understood their genres and audiences, and they understood the rhetorical importance of their release notes. While the books might now be older, the release notes are still valid. Not recognizing why these marketing genres work in their used book section, Powell's Books produces its rare books pages from the same template—even including the same "Be the first to comment on this book." Since there are no release notes, they simply import a physical bibliography, or cut and paste content they may have on hand. This is the case with virtually all of their rare books—page after page with one small photo of the book, a physical bibliography, occasionally something pasted from somewhere else, and the injunction "Be the first to comment on this book." The occasional pasted content sometimes includes an old review, but rare book sales are based on provenance and not original reviews. The first book by many of our most important authors will have been in a very limited edition. That a book is one of only twelve existing first printings flat-signed by an important author is much more important than the quality of its plot.

In a different document recently sent to me by the Conference on College Composition and Communication (CCCC; my personalized convention flyer), a similar and similarly obvious problem arises. The document, meant for me to use for publicity, announces that I will be presenting at the 2010 conference. The announcement is obviously made up of snippets extracted from a database and placed into a template. Throughout, the announcement refers to me as "Lastname Jr" as if *that* were my name. Apparently, when I submitted my proposal, the computer was unable to parse my application. Surely, if one person had glanced at it, I would never have seen it in this condition.

NASA Does It Right

In contrast, NASA maintains a running narrative describing the odyssey of the two Mars Rovers, *Spirit* and *Opportunity*. The style of writing is simple enough to be accessible by about any age group. The content serves two purposes: Inform and persuade.

> Homer's Iliad tells the story of Troy, a city besieged by the Greeks in the Trojan War. Today, a lone robot sits besieged in the sands of Troy while engineers and scientists plot its escape....
> As the rover tried to break free, its wheels began to churn up the soil, uncovering sulfates underneath.

> Sulfates are minerals just beneath the surface that shout to us that they were formed in steam vents, since steam has sulfur in it. Steam is associated with hydrothermal activity—evidence of water-charged explosive volcanism. Such areas could have once supported life. (NASA, 2009b)

The "heroic" robot, *Spirit*, is trapped in quicksand, yet perseveres and continues to make important discoveries—six years past its expected life span. There is even more pathos as we learn that *Spirit* will never escape and is unlikely to "survive the bitter Martian winter." Different writers (especially technical writers) might quibble with the level of pathos in the copy, but nobody can deny the copy is designed to meet clearly identifiable purposes and for clearly identifiable audiences. The audience includes some of those suggested by Albers—student researcher and informed amateur. The purpose (as defined by their mandate) is to inform this audience, but it has the additional purpose of engaging their audiences—building a body of support.

Final Point

When Albers presents us with the hypothetical astronomy website as an example of a complex information system, he could easily use some of the examples I have discussed above. In some of the examples, it is clear the people in charge of the content had no sense of the genres they were using and so had no way of knowing when the content failed. If the content for simple websites and flyers, being produced by people who should understand good writing (rare bookstores and the CCCC), can be problematic, what can we expect in the future when information systems are even more complex? I suggest that having an effective system for evaluating content quality is more important than ever.

Reader Take-Aways

- My studies indicate that few professional writers can effectively evaluate existing texts for relevance, quality of writing, and quality of rhetoric. Yet, as information becomes increasingly complex, being able to do such evaluations becomes more and more critical. Having little understanding of the underlying theories of excellent writing, many who produce websites operate under the assumption that any text with excellent spelling and acceptable grammar is a good text. Professional writers working in complex information systems know this is not true and are in a position to very quickly identify ineffective content and improve it.

- The key to evaluating a text in a digital environment is to know what it is supposed to do and to whom. Once the component parts of an action-based genre (exigency, purpose, audience, structure, demands on the author, and text) have been identified, it becomes a simple matter to know what the text is supposed to be doing and describe how effectively it does that. This becomes a tool that professional writers can use that few outside the profession will understand.

- Professional writers have the best backgrounds and education for the demands of the different processes necessary for such evaluation and should have no difficulty understanding the processes I have described.

- Being able to do these evaluations is particularly valuable to the professional-writing community because, although so many people think they can write, it is easy enough to demonstrate that only professional communicators have the skills to evaluate existing texts for quality (especially quality of rhetoric); once revealed, these skills are valued. On many occasions, when I have been able to point to problems, explain them, and repair them, my contributions have been rewarded. My graduate students have expressed similar experiences in their workplaces. They are rewarded when they apply these ideas.

References

Adobe. (2010). *Adobe TV*. Retrieved from http://tv.adobe.com

Albers, M. (2004). *Communication of complex information: User goals and information needs for dynamic web information*. Mahwah, NJ: Lawrence Erlbaum.

Anderson, P. (1987). *Technical writing: A reader centered approach*. Orlando, FL: Harcourt Brace Jovanovich.

Ash, T. (2008). *Landing page optimization: The definitive guide to testing and tuning for conversions*. San Francisco: Sybex.

Bauerlein, M. (2008, September 19). Online literacy is a lesser kind: Slow reading counterbalances Web skimming. *Chronicle of Higher Education*, sect. B, pp. B10–B11.

Bazerman, C. (2003). The production of information in genred activity spaces: Information, motives, and consequences of the environmental impact statement. *Written Communication, 20*(4), 455–477.

Bledsoe, M., & Hitch, T. (2007). You suck at Photoshop. *My Damned Channel*. Retrieved from http://www.mydamnchannel.com/You_Suck_at_Photoshop/Season_1/1DistortWarpandLayerEffects_1373.aspx

Bloom, B. (Ed.) (1956). *The taxonomy of educational objectives, the classification of educational goals: Handbook 1. Cognitive domain*. Chicago: Susan Fauer.

Career opportunities. (1998). Retrieved from http://imrl.usu.edu/techcomm/ undergrad/careers.htm

CBS. (2010). [Homepage]. Retrieved from http://www.cbsnews.com

Freadman, A. (1994) Anyone for tennis? In A. Freedman & P. Medway (Eds.), *Genre and the new rhetoric* (pp. 43–66). New York: Taylor & Francis.

Gregory, J. (2004). Writing for the Web versus writing for print: Are they really so different? *Technical Communication, 41*(2), 276–285.

Hailey, C. E., & Hailey, D. E. (2003, October). How genre choices effect learning in a digital environment. *Journal of Engineering Education*, 287–294.

Krug, S. (2006). *Don't make me think*. Berkeley, CA: New Riders.

Miller, C. (1994). Genre as a social action. In A. Freedman & P. Medway (Eds.), *Genre and the new rhetoric* (pp. 23–42). New York. Taylor & Francis.

Mission and philosophy. (1998). Retrieved from http://imrl.usu.edu/techcomm/undergrad/philosophy.htm

National Aeronautics and Space Administration (NASA). (2009a). *National Aeronautics and Space Administration President's FY 2010 budget request.* Retrieved from http://www.nasa.gov/pdf/345225main_FY_2010 _UPDATED_final_5-11-09_with_cover.pdf

National Aeronautics and Space Administration (NASA). (2009b). *Sandtrapped Rover makes big discovery.* Retrieved from http://science.nasa.gov/headlines/y2009/02dec_troy.htm?list46156

National Aeronautics and Space Administration (NASA). (2009c). *2008 performance and accountability report.* Retrieved from http://www.nasa.gov/pdf/291255main_NASA_FY08_Performance_and_Accountability_Report.pdf

National Aeronautics and Space Administration (NASA). (2010a). *FAQ: Questions from / answers to potential contest entrants.* Retrieved from http://www.nasa.gov/pdf/426982main_KIDS_IN_MICRO_G_Q_A_021610.pdf

National Aeronautics and Space Administration (NASA). (2010b). *For students.* Retrieved from http://www.nasa.gov/audience/forstudents/index.html

National Aeronautics and Space Administration (NASA). (N.d.). [Homepage]. Retrieved from http://www.nasa.gov

National Public Radio (NPR). (2008). *Guerrilla ad campaign pushes boundaries.* Retrieved from http://www.npr.org/templates/story/story.php?storyId=93661794

National Research Council. (2000). J. Bransford, A. Brown, & R. Cocking *How people learn.* (Eds.), Washington, DC: National Academy Press.

Nielsen, J. (2000). *Designing Web usability.* Berkeley, CA: New Riders.

Powell's Books. (2009). *Synopsis and review of* Beloved. Retrieved from http://www.powells.com/cgi-bin/biblio?inkey=1-9781400033416-1

Redish, J. (2007). *Letting go of the words: Writing Web content that works.* New York: Morgan Kaufmann.

3

Language Complexity and Usability

Nathan Jahnke

Texas Tech University

CONTENTS

Introduction 48
Language as a Complex System 48
Language Complexity in Writer's Help... 49
    Tasks 50
    Test Environment and Procedure 52
    Results 53
        Task 3 54
    Discussion 56
Formal Written English 57
    Language Acquisition 58
    Comprehensible Input 59
    Cognitive Load Learning Theory and Language Acquisition.............. 60
Composition Instruction at Texas Tech... 61
    Course Redesign 62
    Usability and the Complex System 63
Conclusion 64
Reader Take-Aways 65
References.. 65

Abstract

All human languages, as complex systems, require intuitive response from their users; conscious application of heuristics is ineffective. Existing writers' help software tools fail to help novice writers acquire Formal Written English because another language, that of their user interface and of current-traditional composition instruction, stands between users and the system's content. In other words, these writers' help tools are cases of language complexity (e.g., users' ignorance of the technical definition of a *run-on sentence*) obscuring yet more language complexity (the vastly different and largely unknown grammar of Formal Written English as the distinct language the user is meant to acquire). In designing a better system to route around these problems of language complexity, we must look beyond traditional usability testing toward context-independent, ubiquitous access to tailor-made, multidimensional content whose usability is constantly and automatically reassessed.

Introduction

The freshman composition course and the study of rhetoric in general have a long and colorful history (Berlin, 1987), but, to date, usability engineers have paid little attention to the role of language in designing systems to help students improve their writing. My aim in this chapter is to discuss the relationship between language complexity and software usability in the context of composition instruction. In doing so, I hope to shed some light on the role of language complexity in usability in general.

I begin by discussing a recent usability study I helped conduct in which test participants, unbeknownst to them, largely failed to use a system designed to help them improve their writing; I show that, ironically, these task failures appear to have arisen from the same language complexity problem the software was meant to help correct. In order to explain this result, I then turn to a theoretical discussion of language complexity in the context of composition instruction. I take it as the goal of composition instruction to facilitate acquisition of Formal Written English, a language I claim to be no different from the various spoken dialects of English. Finally, I offer a report on some research currently underway at Texas Tech University that promises to help mitigate the problems I discuss in helping students acquire language. In discussing this research, I also speculate on how such a complex system might be tested for usability.

First, though, it is important to understand how I define concepts such as *complexity* and *language,* and to explain why I consider language a *complex system* in the sense of Albers (2005).

Language as a Complex System

"Complex" situations differ from "simple" and "complicated" ones in fundamental and significant ways. Albers (2005) established five criteria for identifying a complex situation in the context of information architecture:

No single answer: The required information does not exist in a single spot and a single answer, "the correct one," does not exist.

Open-ended questions: An open-ended question has no fixed answer that can be defined in advance, and it can be difficult, if not impossible, to even state when the question has been answered completely.

Multidimensional strategies: A person needs to use an evaluation strategy that takes into account multiple factors simultaneously, because the information normally comes from (multiple) sources and can be looked at in different ways.

Has a history: Complex systems have a history, which implies the system has a future. Something has happened in the past that caused the current information to obtain its current values and can affect the future evolution.

Dynamic information: None of the important information has a fixed value; rather, it changes continually. Also, the user's goals and information needs are highly dynamic and are continually refined as new information is revealed.

Nonlinear response: The response of a complex situation to interactions is often nonlinear, meaning it is very sensitive to the initial conditions and influencing factors. Minor changes in starting conditions can result in large differences as the situation evolves, and how the changes will affect the situation is often unpredictable. (Albers, 2005, p.23)

He then went on to assert that, even though complex situations are "the real-world norm" (p. 37), they remained a major design problem in usability engineering (p. 18).

In Table 3.1, I apply Albers's criteria to human language in order to demonstrate that composition instruction represents a "complex situation," as Albers defined it. Any system attempting to address human language problems must grapple with these challenges.

Albers argued that a special perspective is needed to ensure usability of software or other artifacts used in complex situations. In what follows, I demonstrate that language complexity did indeed affect the usability of a writer's help tool I helped test, but I deviate from Albers in that I do not believe said complexity must be dealt with head-on in designing a more usable version of that system. Rather, using a system we are currently developing at Texas Tech University as an example, I demonstrate how one might instead route around language complexity and enable instructors to deliver tailor-made content directly to students.

Language Complexity in Writer's Help

In June 2009, the Usability Research Laboratory (URL) at Texas Tech University conducted a test of an online writer's help resource. Twenty current or former students of composition were recruited to perform nine tasks over a forty-five-minute period. This usability test represented a unique opportunity to

TABLE 3.1

Application of Albers's Criteria to Human Language

No single answer	A student of composition may correct an error in his or her writing in an unlimited number of different ways. With human language, as the saying goes, "There's more than one way to do it."
Open-ended	All language use is discursive. The instructor may not understand what a student meant; the student may not understand what the instructor meant. Interested parties must reach an understanding.
Multidimensional	Register (essentially politeness level), the writer's voice, audience, and many other factors compete for the writer's attention, complicating the writing task. In addition, human language syntax is probably represented multidimensionally in the human brain (e.g., Chomsky, 1965).
Historical	The writer's background influences his or her writing in countless ways. Importantly, some dialects of English are closer to Formal Written English than others. Students need individual attention. Good writing is also reflection: Writers understand who they are now by examining what they wrote in the past.
Dynamic information	All writing is situated in an ever-changing context. Current events change what ought to be written and how we respond to what has been written. Additionally, changing even one word may change the meaning of the entire sentence.
Nonlinear response	Because how human language works is not fully understood, heuristic remedies for formal writing problems cannot be consistently applied. Even if a reliable heuristic could be established, due to the complexity of the underlying language system, transformations could unpredictably (from the point of view of the writer) cascade through the document, rendering technically correct output that is now devoid of its intended meaning.

operationalize complexity in the context of language learning. In the following sections, I explain the methodology used for the test as well as the test environment and our results.

Tasks

The tasks were written to represent different levels of abstraction away from the writer's help interface. In other words, in some tasks, participants simply had to click on the interface hyperlink that included some of the same words as they were given in the task prompt, while others were so abstract that participants had to possess some fundamental knowledge not present in the task prompt in order to arrive at the pertinent information. All tasks are given in Table 3.2.

For example, in Task 3, participants were asked to use the writer's help product to help them revise a sentence that "uses a lot of words to say something pretty simple." In designing this task, we first located a capability of

TABLE 3.2

Tasks Sorted by Language Complexity

Low-Language-Complexity Tasks

4. You don't have a problem with commas, but one of your friends in class does, especially using commas for items in a series. Figure out the best way to alert your friend to help available on writer's help for comma issues related to items in a series.

(relevant menu item: **Punctuation: The Comma**)

6. You need to create an end-of-text, bibliographic citation for the following source using MLA. Use the writer's help to do so correctly:

John Smith wrote the article "10 Steps to Financial Success" on the CNN Money website (http://www.cnnmoney.com/smith_success.htm) 2-12-2008.

(relevant menu item: **MLA Papers**)

7. How would you format both items as bibliographic entries in APA style?

(relevant menu item: **APA Papers**)

High-Language-Complexity Tasks

2. There's just something wrong with this thesis statement: "The Web has had considerable influence on the world." Use writer's help to diagnose and then fix this thesis.

(relevant menu item: **Composing and Revising: Drafting: Revising a Thesis That Is Too Broad**)

3. This sentence uses a lot of words to say something pretty simple: "Basically, in light of the fact that Congressman Fuenches was totally exhausted by his last campaign, there was an expectation on the part of the voters that he would not reduplicate his effort to achieve office in government again." What help does writer's help provide you to improve it?

(relevant menu item: **Word Choice: Concise Language**)

5. If your instructor indicates you need to collect evidence to support your argument, what advice does writer's help offer about what sort of evidence you need to gather for an English composition assignment?

(relevant menu item: **Research: Locating Articles**)

8. On an example paper, the instructor has written "flawed argument" next to the following sentence: "If we can put a man on the moon, we should be able to cure the common cold." What does the writer's help tell us is wrong with this sentence and how it can be fixed?

(relevant menu item: **Academic Writing: Evaluating Arguments**)

the product—in this case, its advice on how to revise "wordy" sentences or to help users write more "concise" ones—and cloaked it in what we considered less specialized language ("a lot of words"). This change qualified Task 3 as a "complex" task, since participants would have to perform some translation of the language of the task prompt to the language of the system being tested. We tested using three low-complexity tasks (in which the language of the prompt was taken directly from the system) and four high-complexity tasks (in which the language of the prompt did not match the language of the system). Regardless of the complexity of the task, the information needed to complete it was confirmed present in the software before the testing began: We were testing only whether people could get to it upon reading the task prompt. In this way, we sought to operationalize the complexity that Howard (2008) found unexpected in his similar test of a writer's help product.

The tasks employed in the study grouped by level of abstraction are given in Table 3.2. (We did not include our first or final tasks in this rubric because they were open-ended, with different participants looking in different areas of the software for help with their own individual problems.) The name of the correct menu choice in the writer's help system is also given below each task.

By giving participants different types of tasks in this way, we sought to relate success on task to correspondence between task language and interface language. In other words, we wanted to answer the question "How does language complexity affect usability?"

We used the popular Morae software package from TechSmith to annotate videos of tests with qualitative observations and to generate reports using quantitative data collected by the software during the tests.

Test Environment and Procedure

The Usability Research Laboratory at Texas Tech is made up of two adjacent rooms with large, one-way glass windows between them. Behind the windows is our observation room, where the observer sits and records qualitative annotations on a Windows PC with Morae software. The observer has the choice of several camera feeds from the test room; one camera is usually placed immediately below the participants' computer screens in order to capture their changing facial expressions and other gestures as they attempt to carry out tasks. Morae also records video of participants' actions on the computer screens along with other data such as mouse clicks, time on task, and so forth.

For this test, participants sat in the test room in front of the one-way windows at a workstation with a Windows XP PC. They received tasks on slips of paper from our facilitator. Participants were asked to read each task aloud as they received it and before they began to carry it out. Then, during the test, the facilitator prompted the participant with questions such as "What do you think that does?" and "Would you have given up by now?" Thus, the facilitator effected active intervention in addition to a more traditional talk-aloud protocol.

There was a five-minute time limit per task: If time ran out before a participant completed the task, our observer recorded a task failure, and the participant was asked to move on to a new task.

After all tasks were complete, we administered the standard System Usability Scale (SUS; Bangor, Kortum, & Miller, 2009) questionnaire through the Morae software for participants to fill out.

Results

Participants failed the more "abstract," language-complex tasks more often than the simple ones, in some cases as often as 55 percent of the time. Language complexity of a given task was therefore a good predictor of task failure. These data are summarized in Table 3.3 and plotted in Figure 3.1.

Contrasting the high failure rate on some tasks was the high average score of 77.35 on the SUS questionnaire we administered after each participant's test was complete.

TABLE 3.3

Failure Rate by Task Language Complexity (Lower Is Better)

Low-Complexity Task	Failure Rate (%)
6. Format MLA citations.	40
7. Format APA citations.	20
4. Help friend with commas.	20

High-Complexity Task	Failure Rate (%)
5. Gather evidence.	55
2. Fix thesis.	50
3. Solve problem with conciseness.	50
8. Correct a flawed argument.	55

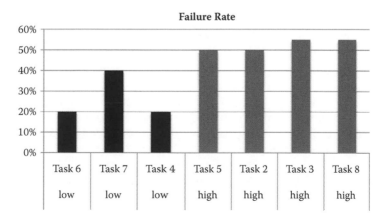

FIGURE 3.1
Failure rate by task language complexity (lower is better).

My team interpreted the high subjective scores for the system's usability to mean that participants were not aware that they were failing the tasks (at no time during the test were they told, "You did this wrong"). We reasoned that if participants knew of their failures, they would have, in turn, failed the system on the SUS questionnaire—or at least not rated it so highly (Bangor et al., 2009).

We were especially interested to discover that, for Task 3, participants misinterpreted the task prompt in a remarkably consistent way.

Task 3

This sentence uses a lot of words to say something pretty simple: "Basically, in light of the fact that Congressman Fuenches was totally exhausted by his last campaign, there was an expectation on the part of the voters that he would not reduplicate his effort to achieve office in government again." What help does writer's help provide you to improve it?

The jargon that composition instructors attach to the above example sentence is "wordiness." They say the example is too "wordy" or lacks "concision." But we didn't include these key terms in the task prompt precisely to test whether participants (1) knew the terms already (remember, they were all current or former students of composition) or (2) would discover the terms during their attempt to use the system to diagnose the example. If the system were truly usable, we reasoned, users wouldn't have to know jargon in order to complete tasks with it (i.e., improving their writing).

The high failure rate on this task speaks to users' inability to find content marked with the key word "concise language" when they either (1) were not aware that the example lacked concision or (2) were aware that this was a wordy sentence, but did not know that the system wanted them to search for "concise" to find help on the problem (in other words, they didn't know the jargon). In this way, the complexity of language doomed participants twice over: In their attempts to find information on how to correct the wordy sentence in the task prompt, they were foiled by terminology.

More interesting, though, was how nine out of twenty (or almost half of all) participants decided that the example was a "run-on sentence" at some point during the task and sought help for run-on sentences in the writer's help system. It is important to know that the example does not represent a run-on sentence. Thus, not only were participants not able to find the correct information (on concision), but also participants actually located incorrect information (on run-on sentences) and believed that they had found the correct information. This is another reason we believe the SUS scores given by the participants were inflated: Participants believed that they had found the correct information when they had actually failed the task. Howard (2008) also experienced this disconnect between success and perceived success in testing a similar product (a writing textbook) set in a similar complex situation (writer's help).

Participants' correct "first clicks" (or whether they initially went to the right place for each task) are given in Table 3.4 and plotted in Figure 3.2.

With one exception, which I will discuss in a moment, participants had trouble locating all and only information associated with the high-language-complexity tasks (2, 3, 5, and 8) and mostly went directly to the correct information for the low-language-complexity tasks (4, 6, and 7). (Again, we expected participants to go off in different directions for the first and final tasks, 1 and 9, which were open-ended.)

It is interesting to note that participants initially did not know where to find information for Task 4 ("Help friend with commas")—even though we categorized this as a low-language-complexity task. Though the failure rate on this task was the lowest of all—20 percent—participants nonetheless started out looking in the wrong places. We attribute this directly to the lack of a top-level menu item called *commas* (participants instead had to

TABLE 3.4

Correct First Clicks by Task Language Complexity (Higher Is Better)

Low-Complexity Task	Correct First Clicks (%)
6. Format MLA citations.	90
7. Format APA citations.	90
4. Help friend with commas.	50

High-Complexity Task	Correct First Clicks (%)
5. Gather evidence.	60
2. Fix thesis.	55
3. Solve problem with conciseness.	35
8. Correct a flawed argument.	30

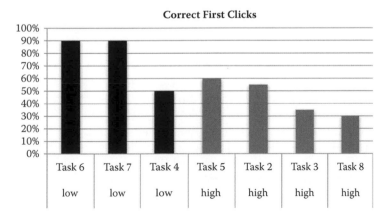

FIGURE 3.2
Correct first clicks by task language complexity (higher is better).

find and click on "Punctuation" to reveal the subheading "The Comma"). Once participants saw and understood the "Punctuation" menu item to be relevant to their interest in completing the task, they had no trouble locating the information on commas below it. This was not the case for the "Format MLA citations" and "Format APA citations" low-language-complexity tasks, for which there were top-level menu items labeled "MLA papers" and "APA papers." We conclude that the existence of a menu item (at any level of navigation) whose label closely matches the language of the task description is a predictor for overall task success, while the existence of a *top-level* menu item whose label closely matches the language of the task description is a predictor for task efficiency. We are left with a very precise sense of what is needed to overcome language complexity in the user interface—a key word from the task prompt must appear at the top level of navigation. Yet this is not always practical to implement, as we will see in the next section.

Discussion

Clearly, a system that leads to incorrect choices—especially without users' awareness that they are incorrect—is not one we would term *usable*. The question, then, is how we ought to improve it. As usability engineers, we need to change the system so that it leads users to all and only correct diagnoses for common writing problems.

The most obvious course of action would be to discover all common writing problems, sort them, and provide top-level navigation links in the system for each one, labeled in ways that users understand. This method breaks down, however, when we take stock of the number of problems that novice writers have; as Albers (2005) reminds us, the whole is greater than the sum of the parts when it comes to complex systems. In the writer's help system we tested in our lab, for example, there were over one hundred subheadings revealed when users clicked the top-level navigation links. We simply cannot call a system "usable" that presents users with a glossary-like, mammoth list of "catchphrases," even if these were carefully selected to avoid the use of terms unfamiliar to most users. We can also imagine that the relative importance and even the very names of these "catchphrases" would change over time for individual users as their writing skills and knowledge of composition jargon improve.

The solution to a complex problem like writer's help has to be different in a fundamental way from the solution to a simple one. In short, we have to redesign the teaching of composition, and then we have to validate our new system, making sure that it is usable. In order to do this, though, we first need to have a better understanding of the nature of human language as a complex system. In this, we will make use of the second language acquisition theories of Krashen (1981, 1985) and others. We must first ask ourselves

"What is the writer's help product trying to do?" And, only after we have an answer to that, "How can we make it more usable?"

Formal Written English

Composition, at least in its most common form, the so-called current-traditional pedagogy (Berlin, 1987), is the teaching of Formal Written English. I use the term *Formal Written English* to mean the special variety of English cultivated by educational institutions and appearing in publications such as *Time* magazine, the *New York Times*, the *BBC News* website, and so on. Formal Written English differs in significant ways from the many spoken dialects of English. Unfortunately, the long history of standardization of the English writing system has obscured many of these distinctions. As a result, most writers are not aware that they are, in fact, writing in a different language than they speak.

This distinction is not limited to English spelling, whose poor representation of the sounds of words is well known; rather, it is a highly complex web of nuances of which we are seldom, if ever, aware. To draw on just one example I have studied, certain speakers of English in the American state of Pennsylvania (or whose parents hail from Pennsylvania) will use what appears from the perspective of Formal Written English to be a truncated form of the passive voice, for example,

    1. This car needs washed.

This construction was shown to have arisen under the influence of another language, Scots Gaelic, when the ancestors of those who settled in Pennsylvania still lived in Ulster in northern Ireland (Murray & Frazer, 1996). Formal Written English would prefer

    2. This car needs to be washed.
or
    3. This car needs washing.

However, English grammarians remain largely unaware of this nonstandard construction (nonstandard in the sense that it is not permissible in Formal Written English). For example, Microsoft Word 2008 version 12.2.1 for the Macintosh allows "This car needs washed" as a grammatical sentence, even though human readers probably would not, especially now that I have singled it out. And this is just one example. Because human language is a complex system, it is impossible to know how many of these nonstandard constructions "fall out" when one moves between his or her dialect of spoken English and the Formal Written variety. Further, even if it were possible to keep track of and to design for all of them, the language system changes with such regularity that such a simple solution would be perpetually out of date.

Clearly, teachers of composition cannot hope to transmit competence in Formal Written English in a single semester, especially when they are not even aware of many of the differences between each student's own dialect of English and the Formal Written variety. All is not lost, though, as there is much insight on this problem to be found in second language acquisition research.

Language Acquisition

Stephen Krashen, an influential human language acquisition theorist, draws a distinction between language *learning* and language *acquisition*. This distinction is fundamental for understanding how adults acquire competence in any language, including Formal Written English.

Specifically, he argues that language *learning*, which involves conscious memorization of rules, is much less helpful to the student than is language *acquisition*, the unconscious development of linguistic faculties in the brain (Krashen, 1985).

Krashen (1985) argues that while certain rules (e.g., "don't end a sentence with a preposition") can be *learned*, languages in general are best *acquired*. The reasons for this in the context of composition instruction are many, but one draws on the discussion in the previous section: If teachers of composition are not aware that "This car needs to be washed" is preferred in Formal Written English over "This car needs washed," then they cannot teach their students this distinction. The students must simply acquire it on their own.

To help us visualize this problem, Krashen (1981) offers the diagram in Figure 3.3. In this diagram, each circle represents language information to be acquired: The smaller the circle, the smaller the amount of information.

Even the most talented theoretical linguists today do not understand completely how the human language system works. In other words, they are not able to explain every nuance of language using current theory. This means that there are parts of every language, including Formal Written English, that are impossible to teach. This assertion comes as no surprise to the expressivist instructor who believes that writing talent cannot be taught (Berlin, 1987), but I am not addressing talent here, only objective correctness. Formal linguists' limited knowledge is represented by the first circle inside the outer circle in Figure 3.3.

The next inner circle represents composition teachers' understanding of current linguistic theory. This understanding is more limited than theoretical linguists', if for no other reason than how fast the field of linguistics is currently moving.

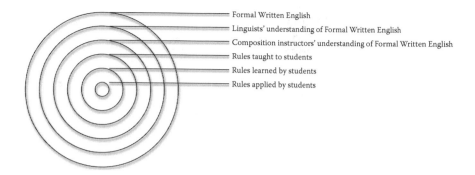

FIGURE 3.3
Little linguistic knowledge can be successfully taught. (Adapted from Krashen, S. (1981). Principles and practice in second language acquisition (pp. 92–94). New York: Pergamon.)

One semester is not long enough to teach every student every rule he or she lacks; the next inner circle represents the number of rules an average student is actually taught in composition.

Of course, students do not learn everything they are taught; the amount they actually learn is represented by the next inner circle. (Some students even learn rules incorrectly, but that is a topic for another time.)

Finally, the few rules that students do learn are applied unevenly to their writing; students are not perfect, and will not catch every violation of every rule. This amount of applied formal knowledge is represented by the inner-most circle in the diagram.

Comprehensible Input

Comprehensible input is the engine of language acquisition. It is defined as any message delivered in the target language that is understood by the language acquirer (Krashen, 1985). Given enough comprehensible input, any mentally competent human being will unconsciously acquire the language in which the input was delivered.

Indeed, this means that one must already understand a language in order to acquire that language. But this apparent paradox breaks down when we remember that it is possible to understand parts of a language (e.g., certain words or phrases) but not others, and that it is often possible to understand what is meant by context alone.

Thus, if a composition instructor is able to offer a student individual attention and instructs his or her student to "correct this run-on sentence by breaking it up into two smaller sentences," while demonstrating what those two smaller sentences look like, and the student does his or her part by reading and understanding the meaning of what the instructor has written, then the student will come to associate this task with the term "correcting a run-on sentence." Crucially, second language acquisition theory holds that *this happens with no conscious effort on the student's part whatsoever.* The linguistic tradition of the latter half of the twentieth century assumes that language is an innate ability of the human species; therefore, no conscious effort is required to acquire or use language. As I have hinted at before, the problem of composition instruction is a particularly thorny one, because the student always deals with two layers of language complexity: The first is concerned with communication between the instructor and the student (e.g., what is meant by a "run-on sentence"), while the second involves the actual correction of the run-on sentence to achieve grammaticality. Thus, the student is always acquiring one language, that of the composition profession, in order to acquire another, Formal Written English. Understanding this distinction is crucial before making any attempt to redesign composition and, by extension, the writer's help system to be more usable.

Language acquisition is particularly efficient when one already understands enough of the target language to make out what is being said most of the time, thereby providing the acquirer with plentiful comprehensible input; this is also the case with Formal Written English. To continue with the example from the previous section, thanks to context clues, speakers of Pennsylvania English understand instantly that "This car needs to be washed" means the same thing as "This car needs washed" upon seeing the former in print, even though they have been using the latter exclusively in speech since they were young children. Then, as the theory goes, after seeing the standard form in print many times, they begin to write, "This car needs to be washed" instead of "This car needs washed," correctly writing a different language than they speak. Again, at no point in the acquisition process are they consciously aware of the changes taking place in their minds that facilitate this transformation from use of the nonstandard spoken form (by default) to the standard form in writing. In fact, second language acquisition theory holds that it is impossible (or at least wildly inefficient) to consciously learn language. The reason this must be true will be explained in the context of usability in the next section.

Cognitive Load Learning Theory and Language Acquisition

Feinberg, Murphy, and Duda (2003) applied cognitive load learning theory to information design and usability engineering. I am interested here mainly in the distinction that cognitive load learning theorists draw between sensory memory, working memory, and long-term memory. Of the three, long-term memory is where we find our unconscious ability to perform complex tasks such as walking and talking. For the purpose of teaching languages, including Formal Written English, use of long-term memory is effectively synonymous with skill in a language: Error-free performance is impossible without the help of long-term memory due to the complexity of the language system. In other words, in the terms of cognitive load learning theory, to acquire a language means to commit its vocabulary and formal structure to long-term memory.

Any system concerning itself with language acquisition must provide sufficient comprehensible input to foster the growth of what cognitive load learning theorists call *schemas*. A schema is much like a mental model in that it allows people to quickly accomplish tasks (such as how to form a complete sentence) under novel circumstances.

Albers (2005) gives the example of schemas for different types of restaurants: Even if you have never been to a particular restaurant before, as long as you know what kind of restaurant it is (fast food, buffet, formal ...), you will be able to place your order and receive your meal, even though the nature of those tasks varies widely depending on the type of restaurant you have chosen. You have a schema for each type of restaurant, and so you don't even have to think about what you have to do to get your food.

This definition of schemas or mental models is significant to the problem of language acquisition because to be *competent* in a human language means to be able to form novel—yet fully grammatical—constructions at will (Chomsky, 1965). Modern linguistics is the study of these schemas in the context of human language.

These concepts are summarized in Table 3.5.

In this section, we saw how the problem of language complexity applies to current-traditional composition instruction, and we saw how, using second language acquisition theory, we might get around said complexity. It remains to be seen how one could design a usable system to assist in this formidable task of language acquisition. In the next section, I offer some anecdotes on just such an experimental course redesign that is currently underway at Texas Tech University, pointing the way toward a very different form of writer's help.

Composition Instruction at Texas Tech

Texas Tech University received a series of large grants from the Texas Higher Education Coordinating Board to redesign its introductory-level composition classes. Together with our partner, Dallas County Community College

TABLE 3.5

Unconscious Performance Is Superior

	Conscious	Unconscious
Cognitive Load	Sensory, working memory	Long-term memory
Human Language	Learning	Acquisition
Efficiency	Low	High

District, we experimented with several new systems to help with the teaching of composition. Whether these systems are usable is the single most important question we must ask ourselves.

But before I describe those systems, it's important to note that Texas Tech had already been at the forefront of innovation in composition instruction before receiving the grants. For example, Dr. Fred Kemp's TOPIC-ICON system had been in operation since 2001. In the ICON system, classroom instruction and grading were separated: Classroom instructors (CIs) would meet with students in person once a week, and document instructors (DIs) would grade students' work anonymously online. It was possible in this online software system, called TOPIC, to include links to core composition concepts in commentary on student writing; for example, if a student had a problem with run-on sentences, then a pre-generated link to specific help with run-on sentences could be included by the DI grading the work. However, these links led to short, text-only descriptions of the most common errors, and this system worked better for formal problems such as comma splices and the aforementioned run-ons than it did for holistic issues such as the importance of doing proper research before writing a research paper.

The team charged with implementing the new course redesign included Dr. Kemp as well as Dr. Rich Rice and me. Development of the experimental systems detailed in the following pages should not by any means be attributed solely to me!

Course Redesign

Basically, the Department of English at Texas Tech was given a license to explore various improvements to its already innovative freshman composition program. I describe these changes here in order to illustrate what a system might look like that gets around the complexity inherent in the task of accepting feedback and addressing issues in writing (from the perspective of the student). This section is not meant as a panacea for addressing language complexity in usability engineering, but rather as a proof-of-concept of how complexity might be partially avoided in one particular context by taking it into consideration in the early stages of design.

We decided to combine our strength in online assessment with the LeCroy TeleLearning Center's two video series on composition intended for use in distance education, titled *The Writer's Circle* and *The Writer's Odyssey*. All together, these video programs constituted twenty-six "Lessons," each thirty minutes long. But we were more interested in delivering students short, targeted help on specific writing problems in order to avoid overloading them with useless information which, as second language acquisition theory claims, they would never use to acquire Formal Written English. To that end, we "chunked" or segmented the twenty-six Lessons into about 500 three- to four-minute "chunks." We then tagged these chunks with one or more key words; for example, a video might be tagged "the writer's voice" and "prewriting," while another might be tagged "run-on," depending on which issues those individual videos addressed.

By serving the videos on demand from a streaming server, we were able to allow DIs to link to them from any feedback they might wish to give students on their writing. We were also able to address the often overlooked issue of user environment by buying every student in our trial his or her own iPod, which we preloaded with every one of the 500 or so video chunks. In this way, students could look up videos by name or number (e.g., the chunk we titled "Decision-Makers," or Chunk #192) on their iPod without having to be in front of a computer to receive and understand their instructors' feedback. By giving students the freedom to use our system in any environment they chose, we eliminated the problem of environmental complexity from our design.

Originally, we were only able to allow DIs to provide feedback on student work holistically; in other words, they were not able to comment on specific passages of student writing without copying and pasting said passages into their commentary. Later, however, we integrated the system with Microsoft Word, so that DIs could use the much more efficient Track Changes commentary feature of that software. Submitting and assessing work became a simple matter of uploading or downloading a single file, and students received commentary on their work just as they would had they submitted it in paper format, with notes and helpful links provided in the margin next to the target areas.

It was primarily with targeted commentary delivered in this way that we were able to overcome the problems with terminology (e.g., "What is a 'run-on' sentence?") that plagued the online writer's help system whose usability testing I discussed earlier in the chapter. By linking directly to appropriate content, including video clips, instructors established one-to-one dialogue with students and eliminated problems with language complexity and navigation.

If students were viewing these video clips online, they also had the option of skipping instantly to other clips tagged with the same key word(s). In this way, we also allowed students to tailor delivery of content to their own personal background and learning style: If a student didn't understand a particular clip, then he or she could simply try the next one on that particular topic. This method is well in line with the communicative approach to language instruction advocated by Krashen and others, the goal of which is to provide the student with as much comprehensible input as possible in the right context.

Usability and the Complex System

How might usability be assessed in such a system, whose use is so closely tied to the needs of the individual student? Attempts to bring students into the usability lab for traditional testing would probably result in invalid data because the context would be too artificial (Genov, Keavney, & Zazelenchuk, 2009).

Rather, we would like to tie use of specific video clips to students' performance on subsequent assignments. We have not yet had time to implement this feature, but we suspect it would help further differentiate helpful chunks from unhelpful ones. In other words, we would implement a ranking system whereby chunks are valued based on how well students perform in assessments after watching them. Then, the more helpful chunks would be sorted first as students click through chunks related by a particular key word. Importantly, assuming the ability of the grading system and the video system to talk to one another, this feature could be totally automated, thereby solving the problem of testing, in a traditional way, a writer's help product masked by language complexity.

Automatic reconfiguration of the system to respond to changing users' needs would satisfy Albers's (2005) requirement that a system made to address a complex task also address the history of said task. More conjecturally, it may also be possible to predict students' future needs based on past ones. There is some evidence for a "natural order" of acquisition of formal features of languages (Krashen, 1981); if true, examining which videos students viewed in which order on a mass scale should reveal this order for Formal Written English for the first time, enabling us to predict with better-than-chance accuracy what students will need help with next.

Another planned feature is that of instructor- and student-submitted videos. While we were pleased with *The Writer's Circle* and *The Writer's Odyssey* from the LeCroy TeleLearning Center, we also acknowledge that even five hundred video clips will not cover every conceivable issue in teaching Formal Written English. We are also very interested to see how user-submitted content would "fare" when pitted against the professionally produced LeCroy clips; in other words, do students learn better from professionally produced teaching videos, from "amateur" videos with low production values made by their own teachers, or from peer-submitted content? What combination of dialogue between stakeholders makes for the most usable system?

Conclusion

Human language is too complex to teach successfully, and the writing situation is too complex to treat with traditional software navigation methods. I have shown that, for a variety of reasons, students cannot be expected to apply rule-based diagnoses to improve their writing; they must instead be exposed to plentiful comprehensible input in the target language—Formal Written English.

In order to ensure that students are receiving this comprehensible input, their instructors must be able to link them directly to it, bypassing the second layer of language complexity of the contemporary Web-based user interface. Taking this paradigm one step farther, use of videos and other material should be tied to the ability to demonstrate competence in the target areas. Content can then be automatically prioritized based on its effectiveness for a particular purpose.

We are still on the first page of the book of usability and language complexity, but the story promises to be a good one. With carefully designed tests and industry partners, we can learn what makes a system usable and apply knowledge from our backgrounds in other disciplines to unlock the world-changing potential of a new generation of software systems.

Reader Take-Aways

- Human language is too complex and not sufficiently well understood to interact with efficiently on a conscious level. Focus on making sure that students of composition get models of correct usage so that they unconsciously acquire Formal Written English and become intuitive users of it.

- When asked what he was looking for in the writer's help interface, one usability test participant replied, "I don't know yet." Avoid isolating users on the other side of a language-laden interface by instead linking them directly to the content they need.

- By the same token, when testing products for clients, do not neglect the issue of language complexity, especially when formulating task prompts.

- People's language and needs are constantly changing. Design systems that self-correct based on constant sampling of user interaction and success rates with specific content.

References

Albers, M. J. (2005). *Communication of complex information*. Mahwah, NJ: Lawrence Erlbaum.

Bangor, A., Kortum, P., & Miller, J. (2009). Determining what individual SUS scores mean: Adding an adjective rating scale. *Journal of Usability Studies*, 4(3), 114–123.

Berlin, J. A. (1987). *Rhetoric and reality: Writing instruction in American colleges, 1900–1985*. Carbondale: Southern Illinois University Press.

Chomsky, N. (1965). *Aspects of the theory of syntax*. Cambridge, MA: MIT Press.

Feinberg, S., Murphy, M., & Duda, J. (2003). Applying learning theory to the design of Web-based instruction. In M. J. Albers & B. Mazur (Eds.), *Content and complexity: Information design in technical communication* (pp. 103–128). Mahwah, NJ: Lawrence Erlbaum.

Genov, A., Keavney, M., & Zazelenchuk, T. (2009). Usability testing with real data. *Journal of Usability Studies, 4*(2), 85–92.

Howard, T. W. (2008). Unexpected complexity in a traditional usability study. *Journal of Usability Studies, 3*(4), 189–205.

Krashen, S. (1981). *Principles and practice in second language acquisition*. New York: Pergamon.

Krashen, S. (1985). *Inquiries & insights*. Hayward, CA: Alemany.

Murray, T. E., & Frazer, T. (1996). Need + past participle in American English. *American Speech, 71*(3), 255–271.

4

Innovation and Collaboration in Product Development: Creating a New Role for Usability Studies in Educational Publishing

Tharon W. Howard

Clemson University

Michael Greer

Pearson Higher Education

CONTENTS

Introduction 68
Background of the Product...68
Finding Opportunity in "Failures"..70
Unexpected Complexity...73
    Test Methodology 74
    Selected Results..76
    Implications 78
Collaboration and Rhetoric..79
Driving Innovation in Product Development...82
Conclusion 84
Reader Take-Aways 85
References..86

Abstract

In July 2006, we began working together on a series of studies designed to assess the usability of a new college writing handbook. Our purpose here is to report on these usability studies and their implications concerning the methods for bringing user experience design principles into the product development cycle in educational publishing. Our successful collaboration has demonstrated how a pilot project which was originally intended only to demonstrate the value of usability testing to the revision of textbooks eventually impacted the product development process for a major publisher. Specifically, we found that exploring the limits of complexity in the handbooks we tested helped us move from what we call an *accommodationist model* of textbook design to a *constructivist model.*

Introduction

In July 2006, we began working together on a series of studies designed to assess the usability of a new college writing handbook. One of us (Howard) was commissioned by the handbook's publisher to conduct the usability studies. The other (Greer) is a development editor who had been working since 2004 on the development of the new handbook. Our purpose here is to report on these usability studies and their implications concerning the methods for bringing user experience design principles into the product development cycle in educational publishing. We also intend to reflect more broadly on the lessons we have learned over the course of this multiyear collaboration—lessons about how to leverage usability testing to drive innovation and about how to build collaborative relationships that support user testing as an integral, ongoing part of the development process. More importantly, this chapter describes how we learned that we didn't have to use usability testing merely as a means of discovering how to "accommodate" users by locating and recommending ways to "fix" the usability errors in the texts we were designing. Instead of using usability testing as a means of accommodating our texts to users, this chapter explains how we found that exploring the complex problems in a familiar, well-established product helped designers, authors, and editors realize ways that they could constructively adapt users to the texts.

Background of the Product

College writing handbooks are in many ways analogous to software documentation and product user manuals. Handbooks focus on practical, how-to information (e.g., how to edit and correct grammatical errors, and how to cite sources); they are intended for novice users with scant or fragmentary subject matter knowledge; and they are commonly consulted by users who are under stress, encountering problems, or experiencing other external pressures. Ease of use and accuracy of information are of paramount importance. Users of writing handbooks, like users of product manuals, are often not aware of the nature of the problem they are experiencing, how to name that problem, or, often, that they have a problem at all. As our studies illustrate, many student handbook users thought they had found a correct solution to their problem when in fact they had not. One of the greatest challenges for any handbook publisher is finding a way to help students find solutions to problems they do not know they have.

Historically, writing handbooks and user manuals have been understood and treated (by authors and publishers) in much the same way, as containers for the delivery of information. It is assumed, and implicit in the architecture and design of both handbooks and user manuals, that the document itself (the handbook or manual) functions as a conduit for the transfer of information from the expert (author) to the novice (user). This content delivery model is at odds with what usability professionals know about how users make sense of texts. Users will make sense of a document in whatever ways best suit their needs and purposes at the time; they will usually settle for the first solution they find that satisfactorily answers their particular problem at that point in time. The determining factor in whether a user is successful in using a handbook or manual is, therefore, not the quality or accuracy of the information contained in the document, but the user's effectiveness in finding, making sense of, and applying that information.

The distinction between a content delivery model and a user experience model is crucial. A content delivery model places primary emphasis on the accuracy and quality of the information contained in the document itself. It assumes that users will be able to extract that information with relative ease as long as the document is accurate and clear. In other words, the goal of testing in this model is to discover where information is unclear to users and then accommodate the users by changing the delivery system to meet their needs. Based on what we have learned over the course of these studies, we would argue that a content delivery model can result in an effective and usable document only when the conventions of document design and document literacy are stable and shared by the author and user alike.

In the case of college writing handbooks, however, the conventions are neither shared nor stable. While authors and teachers may be well versed in the conventions of handbook architecture and design, the student users most certainly are not. Students do not have the basic grammar knowledge necessary to make sense of even the tables of contents. A student may know, for example, that she needs to repair misplaced commas in her paper; but she usually does not know whether to look under subordination, sentence structure, or run-on sentences, among many other possibilities. Thus, the conventions of handbook design require a working knowledge of grammar terminology that student users simply do not have.

Further, student reading and literacy practices in general are undergoing a radical and wide-ranging shift from a print-centered to a multimodal-visual literacy. In fact, we have observed a wide gap between teacher and expert reading practices, grounded in years of experience with printed texts and a familiarity with established textbook conventions, and student and novice reading practices, grounded in years of experience with multimodal, multimedia, and visual texts. In short, conventional handbook design is built for teachers but not for students in the same ways that many software manuals are built for software engineers and not actual users.

This divide between producers and consumers is not only deep in terms of content, but also widened by changing expectations of literacy. Gunther Kress (2003), among others, argues that literacy in the new media age is characterized by a shift from the "dominance of writing to the new dominance of the image, and ... from the dominance of the medium of the book to the dominance of the medium of the screen" (p. 1). Clearly, a fundamental shift in the very definition of literacy would have an impact on the way educational texts are designed and developed. As yet, however, the dominant models and conventions used in the production of handbooks (and most other educational texts) continue to be grounded in traditional print-based reading practices and have yet to fully understand or embrace emerging new media literacies. As a result, as we began to understand and interpret the data from our usability studies, we began to see ourselves as advocates for the student users. We wanted the stakeholders in the publishing process (the authors, designers, editorial management, and marketing team) to understand that students are not "bad readers" who needed to be accommodated by making the design more direct, simpler, and less engaging. Instead, we argued, they simply read differently. Understanding how they read differently would be a key in the process of designing a more effective handbook for them.

This shift from a content delivery model to a user experience model was a key component in the rhetoric of the reports and memos we were composing for decision makers and stakeholders in the publishing process during the course of the new handbook's product development cycle. One of our more significant, if easily overlooked, observations is that usability studies represent not just a methodology but also a language. A usability study is more than a set of data about user experiences with a document; it is also a language and a discourse for framing, describing, and understanding those experiences. By beginning to write and talk about handbooks as user experiences (rather than as books, containers, or products), we discovered that we were able to persuade key players in the publishing process to entertain and even embrace innovations that represented significant departures from existing conventions of handbook organization and design.

Finding Opportunity in "Failures"

Ironically, one of the motivations for the original usability study in 2006 was the publisher's concern that the new handbook was perhaps "too innovative."

College writing handbooks are marketed in a highly competitive landscape, and are promoted to "legacy users"—that is, college writing teachers accustomed to a long-established set of conventions for the presentation, organization, and delivery of information in handbooks. Because the new handbook we were producing departed from many of these conventions, most notably in its highly visual layout (Wysocki & Lynch, 2009), the publisher decided to commission Howard to conduct the study as a side-by-side test against the current market-leading handbook, by then in its sixth edition (Hacker, 2007). As Figure 4.1 illustrates, the new handbook sought to visually represent "patterns" that exist in sentence structures. Rather than merely providing a verbal label for a pattern such as *appositive* or *relative clause*, which has been the dominant approach in the handbook market for years, this approach sought to break with tradition and to help student-users to recognize patterns visually.

The results of the study were striking: While the new visual handbook was judged by users to be significantly more usable than the existing market leader in the head-to-head test involving traditional usability measures, the results of users' performance as judged by experts demonstrated that neither handbook was particularly usable for today's college students (Howard, 2008). In other words, users thought they had successfully completed simple tasks, when in fact they had failed to perform satisfactorily. This finding confirmed our intuition that handbooks had become bound by a set of outdated conventions that were neither recognized nor understood by their intended audience of first-year college composition students. But it also showed that we hadn't gone nearly far enough in what many feared was a far too radical redesign.

Rather than being discouraged by the low usability scores on both handbooks, we began to see the results of the initial study as valuable data we could use as leverage to persuade the handbook's publisher to embrace

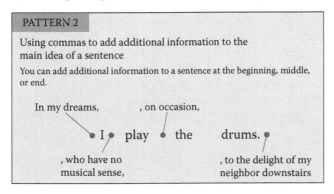

FIGURE 4.1

A visual approach to grammar instruction: The new handbook sought to visually represent patterns in sentence structures. (From Wysocki, A.F. Lynch, D.A. (2010). *The DK handbook*, 2nd ed. Copyright 2011 Pearson Education, Inc. Reprinted with permission.)

innovation as a necessary and viable market strategy. This insight was critical to our long-term success at making usability studies a routine part of the development process and encouraging the company to shift from an accommodationist, content delivery product development model. Adding to the layers of rhetorical complexity already visible in the scenario surrounding this usability study is the fact that the 2006 student-user test marked the first time this publisher had commissioned a formal usability study as part of its textbook production process. Educational researchers have tested textbooks independently of publishers and post facto, but in terms of the actual production of texts, market and user feedback in college textbook publishing has traditionally come from editorial reviews (written by teachers, not students) and "class tests" (where a teacher uses sample chapters of a new textbook in a classroom setting). The publisher's decision to seek and obtain direct, scientific feedback from the students who were the intended end users of the product marked a significant departure from its traditional product development practices. It is possible, even probable, that many publishers would have dismissed the results of the 2006 study, or been defensive or even hostile toward the report's findings. After all, the study did conclude that, from a usability standpoint, the prevailing product models in the lucrative handbook market were essentially failing to meet the needs of their student users. Instead, because we were part of the production process rather than post facto validation testing, we were reasonably successful in convincing the publisher to see this failure as an opportunity.

By the time the first edition of the new handbook was published in 2008, the usability study had become a key part of its marketing story and strategy. A second usability study, based on the published first edition, was completed in October 2008. That study, combined with the insights gained from the 2006 study and the many conversations we have had since its completion, informed the development of the second edition, published in January 2010. Usability studies have become, in this case, an integral part of the product development process. The findings of the 2008 study were fed directly into the work of the authors and the publisher as part of the course of the revision process for the second edition.

Our purpose in what follows is to place the results of the 2006 and 2008 usability studies in the context of our experiences of presenting them to the various stakeholders in the publishing process. Our account has three primary points of focus:

> *Introducing usability testing into an industry that was relatively unfamiliar with such testing.* As was previously mentioned, college publishers traditionally used editorial reviews from teaching faculty as the primary source of user feedback. Student feedback was limited to occasional "class tests" (filtered through the instructor) or focus groups

on campuses, usually conducted for marketing rather than editorial purposes. The 2006 handbook study was to our knowledge the first time that a formal usability study was conducted by this handbook's publisher. This meant that we spent substantial time and energy explaining to management at the publisher what usability testing is and how it differs from other forms of market feedback.

The importance of strong collaboration between the usability tester and the product developer. None of the success we have encountered so far would have been possible without a close collaboration between the usability professional (Howard) and the product developer (Greer). While the role of usability testers has often been an outside "vendor" role, in this case it was essential to build a collaborative approach that enabled us to work together to "sell" the publisher on the ideas we were developing around the results of the study. Because the usability tests so fundamentally challenged the inherent conventions in handbook design, we had to repeatedly reinforce the message that student usability needed to be a driving concern in the product development process.

Innovative product development. The discovery of unexpected complexities in the handbook study enabled us to revise our thinking about the process for handbook development. We were able, at least in part, to move from a content delivery model to a new model informed by user experience design. The fact that students had so much trouble navigating and using conventional handbooks pointed the way to significant new opportunities for redesigning a handbook from the ground up. Because the first edition met and exceeded the publisher's expectations in terms of market success, we are now in a position to continue to advocate a new model of product design based on principles of usability and user experience design.

Unexpected Complexity

For the faculty who assign them and the publishers that sell them, college writing handbooks are a stable commodity. Conventional wisdom in the marketplace is that all handbooks are more or less the same; indeed, the organization and presentation of information in most handbooks are based on a familiar pattern. The tasks for which students are expected to use handbooks represent practices for which most writing teachers have acquired years of experience: evaluating source materials for a research project; citing and documenting sources properly in a research paper; editing and revising

Formal Written English prose; and correcting matters of written style, usage, and convention.

Because teachers are familiar with these practices, and with the conventions shaping typical handbook architecture and layout, they are apt to perceive such tasks as relatively simple. The 2006 usability study revealed, however, that these tasks and practices are in fact substantially complex (Howard, 2008). It is in the confusion between complex and simple problems, and the misapplication of simple frameworks to complex problems, that we begin to discover the gap between teacher and student user experiences, as well as a clue about where to start rethinking the product development process.

Definitions of complex problems vary, but according to Albers (2003), "In complex problem solving, rather than simply completing a task, the user needs to be aware of the entire situational context in order to make good decisions" (p. 263). In doing a comparative usability test of two handbooks for students in freshman composition classes, the 2006 study provided a very interesting case in point for Albers's statement. Students reported that they found the writing handbooks were easy to use, but they chose incorrect solutions to problems because the handbooks did not help them understand the complexity of their task and did not provide guidance on how to choose wisely within that complexity.

Test Methodology

The 2006 study used active intervention, think-aloud protocols that asked users to complete tasks with comparable chapters from both handbooks. Twelve participants from six different introductory composition classes were recruited for the study: six from two-year colleges and six from a four-year institution. We decided to use twelve participants because Virzi (1992) and others have shown that five to six representative users will capture up to 90 percent of the major usability errors in a product . Because our clients were familiar with focus groups and other types of marketing studies, we were encouraged to double this number in order to increase our clients' confidence in the data. All of the participants were 18 and 19 year olds enrolled in their first semester of college, their average high school GPA was 3.4, seven participants were male, five were females, and three were students of color. After going through the informed consent statement agreement and giving their permission to be videotaped, participants in the study were asked to participate in a pre-scenario interview. This interview collected basic demographic information about participants' experiences with high school English classrooms. The questions were intended to help us gauge whether the participants were representative of typical first-year composition students.

In addition to basic information about participants, the pretest interview sought to learn about the students' attitudes toward writing courses they have had in the past and their performance in them in case it became necessary for us to gauge their level of writing apprehension. Since we were also particularly interested in the development of writing handbooks, they were also asked to describe if they used writing handbooks in previous writing courses and, if so, how they have used them and what attitudes they had toward them.

After collecting this basic information about the participants' pre-scenario attitudes, we introduced participants to the scenario. To help with their comfort level, participants were reminded several times that we were not "testing" them; instead, we were testing the usability of the handbooks being examined. They were instructed to talk out loud and to verbalize their thoughts as they attempted to use the textbook to perform the tasks provided. To control for first-use bias, half of the participants used the new visual handbook prototype first, and the other half used the traditional handbook first. Participants were videotaped as they performed the tasks, talking aloud and saying what they were thinking as they worked. Over twenty-seven hours of video were recorded, and the users' behaviors were subsequently coded by two independent raters and analyzed. The accuracy of participants' performances was also assessed during these protocols. Two pilot tests were also conducted to ensure that the data collection instruments were functioning correctly and that the tasks we provided made sense to the participants and yielded the data we were seeking (data from the pilots were not reported).

In the first task, users were asked to assume that they were working in their current composition class on a research paper for a grade in the course. The scenario required that they create a works cited entry using MLA documentation style. The researchers provided books with passages marked in them that users had quoted in the research paper they had written.

In the second scenario, users were asked to identify comma errors in a paragraph and to provide the page numbers from the handbooks that provided information about the correct comma usage. In order to attempt to provide a wide range of comma errors, we provided students with a sample student essay that was littered with comma errors typical of student writing. The grammatical errors we asked students to identify were based on the twenty most common grammatical errors as identified by Lunsford and Lunsford (2008) in order to ensure that we were asking students to identify grammar problems they were likely to encounter routinely.

The third scenario gave users a research-paper prompt that included both a research topic and a specific audience for the paper. Users were then given possible sources for the research paper, and they were asked to use the handbooks and to indicate if the source was acceptable or unacceptable, or if more info was required. Users were again instructed to provide the page numbers from the handbook that enabled them to make their determinations.

Each scenario increased the complexity of the task to be performed. The citing sources scenario essentially asked users to follow a model in order to complete a task; users simply had to imitate the pattern for citing an MLA source. Identifying nontrivial comma errors was slightly more complex because it asked users to apply grammatical rules to a situation and to make a judgment. And the final task, evaluating possible sources for a library research paper, was the most complex because it required an understanding of the rhetorical situation in which the sources would be used. Users had to make a judgment about the appropriateness of a source based on the exigency for the research paper, the audience for the piece, and a wide variety of other environmental factors.

Selected Results

Because we were mainly interested in improving the new handbook prototype, most of our coding and data analysis focused on addressing usability issues that were unique to the publisher's prototype and not of general interest here. Hence, we will focus only on a small slice of the data that illustrates how the unexpected complexity of our tasks revealed itself.

As Redish (2007) has suggested, traditional ease of use measured alone offered no real sense of the complexity involved in the tasks. Overall, users reported that they preferred the new prototype's visual ease of use to the more verbal approach used in the traditional handbook. When asked to rank the "overall" ease of use for the two texts after they had actually used both handbooks, nine of the twelve users preferred the visual prototype, and nine of the twelve indicated that they would recommend it to their teachers for their entire class. Users recommending the prototype also appeared to have a stronger preference for their recommendations than the three recommending the traditional handbook. Users were asked to indicate the strength of their preference on a scale from 1 to 10, where 1 indicated that they thought the text was *slightly better* and 10 indicated that it was *vastly superior.* The nine users recommending the new handbook prototype averaged 7.44 (standard deviation was 2.35), and the three recommending the competing handbook averaged 6.00 (standard deviation was 1.0).

Yet, while users' clear preferences for the visual approach and their overall "ease of use" scores are suggestive, it would be an error to conclude that the prototype's visual approach was more "usable" than a verbal approach used by the market leader at that time. The ease of use evaluations above do not give a complete picture of the usability of the texts because the users' evaluations must also be considered in light of the question of whether users were actually able to complete the task "successfully."

Users may initially give a positive evaluation of the ease of use for a product that they thought had helped them complete a task, but if a text misled them by allowing them to believe that they had finished the task when, in reality, the task was only partially completed, then the users' initial assessments are less valuable as a measure of usability. It is at this point that the issue of complex problem solving manifested itself in our study, and it is by means of content experts' assessments of the quality of the users' performances that researchers who are working with new clients can observe when complex problems may be disrupting the findings in a traditional usability study.

In this study, both the prototype and the traditional handbook excerpts failed the users when it came to helping them successfully complete acceptable works cited entries for the works provided. All twelve users failed to provide a works cited entry that was judged completely satisfactory by college-level composition instructors. For example, users failed to list the authors of an article in an anthology, they failed to list a title of the essay, they failed to include the number of an edition, they failed to provide the page numbers for articles, and so on.

Admittedly, the books that users were tasked with citing were challenging. But a strength of the study was that the task was also realistic. Pilot testing revealed that students wouldn't use the handbooks to cite simple sources like a single-authored book, so the researchers were compelled to use challenging titles. One of the works that users had to cite was a corporate author where the corporation was also the publisher. The other was a book chapter with three authors that was published in the second edition of an anthology edited by four people. It was necessary to challenge the users so that they would actually need to use the handbooks to complete the tasks, and it is the case that all the information needed to cite both of the texts is provided in both handbooks. The books were intended to support precisely this sort of challenging citation, so if the handbooks were to be considered truly usable, it seems legitimate to have expected that users should have experienced more success than the near-total failure we observed.

Furthermore, we saw similar performance issues in the responses to the punctuation scenario. Although the findings were less problematic than the MLA task from a performance perspective, our study found that users consistently failed to correctly indicate when the use of commas was required, not required, or optional, and they also failed to provide the correct page number from the texts where they obtained the information. For example, eleven of the twelve users incorrectly stated that a comma was required rather than optional after short, two-word introductory clauses, and once again, this finding was observed for both handbooks. However, the users were completely unaware of this deficiency in their performance and did not consider this factor when they assessed the "usability" of the handbooks.

So why did users of *both* handbooks do so poorly without realizing it? Obviously, if only one text had poor performance indicators, then this would have suggested that the information delivery techniques may have been at issue. The fact that both failed our users was crucial in recognizing that we were dealing with unexpected complexity. Users were dealing with complex problems which the books were treating as simple imitation problems and which, consequently, the users also thought were simple problems. It turned out, however, that citing an MLA source had unexpected complexity because users lacked sufficient knowledge of their "situational context" (Albers, 2003, p. 263). Users didn't know how to differentiate between an encyclopedia article and the anthology they were actually supposed to be citing. They also lacked an understanding of the term *corporate author* and didn't realize that corporations can also be publishers. It turns out that, if users can't identify the type of source they're citing correctly, they can't use a model to correctly solve a problem. Simply put, if someone tries to create a works cited entry for an article in an anthology using an encyclopedia entry as the model, they will fail to generate a correctly formatted citation.

By way of contrast, users performed more successfully on the more complex task that we gave them when they were using the prototype. All of the users of the evaluating sources chapter in the new visual handbook performed successfully. Unlike the simple modeling approach used in the MLA material or the simple pattern recognition approach used in punctuation, the material provided here mainly used "If … then" scenarios that users could play (Flower, Hayes, & Swarts, 1983). Through the use of mini-stories about fictional students attempting to solve the same kinds of evaluation problems that the handbook users faced, the handbook illustrated the complex variables and situational contexts that needed to be considered in order to solve the problem. The handbook treated the material as a complex problem and *created a clear role for users to play as they used the text through the empathetic storytelling*. Because the users understood the role they were supposed to play, they also understood the logic they needed to follow in order to make the decisions necessary to complete the task at hand. Put another way, the textbook used storytelling to construct the users in ways that provided them with an interpretive framework for success. Through the use of role-playing, users were adapted to the content rather than the publisher attempting to adapt the content to them.

Implications

The 2006 study provides empirical evidence that validates Redish's (2007) observation, "Ease of use—what we typically focus on in usability testing—is critical but not sufficient for any product" (p. 104). The ease-of-use indicators in this study, both quantitative and qualitative, suggested that our users found the prototype handbook and its visuals to be an "easy-to-use solution" to their perceived needs. And the publisher would very likely have been entirely satisfied with the ease-of-use data we provided. However, had the study's design solely focused on what the publishers told us they wanted, we would have failed both them and the users. The true complexity of the tasks would not have been revealed, the lack of "utility" in the handbooks would not have manifested itself, and we would never have seen the need to "think outside the box" of traditional handbook design. We would have lacked the cognitive dissonance that motivated at least some of the creative thinking that ultimately led to the successful handbook published in 2008.

Collaboration and Rhetoric

The 2006 usability study provided us with two essential bodies of data that provided a framework for making revisions to the content and design of the working prototype handbook. First, the study provided empirical data showing specific pages in the handbook where users reported confusion or difficulty. Second, the discovery of the unexpected complexity provided a model for better understanding and describing handbook user behavior.

Once we had shared and discussed the results of the usability study, we were left with the task of finding a way to present it to the authors and key decision makers at the publishing house. Howard's user testing had highlighted the unexpected complexity of the tasks, and in so doing had opened up a model for developing user experience frameworks that might succeed where the existing conventions had failed. Yet we faced a significant rhetorical challenge: how to present potentially bad news (the new handbook did not support students' uses in the way the publisher wanted it to). In an early cover memo to the editorial and management team, Greer wrote, "We may in fact be at a turning point in the evolution of the handbook genre, where we can now see that some of the tried-and-true conventions for presenting information are no longer working for students. So we have the opportunity (and the challenge) of designing something that will work."

Our strategy for framing and presenting the usability study results depended on two key factors. First, it was essential that the two of us were speaking with a single voice, writing as informed observers and advocates for the student test subjects. Second, it was essential to frame the rhetoric of our discussions not as a simple recital of the isolated "errors" in the handbook prototype, but as a new approach to the development and design of the handbook, based in a body of empirical data about actual student experiences.

We used Peter Block's "Balance of Responsibility" approach to achieve this (Block, 2000). As Figure 4.2 illustrates, our goal was to share our responsibility for each decision made during the planning, data collection, and reporting phases of the usability testing as much as possible. This is somewhat unique since typically, when an external usability consultant is brought into a corporate environment and tasked with designing usability studies, the consultant's expertise makes him or her "responsible" for decisions about, for example, what kinds of research methodologies to use in a particular situation. Indeed, frequently a client will adopt the attitude that the testing has been "outsourced" and the usability professional should go away, do the testing job, and report back with results and recommendations when done.

We assiduously sought to avoid this "expert vendor" model and sought instead to share responsibility equally for all decisions about the goals for the study, the research questions to be pursued, the design of data collection instruments, and the reporting of results. In fact, we even wrote into the contract that the publisher would allocate personnel resources to a "liaison" who would work with Howard to ensure that all decisions had a balance of responsibility. This ensured, on one hand, that Howard always had first-hand information about the publisher's needs and could design studies that would yield the highest return on investment, but it also ensured, on the other hand, that Greer and other members of the product development team understood the full range of research methods available to them and how they could best interpret and utilize those results. Balancing responsibility took considerably more time on the front end of the study because it meant that both sides had to educate each other and that means for adjudicating decisions had to be negotiated. However, the time we invested in each other

	Client's Sole Responsibility	Shared Responsibility	Consultant's Sole Responsibility
• Define initial problem	| _____	↓	_____ |
• Decide appropriateness and feasibility of project	| _____	↓	_____ |
• Establish research goals	| _____	↓	_____ |
• Design methodology (user tasks, participant profiles, data collection instruments)	| _____	↓	_____ |
• Collect, code, & analyze data	| _____	↓	_____ |

FIGURE 4.2
Block's "balance of responsibility": In a shared responsibility model, decisions about the goals and design of a usability study are shared equally by client and consultant. (Adapted from Block, p. (2000). *Flawless consulting: A guide to getting your expertise used* (p.36). San Francisco: Jossey-Bass/Pfeiffer.)

paid huge dividends when the data came because not only were the results obviously actionable to the design team, but also the members of the team had sufficient understanding and clarity about the results that they were able to convince other internal decision makers to invest in the recommendations resulting from the work.

The rhetoric that we came to use to frame our understanding of the test's results was based in two primary claims. First was that the success of the new handbook prototype in the evaluating sources task, and the failure of both handbooks on the works cited task, could be partially attributed to an incomplete understanding of users' goals and task environment. It was obvious to everyone involved that improving an author's understanding of the users' needs was likely to result in more usable information products, hence the mantra "Know thy user." This observation was largely the motivation behind the publisher's decision to commission the usability studies in the first place.

What was less fully understood is the role that authors play in actually constructing the use or task environment for users. All too frequently, we tend to think in terms of accommodation of users' needs, and we tend to overlook the important role that authors and designers play in the construction of users' task environment. Indeed, the movement away from theories of "user-centered design" in the 1990s toward "user experience design" is largely a recognition of this complex negotiation between accommodation of users on the one hand and construction of user experiences on the other. Thus, our second primary claim was that successful texts and information products create roles and provide interpretive frameworks that users can deploy in order to successfully complete tasks and achieve their goals for an information product.

The usability studies revealed that handbook users scanned pages for examples that matched the mental models they had for patterns and only stopped to read material when they found patterns that matched those models. In what we are calling the accommodationist model, then, developing a new, more usable handbook would be relatively simple. Isolate those pages where users had a hard time finding the pattern, and make the patterns more visible on those pages. Accommodate the user's needs, and make that the basis of your product development strategy. But what if your users are matching the wrong patterns because of a more fundamental misunderstanding of the context and task environment? Howard's interpretation of the study's results was that this was exactly the problem the test subjects were encountering. The students saw their task as a simple pattern-matching process when in fact it was a far more complex task that required a working knowledge of types of sources (or comma errors)—knowledge the students did not have. In order to successfully build a more usable handbook, we would need to move away from an accommodationist model toward what we are calling a *constructivist* model. In a constructivist model, the role of the author and the text in constructing the user's experience is emphasized.

Howard's summary report on the 2006 test included the following:

> The study further found that the evaluating sources section of the *DK Handbook* did accomplish what the MLA and comma sections did not. Through its judicious use of questions followed by detailed discussion, effective page design and layout, [and] consistent and logical use of fonts, the evaluating sources section of the *DK Handbook* created a user experience that enabled users in this study to complete tasks more successfully and to report greater satisfaction with the ease of use. Perhaps this section was more complete and more polished than the other 2 sections used in this study (to avoid bias, the researchers have not communicated with the authors and do not have information about the status of the manuscript, its proposed directions, scope, etc.). However, if this section represents the authors' goals, then this bodes well for the usability of future iterations of the text.
>
> Increasingly, students are going online and using web-based resources for information about issues related to research paper development. During the pre-text interviews, for example, several of our users indicated that they used sources like Easybib.com, the Purdue OWL, and Citationmachine.net to produce their works cited entries. Many modern students appear to want a "user experience" in their textbooks which approximates what they find on the Web. Because of its use of well-conceived questions, guided decision-making practices, and a page layout that allows for "hypertextual" reading on the printed page, the Evaluating Sources section of the *DK* prototype should serve as a model for other sections of the book.

With the benefit of hindsight, we can now recognize that this was the crux of the argument that ultimately proved successful in shaping the revisions made to the prototype handbook following our discussion of the test results. Without fully realizing it at the time, we had isolated a rhetorical framework that would allow us to bring a student-user experience perspective into the center of the product development cycle, and in so doing, to revolutionize, potentially, the entire product development process. We realized that resolving the issues of unexpected complexity was simultaneously forcing us to move our thinking about the product development process from an accommodationist model to a more innovative constructionist model. In other words, we were no longer simply accommodating problems that users had and fixing page-level errors in the manual; instead, we were constructing user experiences that were creating contexts in which users could solve complex problems.

Driving Innovation in Product Development

In part because it has been relatively slow to adopt user testing and a user experience design perspective, the print textbook industry remains anchored in a relatively traditional product development and production model. The process is linear, and is based on a series of assumptions about how content is developed and delivered.

As Figure 4.3 illustrates, the existing product development process begins with market feedback (expert content reviews, focus groups, and class testing, as described above). Then the content is developed based on this feedback. In practice, this generally means that the authors work collaboratively with an editor to deliver a complete manuscript, which is in turn handed off to a production team. The production team is then responsible for the design and layout of the final product.

Our 2006 and 2008 usability studies have opened the door to a range of new approaches to thinking about the development and design of educational texts. In order to support a new generation of student users, who, our studies showed, read and process textual information in new ways, the product development cycle for print textbooks needs to model that used in industries like Web design and new media development, where user experience design principles are more fully accepted and widely employed.

Through usability testing, we recognized the need to move from a product development model that no longer focused on delivering content simply adapted to users' needs. We recognized that we needed a product development process to design textbooks that constructed user experiences more analogous to the design of video games. The failure of adapting content to what we thought users needed compared with the successful use of storytelling techniques used to help students learn to solve complex problems made us realize that we had to construct experiences that encouraged our users to play productive roles in our instructional texts rather than merely adapt content. We needed to find ways to move students closer to the content rather than vice versa. The constructionist approach we advocate here is in some ways similar to the iterative usability model proposed by Genov (2005, p. 20). Genov applies control systems (or cybernetic) theory to the explanation of complex behavioral systems in a variety of fields, emphasizing, as we do here, the importance of an iterative feedback loop. Figure 4.4 illustrates the new constructionist process model we have been

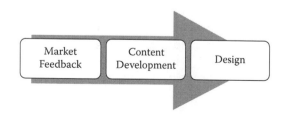

FIGURE 4.3

The accommodationist model of product development: The process is linear, and is based on a series of unexamined assumptions about how content is developed and delivered.

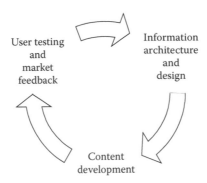

FIGURE 4.4
A redesigned constructionist model of product development: This cyclical model builds user feedback into a recursive development process and encourages users to play a productive role in the construction of the user experience.

advocating throughout the publishing company. This new approach has been attempted for psychology, public speaking, developmental writing, and developmental mathematics textbooks.

Conclusion

Our successful collaboration hopefully has demonstrated how a pilot project that was originally intended only to demonstrate the value of usability testing to the revision of textbooks eventually impacted the product development process for a major publisher. As we describe here, our success can be attributed to three factors. First, our decision to use Block's "Balance of Responsibility" model so that we were forced to collaborate throughout every phase of the project was a key element in our success (Block, 2000). Second, the decision to design a research method that compared a prototype against the market leader and to supplement traditional "ease-of-use" measures with performance measures was critical in helping us to capture the unexpected complexity of the tasks we asked users to perform. Third, the decision to treat the unexpected complexity we found as an opportunity to innovate rather than merely to adapt was also significant. Usability testing helped us realize that we had been victimizing ourselves by trying to adapt the material in order to make problems "simple" for the audience, and when we saw that it was actually more effective to attempt to control the user experience and to change the ways that users approached material, we realized we needed a constructivist process model to ensure that we could repeat the process with future textbooks and instructional materials.

Reader Take-Aways

- Helping designers and authors explore the complexity in their products can enable clients to realize new opportunities for innovation and can help them create new strategies for market success. In this case, helping the publisher to recognize the complexity involved in seemingly simple problems like citing MLA sources and in responding to commas also helped them design product that gained tremendous market share in a market that had been dominated by one product design approach for decades.

- Complex problems for users can masquerade as simple problems for authors, editors, designers, and even expert reviewers with experience in the tasks being performed. This can blind product developers to alternative opportunities for product delivery. We discovered that one methodological way to recognize when this is happening is when users give a positive evaluation of the ease of use for products that they believe help them complete a task but that, in fact, mislead them by allowing them to believe that they had finished tasks that are inappropriately or only partially completed.

- With new clients, usability specialists need to prepare for the likelihood of encountering complex problems masquerading as simple ones and ensure that the studies we design use sound methodological triangulation techniques, including content experts' assessments of the quality of the users' performances and, when possible, head-to-head comparisons with competing products.

- Users in our studies assumed that they were dealing with a simple problem, and once they found what they thought was the simple solution, they didn't look any farther for more complex answers. To avoid giving users a false sense of success which can ultimately lead to poor user performance, designers need to present tasks to users in ways that help them understand where to begin the decision-making process. When users understand the role they are supposed to play, they also understand the logic they're supposed to follow in order to make the decisions necessary to complete the tasks at hand. Our studies found that modeling complex problem-solving behaviors through the use of "scenarios" may lead to more effective performance from users when it is simply not possible to capture or replicate all of the potential variables in a situational context (Flower et al., 1983).

- Exploring the depths of complexity in seemingly simple tasks can break a company out of the accommodationist model of product design where the authors' or designers' role is seen as merely one of adapting technology or delivery systems to accommodate users' needs. Exploring the complexity of seemingly simple tasks can help designers perceive new means of constructing user roles that position users for the successful use of products. Rather than merely tweaking the "usability problems" in a product's delivery system, exploring complexity in user tasks encouraged our company to redefine the designers' role as one that was responsible for *constructing* the users' entire experience and positioning users to the content.

References

Albers, M. J. (2003). Complex problem solving and content analysis. In M. J. Albers & B. Mazur (Eds.), *Content and complexity: Information design in technical communication* (pp. 263–284). Mahwah, NJ: Lawrence Erlbaum.

Block, P. (2000). *Flawless consulting: A guide to getting your expertise used*. San Francisco: Jossey-Bass/Pfeiffer.

Flower, L. J., Hayes, R., & Swarts, H. (1983). Revising functional documents: The scenario principle. In P. V. Anderson, R. J. Brockmann, & C. R. Miller (Eds.), *New essays in technical and scientific communication: Research, theory, practice* (pp. 41–58). Farmingdale, NY: Baywood.

Genov, A. (2005). Iterative usability testing as continuous feedback: A control systems perspective. *Journal of Usability Studies, 1*(1), 18–27.

Hacker, D. (2007). *A writer's reference* (6th ed.). Boston: Bedford.

Howard, T. W. (2008). Unexpected complexity in a traditional usability study. *Journal of Usability Studies, 3*(4), 189–205.

Kress, G. (2003). *Literacy in the new media age*. London: Routledge.

Lunsford, A. A., & Lunsford, K. J. (2008). Mistakes are a fact of life: A national comparative study. *College Composition and Communication, 59*(4), 781–806.

Redish, J. (2007). Expanding usability testing to evaluate complex systems. *Journal of Usability Studies, 2*(3), 102–111.

Virzi, R. (1992). Refining the test phase of usability evaluation: How many subjects is enough? *Human Factors, 34*, 457–468.

Wysocki, A. F., & Lynch, D. A. (2009). *The DK handbook*. New York: Pearson Longman.

Section II

Theorizing Complexity: Ideas for Conceptualizing Usability and Complex Systems

Faced with fundamental problems brought on by an increased awareness of the complexity of usability systems, as well as an understanding that new approaches are necessary to address such complexity, usability practitioners recognize that they need a new theoretical framework for informing their work. The chapters in this section of the book offer innovative ways to conceptualize usability and complex systems.

Brian Still theorizes that we should look to ecology to revitalize the small-scale usability test. Ecologists, Still points out, have long used micro studies of specific food webs to account for the dynamic forces at work in shaping larger ecosystems. Such a mind-set arguably can be useful to usability practitioners as they move away from problem finding and more toward analyzing the many ways that users synthesize various information elements in complex, and changing, environments.

Michael J. Albers argues that a typical approach to usability tests identifies problems but does not effectively recognize how readers figure out what information is relevant and how it connects to the problems discovered. In complex situations, the reader does not gain an integrated understanding of information, which results in a lack of situational clarity that leads to poor decisions and thus errors. Albers focuses on what usability tests can do to help people understand information relationships.

Focusing on health care and also industrial engineering, Vladimir Stantchev asserts that rather than a single test or even a series of iterative tests, we need instead to pursue a continuous evaluation of complex systems that relies on the right mix of empirical and technical methods. Finally, Julie Fisher explores e-commerce websites. Specifically, her review of current research enables her to offer guidelines for employing usability testing of complex e-commerce systems.

5

Mapping Usability: An Ecologial Framework for Analyzing User Experience

Brian Still

Texas Tech University

CONTENTS

Introduction 90
Usability and Ecology 93
    The User Web 93
    Mapping the User Experience 95
    Embracing the Hypothetical 97
    Making the Business Case for Evaluating Complexity 100
    Repurposing the Small-Scale Usability Test 101
Discount Testing versus the Ecological Approach 103
    Other Methods for the Ecological Approach 103
Conclusion 105
Reader Take-Aways 106
References 107

Abstract

How we perceive usability testing, and what we hope to accomplish from conducting a series of small steps as part of an iterative process, must change, especially when we are tasked with evaluating a complex system. This chapter proposes that usability practitioners look to ecology for guidance. Ecologists for some time have relied quite successfully upon small-scale studies, such as the ones we conduct, not to solve easy problems but to engage in the ongoing monitoring of very unstable, very complex ecosystems. Drawing on research from ecology and examples from previous usability testing, this chapter details how an ecological framework for usability would work.

Introduction

During preparation for recent testing of a client's online textbook product, our usability research lab conducted a series of site visits to supplement other work we were doing to determine representative user profiles. Our client had provided us with a fair amount of useful data already about users, including video of interviews it had conducted at university campuses in the United States. In these interviews, students were shown an electronic wireframe of the online textbook and then asked a series of contextualized questions, such as "If you were looking for _____ in the book, how do you think you should be able to find it?"

The only problem with all of these very good data, along with the other data we collected via pretesting surveys, our own interviews, as well as other user analysis methods, was that not a single piece of data was extracted from students in any sort of environment where they might actually use an online textbook. So when we visited their dormitory rooms, the library, or wherever else they might do their work, we were surprised by just how that work was done. Distractions abounded not just from the noise of others around them, but also from what the students generated on their own. Unlike the typical lab setting where the user is walled off from distraction and given a relatively pristine work environment, the places where our users for this study worked were far from distraction free. Cell phones were on and used to take and make calls or, more frequently, text messages. Multiple applications, most having nothing to do with the application in question, ran on the computer, including a variety of popular instant-messaging software. Facebook, an online social community, was also actively used.

Had we decided to forgo this field research and ultimately conduct a typical laboratory test, this user behavior would not have been revealed. We might very well have found the right users, conducted perhaps a cognitive walk-through or another time-tested technique to understand key tasks that users might perform, and think-aloud protocol during testing would have yielded information to supplement other data we collected. But what would have been missing was an understanding of the environment that users occupied when using the product—an environment that clearly had the potential to make simple task performance more challenging, if not more complex.

In a nutshell, this is the problem that plagues the typical usability test in the lab. As Redish (2007) noted, this kind of testing "is too short, too 'small task'-based, and not context-rich enough to handle the long, complex, and differing scenarios that typify the work situations that these complex information systems must satisfy" (p. 106). The fact is that by visiting students, we were able to devise more open-ended testing scenarios. We tested outside of the lab. We didn't give the students specific tasks to perform. We asked them only to use the product to complete work they had to do for the site, such as revise a paper, finish homework, or prepare for an exam. And finally we told them to feel free to do what they always do—answer the phone, text message, watch television, and otherwise engage in any of the distractions that typically interrupt them as they work.

The result was atypical data. Dwell times, task completion rates, and other traditional measures carried little weight or weren't applicable. Think-aloud wasn't natural, so we let them work, take notes, and then carry out retrospective recall at the end—a method that Redish (2007) encouraged us to consider using for analyzing complex systems (p. 107).

Using the System Usability Scale (SUS; Bangor, Kortum, & Miller, 2009) as a post-test measurement of satisfaction also didn't feel like an effective tool. Barnum and Palmer (this book, Chapter 12) already do an excellent job of critiquing the SUS, and I won't repeat its shortcomings here. However, in our past work we often noted the glaring discrepancies between high SUS scores and actual task performance success, especially in student user populations who were tasked with using products in complex system environments. Often these users really liked the product but couldn't use it successfully. Avoiding this problematic contrast, we stuck with a more open-ended interview after testing. Yes, this didn't provide easy-to-graph answers to specific problems. But in dynamic, complex systems there is never a set of finite problems. New problems emerge all the time or change their shape as multiple forces—user, product, environment, and so on—interact with each other.

Although we didn't acquire easily analyzable information, what we did learn was surprisingly enlightening. Users not only were distracted while using the product because of the many other activities that they engaged in, but also ironically relied upon some of the distractions, such as instant messaging or other community networks outside the product, to help them find information. Fascinatingly, the very things that cre-ated noise for them also served as resources. Yes, instant messaging with friends delayed efficiency, but the complexity of the system was such that when students felt stuck they stopped silly banter with their friends and instead asked for serious help, which often came as quickly as the ban-ter. Ultimately, we recommended to the client that they consider add-ing instant messaging as well as a community forum to their product and then test their usability—both suggestions for making the complex system design work that would never have been considered in typical testing.

It is not that there is a problem, in my opinion, with smaller tests. There is also still a value, I believe, in laboratory testing. But how we perceive testing, and what we hope to accomplish from conducting a series of small steps as part of an iterative process, must change, especially when tasked with evaluating a complex system. We cannot assume that iteration alone will get us to a better product, not when that product, the user, and the entire complex system in which they both reside—which influences them and, in turn, is influenced by them—are in constant flux. As Howard (2008), among others, makes clear, we risk failure if we assume simplicity and operate that way when everything around us tells us we are working with something complex.

What, then, should be our intellectual approach for rethinking the purpose of the typical, small-scale usability test so that it still can serve as a viable resource for us when evaluating complex systems? How can we, given the impetus in our field and from clients for affordable and fast testing, and for recommendations that address specific problems with easy solutions, make the case for delivering testing results that might not be as clear-cut but may still, in reality, be more accurate and useful?

It isn't that we should do away with the small-scale test. Limited resources, such as time, funding, or access to user population, necessitate that we still try to be productive with the smallest number of users possible, especially when delivering results to clients. To better manage a product, even if it is complicated in its design, we also must continue to evaluate just particular parts of it and not the entire, overwhelming thing. However, this should not mean that we are stuck somehow with an outmoded evaluation method in which we are damned if we do use it and damned if we do not. Rather, we need only to repurpose it. In fact, I argue in this chapter that if we look outside our field to another discipline, ecology, we see practitioners that for some time have relied quite successfully upon small-scale studies, such as the ones we conduct, not to solve easy problems but to engage in the ongoing monitoring of very unstable, very complex ecosystems, full of multiple elements that are intertwined and interdependent.

Ecologists take a micro approach to examining such systems, just as usability practitioners limit their focus. However, for the most part similarities end there. For ecologists, "microcosm studies," Worm and Duffy (2003) explained, "simplify food-web research into semi-isolated units such as small springs, decaying logs and cowpats. Principals gleaned from food-web microcosm studies are used to extrapolate smaller dynamic concepts to larger system[s]" (p. 628). In other words, small studies are not seen as ends in themselves. They are not isolated from their environment. Rather they are illuminating because of how they are used.

If given the proper context, and seen as a tool not for finding a limited set of fixed problems but instead for extrapolating greater understanding about the overall complexity of the entire system, small-scale usability tests can serve the same purpose. For the remainder of this chapter, I discuss further how an ecological framework for usability would work, I provide examples from previous studies as support, and then I also detail reasons for why usability as ecology should be considered if not adopted as a methodological framework for the usability testing of complex systems.

Usability and Ecology

Begon, Townsend, and Harper (2006), when discussing the study of ecosystems, wrote the following:

> [C]ommunities of organisms have properties that are the sum of the properties of the individual denizens plus their interactions. The interactions are what make the community more than the sum of its parts. Just as it is a reasonable aim for a physiologist to study the behavior of different sorts of cells and tissues and then attempt to use a knowledge of their interactions to explain the behavior of a whole organism, so ecologists may use their knowledge of interactions between organisms in an attempt to explain the behavior and structure of a whole community. (p. 467)

Arguably, the same can be said of usability practitioners. If we see our work, even when carrying out small-scale tests, as an effort to reveal more knowledge about the behavior and structure of an entire community, then the value of the small-scale test is strengthened, not diminished. Such an approach also will lead us to appreciate that just as the small test can illustrate larger issues, larger issues—or elements of the larger community beyond the users and products spotlighted in the small-scale test—can and do influence micro-level issues. At the same time, it will illuminate more clearly the significance of good user-centered design. When we improve a product or make it usable, we do so not just to solve immediate problems discovered during an isolated test. What we fix or don't fix has ramifications beyond itself.

The User Web

Beginning with an ecosystem, ecologists drill down first to food webs, which comprise a number of interconnected food chains. A food chain might be as simple as the wolf that feeds on the deer that feeds on clover. Ecologists know, however, that any impact on this chain will disrupt the ability of one part of it to thrive, causing other parts of it to diminish. At the same time, as Figure 5.1 demonstrates, what goes on in this chain impacts other chains in the food web, causing negative effects to it which, in turn, influence the entire ecosystem.

If there aren't enough wolves to cull off weaker deer, the deer population explodes, resulting in less clover and other natural foods and cover for other species, like squirrels, as the growing deer population consumes larger quantities. Even this fact rebounds to impact the deer, which because of their increased size eat away their food so much that their population cannot be sustained.

Let's extrapolate this process to apply to usability and complex systems. We tested for a client last year developing a comprehensive online reference

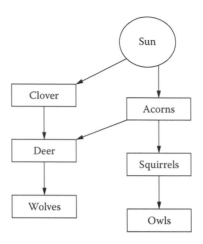

FIGURE 5.1
Deer–wolf food chain.

for university-level composition students. This system was more than an online textbook, although students based on their permitted access could find and use a variety of relevant e-versions of textbooks. Students could take quizzes for a grade, check their grades online for these quizzes and other submitted assignments, and search for examples to help with writing, documentation, and other activities related to composition. The system actually wasn't just one site but instead comprised a number of different sites and really other systems. These often had their own design, including different layout and navigation schemes. And many opened up as separate windows when students searched for them. In short, this system by its very design would qualify as complex.

What allows us metaphorically to consider it an ecosystem is the fact that multiple kinds of users or creatures could live in it, many with different goals for being there, living, as it were, differently; feeding on different things; and contributing in their own ways differently. Interestingly, had the designers approached this very convoluted system in this way, perhaps, arguably, they would not have been so dismayed when they encountered such a wide variety of usability problems during the initial phase of testing we conducted for them, after the project went live! I actually remember hosting the clients at our facility and one developer, after the first user was tested, walked outside the observation room and called back to the main office, saying, "Well, you're not going to believe this but it's a disaster. We need to shut it down. We need to shut the whole thing down and bring back the old site until we can fix everything."

There were many problems that testing revealed, but ultimately the primary reason or problem driving all these was that the designers didn't understand the ecosystem surrounding their software. They had made the assumption that a simple system was at work, one that Figure 5.2 illustrates.

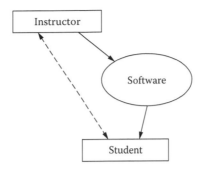

FIGURE 5.2
Simple system diagram.

They thought that users of the software would have the same relative knowledge and motivation to use it. They thought that students would not interact with each other, and that students would interact only with instructors, perhaps directly or via the information that instructors posted through the software. In other words, they thought there was only one kind of food and everyone would like it, would know how to find it, and could live on it. In fact, a careful analysis would have shown, driven from an ecological perspective, a great deal of complexity at work.

Mapping the User Experience

The diagram in Figure 5.3 lays out the client's software ecosystem. It is a snapshot taken with the software that was already part of the environment. A mapping could occur before this in an early design phase, before the software was introduced, providing developers an understanding of the ecosystem so that when they introduced the software to it, the design would be more usable.

I would encourage all designers of complex systems, as well as usability evaluators, to consider employing and maintaining a map like Figure 5.3. A usability ecosystem diagram can be as simple as a sketch done on a whiteboard. To do even this, however, especially as an activity at the very beginning of the testing or design process with all team members present, serves to conceptualize visually the situation, revealing what really is happening: the barriers, participants, and processes involved. Then, as the testing and design begins, the map can be updated as more detailed, focused analyses help to crystallize the team's knowledge of the ecosystem. If new barriers are found, they are added. If user goals change, they are modified. Somewhat akin to the excellent visual model that Albers (2003) encouraged developers to adopt when designing for complex systems (p. 229), such a diagram also takes away any sort of messiness, filling what in my opinion is a glaring hole in alternative ways to visualize the design and evaluation process that don't involve waterfalls or other ineffective hierarchical renderings.

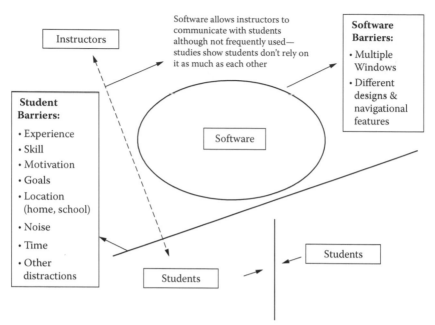

FIGURE 5.3
Ecosystem map of software.

As Figure 5.3 illustrates, there are, in fact, multiple barriers that affect user interaction with the software. Depending on student users' interests or goals, they are looking for different things. Other influences are at work as well, including writing experience and skill, location, and motivation, which all serve as barriers hindering users from getting to the food they need. We already know, for example, that students like to get help from each other, but as the diagram shows they are isolated from any peer support by the software. Maybe had the designers understood how students live in such an ecosystem, they would have made modifications to the software. Some students, as the diagram shows, work from home, and some at the library. Some self-limit the time they commit to the work, increasing its difficulty for them. Some students have considerable online experience, or bring with them greater knowledge of how to write, and maybe more confidence as well.

Instructors too are part of this ecosystem, as both predators and prey. The software, in other words, is designed so that instructors receive information from it, such as gradebook reports and resources for teaching, as well as contribute information to it, such as by creating quizzes and uploading assignments and lecture notes. The software's content and navigational pathways to such content are never fixed. The ecosystem has many environmental factors contributing to its flux, influencing the many different but interconnected participants within it.

An ecological approach, such as that which Figure 5.3 presents, would have encouraged a more comprehensive "examination of this. Begon, Townsend, and Harper (2006) wrote that "community ecology seeks to understand the manner in which groupings of species are distributed in nature, and the ways these groupings can be influenced by their abiotic environment ... and by interactions among species populations" (p. 469). Small tests make this happen, just as they would for the composition reference software. However, such tests would happen sooner and would be focused less on problem solving and more on the discovery of information about the software's ecosystem. As such, the tests would likely be in the field, or key elements of the environment would be included in laboratory evaluation. In addition, every effort would be made to incorporate the testing into the design process so that it was ongoing. Morgan (2004) made the case for how this longitudinal evaluation can be done when he detailed the efforts of eBay to track long term a representative sampling of users from different countries. To do this need not be too costly and involve large user samples. Since 1958, ecologists have studied a simple food chain, moose and wolves, on Isle Royale in Lake Superior. Other similar studies, of finches in the Galapagos Islands or of fruit flies in Hawaii, serve as microcosmic illuminations beyond themselves. They enable ecologists like Begon et al. (2006) to deal with what they called the "daunting problems" posed by communities that are "enormous and complex" (p. 470). Developers of the composition reference software could isolate, like ecologists, an island of users—say, one group in one location—and use them as a continual resource. They also could change that island to see if results varied. As long as they maintained focus on how users interacted with each other and the product and how that interaction was influenced by the surrounding environment, they could keep track of "patterns in the community's collective and emergent properties." These "patterns are repeated consistencies," according to Begon et al., "or repeated trends in species richness along different environmental gradients. Recognition of patterns leads, in turn, to the forming of hypotheses about the causes of these patterns" (p. 470). And from these hypotheses, usability practitioners as well as product designers learn what changes are needed in the system, and what methods are ideally suited for discovering these changes, to accommodate the improvement of the user's experience.

Embracing the Hypothetical

Of course, hypothesis-driven usability testing is not something that really works when delivering results to clients. Clients aren't interested in the scientific method. They want to know how to make their product immediately better. Often too, as Molich, Ede, Kaasgaard, and Karyukin (2004) noted, even when given a wide range of issues to address, they change, for whatever reasons, only a few. Still, since we are operating in new spaces, examining new methods for addressing complexity in user-centered systems, a certain amount of hypothetical thinking is absolutely necessary. Otherwise, we end up testing the same way with the same results that miss the mark on what really is unusable about a product. How then do we reconcile this dilemma? How do we embrace science and business so that we're doing our jobs the right way and clients are getting what they want? We begin first by arming ourselves with stories like the following. No business wants to repeat the mistakes of another.

A local software company contracted with us last year to test a redesigned touch screen smoking cessation kiosk it had removed from the market not long after its deployment. The company had paid for a series of small-scale usability tests. Novice users, defined simply as people who had not used a touch screen kiosk, were recruited for each round of testing. To insure accurate representation, each user was a smoker and of an age similar to the targeted audience who would more than likely encounter the kiosks at public health centers, hospitals, or other similar locations.

After each stage of testing, recommendations to address specific issues discovered were made, and the software company dutifully attempted to fix every one. Different versions of the kiosk were tested, beginning first with a wireframe so that any big problems encountered could be caught early enough to be corrected. In short, both client and usability evaluator did everything right (as they understood it) to deliver a usable product to the market. Consistently improving task performance for each phase of testing, in terms of both efficiency and error rates, along with a very high SUS rate for the final test, told them that their hard work would pay off.

Unfortunately, customers, such as hospital or clinic administrators, began to call the company with complaints almost as soon as the kiosks were rolled out to the public. When the company contacted those who didn't complain to find out if the problems they were hearing were consistent among all customers, what was heard didn't help to soothe concerns: It seemed that in those places where the kiosks were installed but there were no complaints, it was because the kiosks weren't being used at all.

So the company's first question for us was simple: Why? The company was torn even to go this far because many of its representatives felt burned by the usability process. It had done everything the perfect usability test client should do, and then this had happened. In truth, operating within the parameters of the discount usability testing approach, one would have trouble faulting the usability evaluators. They had recruited users matching the client's suggested profile. They had conducted low- and high-fidelity-prototype, iterative testing. They had found errors, evaluated for efficiency and satisfaction, and made available comprehensive reports, complete with video and audio clips of users in action, for support. What then could have gone wrong?

Working with the story we had been given, as well as any data the client had to share, we let them know first that it was our intention to conduct a thorough, due diligent assessment of the situation. The argument we made was that we needed to understand everything even beyond the software's functionality before we could make any hypotheses. Yes, we said "hypotheses." To buffer any hesitation on the client's part to embrace what seemed like a time-consuming approach, we made the rhetorical case that users and the software they use cannot be isolated. Without attacking the evaluation, we made it clear that we took an ecological approach to testing.

Now, it wasn't that we had figured out all of this ahead of time. It's just that we were struggling with a way to describe to the client a methodological framework that made sense to us and to the company, and ecology, more than any other science, gets emergence. And a basic hypothetical assumption we made about the smoking cessation kiosk was that the failure of it, even without yet knowing the specific reasons, was promulgated by a yet-to-be diagnosed complexity. In ecology, as we told the client, complexity is always assumed, even when just studying the interaction of a single predator and its prey. If anything happens to the prey, the predator is impacted, and vice versa. And this change impacts other species, other aspects of the environment, just as these things can impact it. As good as the testing was that the client paid for, it was too isolated. In focusing just on the users and their interactions with the software, it had walled itself off from the rest of the system and the potential for emerging factors to be at work in the system, affecting user and software interaction.

The software company, accepting this approach, wanted to know what it meant in terms of work and payment. We indicated that the first thing that needed to be done was an ecological mapping of the surrounding system, informed by site visits and other fieldwork. All of this would be included as part of the cost of discovery, and the mapping would be something that would continue, informed later by testing. The testing would be small again, with just a few users, but it would be done remotely, and it would be ongoing, even if it meant the company (or us, contracted through them) would have to go on-site to solicit real users to participate in brief walk-throughs of the "live" kiosks while at particular clinics or hospitals.

After we got started, we learned quite quickly, through on-site observation as well as user walk-throughs and interviews, that the kiosk software, on its own, worked well. The problem was that many users didn't understand its purpose. It didn't seem to fit in the environment where it was placed—in other words, it wasn't a food the users were familiar with.

The target users also were more than just novices. Given the location of the kiosks at public health care facilities, many of the users were uninsured and, therefore, were indigent and had never used any kind of digital electronic device before for educational purposes. In addition, they were often illiterate and couldn't read the English-only touch screens, or, if they were literate, they spoke and read Spanish, not English. Efficiency experienced in the lab had no basis in the environment where real users dwelled.

Ultimately, many other issues about the environment emerged that showed a significant (or the potential for) impact on the usability of the smoking cessation kiosk. The ecological mapping we conducted and presented to the client clearly demonstrated this and rhetorically made the argument for continuing to do it our way more palatable. Usability as ecology, therefore, worked not just to inform our own practices but also to make the business case to the client because the mapping showed to them the forces at work in the system, as well as the need for other potential, emerging factors to be monitored with ongoing testing.

Making the Business Case for Evaluating Complexity

The business client will accept that complexity must be addressed if the approach to understanding, evaluating, and designing for such complexity is not itself unnecessarily complex and seemingly open-ended. Yes, the ecological map is dynamic in that it suggests that it can and will change, and this sort of open-endedness can make people squeamish. But it provides also to the client snapshots with which they are familiar given how marketing trends and other measurements capture moments in the business cycle. For evaluator and client, the ecological framework provides a plausible alternative to the discount, small-scale testing process in which, at the end of either one or a series of tests, there is a faulty assumption that all problems are discovered and the product is ready for use. No product ever experiences conclusive stasis in a complex adaptive system. Dynamic emergence must be assumed in its design and assessment. Paul du Gay et al.'s analysis (du Gay, Hall, Janes, Mackay, & Negus, 1997) of the Sony Walkman's introduction to the consumer in the 1980s demonstrates this. Designers had intended the Walkman to be used by two people at once. Not only was a dual headpiece provided that would allow two people to listen together to music from the Walkman, but also corresponding advertising appeared in magazines and in other media that showed couples enjoying the Walkman together.

Of course, the only problem with this is that when the Walkman hit the market, as du Gay pointed out (du Gay et al., 1997), people didn't want to share their Walkmans (p. 59). There might have been some who emulated the advertising, but for the most part the Walkman became an individual tool. Consumption, not just production, as part of the circuit of culture surrounding the Walkman determined its ultimate use. Designer intention, therefore, was trumped by actual use, as is so often the case. There is so much more going on in the surrounding environment that must be understood and made (as much as possible) a part of any testing process.

Doing this is not without confusion. Evaluating complexity is problematic, even for researchers working in other fields. Adami's (2002) article on evolutionary biology shed light on this. He wrote that "nobody knows precisely what is meant by the word 'complexity.'" In fact, Adami continued, "[C]omplexity is so general a term that it seems to mean something different to everyone" (p. 1085). There is statistical complexity, which for Adami isn't especially helpful because even though it "characterizes the amount of information necessary to predict the future state of a machine ... it fails to address their meaning in a complex world" (p. 1086). Hierarchical complexity looks at the "levels of nestedness" (p. 1086), but this approach cannot be universalized, in Adami's opinion, given the diversity of creatures and systems. Structural and functional complexity have also been examined, but Adami regards both, like the others just mentioned, as not effective for getting at the dynamism of complexity found in biology.

What he turns to for defining complexity, and what is useful also for the usability practitioner in attempting to explain the complexity at work in shaping the ecosystems surrounding users and products, is a more physical approach, reflective, interestingly, of an ecological understanding of how everything functions within a particular context. In other words, Adami (2002) argued for an understanding of complexity that "refers to the amount of information that is stored ... about a particular environment" (p. 1087). Typically complexity, when applied to some sequence, often reflects what he called "unchanging laws of mathematics." But physical complexity focuses on information in context, meaning that any action or sequence examined to understand its complexity takes into consideration that any measurement of such complexity is "relative, or conditional on the environment." In a nutshell, therefore, "physical complexity is information about the environment that can be used to make predictions about it" (Adami, p. 1087).

Repurposing the Small-Scale Usability Test

Harvesting such information, even in the evolutionary biology setting Adami occupies, does not mean an examination of the entire system. Doing this isn't possible in that field, according to Adami, because it lacks the tools. We do as well. However, one tool we do have, one this chapter has argued needs to be repurposed and not done away with, is the small-scale test. As Adami (2002) stated, "[A]n increase in complexity can only be observed in any particular niche" (p. 1092). In biology, this may mean a focused examination of just one genome. In ecology it may mean examining a particular food chain. In usability, it means looking at one user group, maybe even a very limited number, and maybe, should the software be complicated, it means examining just a single process or function. But in context, informed by an understanding if not an active incorporation of the physical environment that influences, and is influenced by, that smaller user group, such an examination can be fruitful.

We carried out testing just recently of software used by colleges and universities to maintain their academic and financial records. The test was very small, and unfortunately for the client and users of the product it came far too late in the process to be helpful, at least in any ways that could be incorporated as immediate, positive changes.

To say that end users were frustrated with the new software was an understatement. They had used for some time another suite of software to carry out their work, and although this suite had usability and functionality issues (one reason why it was being replaced by the new software), it was a known entity. Over time, its end users had developed an informal support network for working with it. Experienced staff trained new staff on using it, or could always be contacted by phone or e-mail for the answer to a quick question about a particular issue. Everyone had more or less developed mental models too that enabled them to work efficiently despite any obstacles the legacy suite might have posed.

However, all of this history, for the individual end user and the group as a whole, was basically ignored during the rollout phase for the new software. Even though developers indicated that customizations could occur, those who would need them most and would know also where best to include them, the staff end users, were not consulted. As subject matter experts, they were not represented on the new software's implementation team. So when the change to it came, it hit hard. Some end users openly cried at introductory meetings discussing the new software. A few literally retired early rather than deal with the problems they thought would be inevitable. Most demonstrated an immediate resistance that only grew in intensity and scope, eventually motivating supervisors of staff to chime in and also express concerns. Today, whenever anything goes wrong with a student's record or a financial transaction at the university, human error is never blamed. People just say the name of the software and roll their eyes or mutter something negative under their breaths.

Now what our testing revealed was that when it came to functionality, the new software actually worked fine, if not better, than the complicated suite (comprising sometimes different software for different stages of the same process) that had been in use for some time. However, its look and feel were unappealing to users. Alert popup boxes came up even when tasks were completed successfully. This could have been turned off, but no end users were ever tested during rollout to see if they regarded this as annoying or distracting. Still further, users found the navigation labels, all of which could be altered, nonintuitive. Not being able to input help documentation, something they could do for their last system, was also challenging—as was the fact that all previous help documentation they had created and stored, either mentally or physically (in Post-it® notes on computers or on resource files in Microsoft Word—complete with screen captures), no longer had any benefit.

Had designers incorporated staff end users in the process, something Redish (2007) suggested as a valuable method for evaluating complex systems, many of the problems that impacted the adoption of the new software might have been avoided. According to Redish, "domain experts," like the staff end users with combined centuries of experience managing the university's records, "must be partners in the evaluation, just as they must be partners throughout the planning, design, and development of the systems" (p. 105). Collaboration is the key. Neither should we as usability experts wall ourselves off from clients or other domain experts, trusting just our own experience and the previous validity of the typical test, nor should we remove from the process (or make inconsequential by our treatment of it) the environment where users and product interact.

Discount Testing versus the Ecological Approach

At this stage, it is useful to summarize the argument I have tried to make to this point. Because complex systems by their very nature are dynamic and open-ended, and because the environment of such systems plays a crucial role in how users interact with products, the typical small-scale usability test must be repurposed. An ecological framework for understanding how to reconsider and retool the small-scale test is useful for us to consider, if not implement. To visualize how that might look in comparison to what is typically done now in discount testing, see Table 5.1, which offers a comparison.

Other Methods for the Ecological Approach

Certainly other methods can be incorporated as tools of discovery in the ecological approach to usability. Apart from the ones discussed above, Redish (2007) suggested a few in addition to using retrospective recall or testing outside the lab. One interesting idea she offered is to build simulations (Redish, p. 107). Such simulations or models, assuming they match closely the complexity of the actual system being simulated, would work well as an experimental site for evaluation. Interestingly, models like this have always been highly prevalent in ecology (Radcliffe, 2007), allowing researchers to introduce variability, such as hypothetical new properties or other factors, to see how it might impact the system. In fact, I remember many years ago coming across a very basic but nonetheless instructive Moose/Wolves software game. Students could change either population, make weather more or less harsh, and modify other factors as a means of understanding, in a simulated environment, what might happen if things were different. Alstad (2007) at the University of Minnesota employs a much more sophisticated form of this. Populus software offers an array of simulations that students can manipulate as they learn hands-on key principles of biology and evolutionary ecology.

TABLE 5.1

Comparison of Usability Approaches

Typical Discount Testing	Ecological Approach
Limited user and task analyses (dependent upon time frame and stage of testing, the thoroughness of analysis could vary); however, no requirement that evaluation of user environment be carried out or made part of the test environment.	Exhaustive discovery process that results in ecological mapping of the system; extensive use of site visits and other tools for understanding how users work in the actual environment where the product will be used.
Specific tasks for users to perform that result in specific measurements.	Open-ended, more realistic scenario(s).
Although not always the case, laboratory testing (or testing in an environment without distraction) is favored.	Testing if possible done where users actually use the product.
Somewhat homogeneous user population, often broken down into simple user experience categories such as *novice*, *intermediate*, and *expert*.	User profiles based on factors other than just experience with product.
Five users per test; if iterative testing is employed, then three tests or fifteen users total.	Given that there is never assumed to be a finite set of problems that can be discovered regardless of the number of specific tasks performed or tests executed, user test population size is not as important as who is being tested and where. Also important is that the testing process never ends. Ongoing testing should continue even with a small sample of users.
Concurrent think-aloud protocol to solicit feedback from users.	Concurrent think-aloud protocol not natural and can be distracting; instead, retrospective recall is used.
Different data collected but focus, by way of specific tasks used in testing, placed on understanding user efficiency, error, and satisfaction.	Focus is on gathering information that helps evaluators understand the system better so that the product's design works for users in that system. As Albers (2003) wrote, "[T]he design analysis must focus on determining and defining user goals and providing sufficient information to address the users' informational and psychological needs" (p. 38).

Such models like this allow for experimentation, for proactive research that enables practitioners to stay a step ahead. Since the focus for a complex system is not on solving all the problems but instead on providing information to designers that allows them to make ongoing, effective modifications, a model of some kind, if sufficiently like the real thing, might prove very useful. To date, some work has been done on this, but more is needed to explore its potential.

Closely akin to modeling, although less concrete, is early-stage paper prototyping. It has proven to be a valuable tool, as Snyder (2003) and others have argued, for usability evaluators. I believe that reinvigorating it so that it is more a part of testing would prove valuable, especially for complex systems evaluation. Paper prototyping encourages collaboration at the earliest possible stage of design, bringing together domain experts, developers, and many other representatives from other groups that during later-stage testing often are not sufficiently equipped, or are not typically part of the process then. It also serves well, regardless of the product's stage of development, to encourage users to offer creative feedback, since they can use paper to extend what the prototype may not offer, sketching out ideas for navigation and other design elements, or even writing down what they are thinking. Redish (2007) argued that "creativity and innovation" should be new factors we consider when assessing the usability of complex systems (p. 104). I just recently coauthored a study with John Morris (Still & Morris, in press) of a university library website in which we married low-fidelity paper prototyping with medium-fidelity wireframe prototyping. When user navigation led to nonexisting pages or dead ends, users were encouraged to create what they thought should be where there was nothing. By employing this blank-page technique, we acquired insights into users' mental models regarding site content and design, providing developers with useful data about how users conceptualize the information they encounter. It also encouraged just the sort of positive, proactive user involvement that Becker (2004), Genov (2005), and others suggest is crucial for effective usability testing and, as a result, equally effective user-centered design.

Other things, including experimenting with the inclusion of noise or other distractions to measure impact on user effectiveness, especially when compared to testing without such interference, should also be considered.

Conclusion

Experimentation, given where we are now, is key. We are really experiencing a new paradigm shift in usability evaluation, and the purpose of this book has been to make known particular studies or methods that usability professionals have experimented with to address complex systems. Although hypotheses may seem, as previously noted, not workable when testing for clients who want actionable results, we cannot continue to do the same kind of testing, not when examining complex systems. To go forward, we have to experiment. This may mean we recommend less, but Spool (2007) asked, "[W]hat would happen if researchers stopped delivering recommendations altogether?" (p. 159). "If," Spool wrote, "we're no longer responsible for putting together recommendations … we can devote some of that 'found' time to innovating new techniques" (p. 160)—techniques that do not just focus on how to analyze users, but also help clients, developers, and collaborative "teams," according to Spool, "focus on user needs" (p. 160).

Collaboration like this, not just among a team of different professionals at a single workplace but also across the usability field, is another idea for developing and sustaining an ecological approach to analyzing complex systems. Scientists share data. Usability researchers, often working for clients who want nondisclosure agreements, do not, or if they do then they do not nearly as much as they should. Imagine how successful micro studies of a particular systems environment would be if all those carrying out such studies collaborated with each other. Kreitzberg (2006) argued for usability to have a collaborative knowledge space similar to that of other sciences, like ecology. The advantages he saw for doing this include the "potential to tap the great store of usability-related experiential and tacit knowledge that exists among usability practitioners" (p. 110). According to Kreitzberg, it also "will encourage the integration of research and practice." At the same time, if it is structured similarly to Wikipedia or another similar social knowledge software, "it can be updated quickly to reflect the dynamism of the field" (Kreitzberg, p. 110).

This book has contributed, in part, to opening up such a collaborative space by providing new approaches to evaluating the usability of complex systems. Now, Albers (2003) makes clear that no single approach or method, however effective, can work (p. 223). For too long "most document and web design," Albers wrote, "assumes a single answer exists and the emphasis is to help the user find it" (p. 223). But to espouse an ecological approach is not to offer up a single method, to argue that one way of testing should replace another. Rather, it is to offer an intellectual framework that offers good reasons for embracing a micro-analytical approach to macro-sized issues. Usability as ecology is about openness without disarray. It offers intellectual justification for usability evaluators to be open to experimentation, open to the possibility of nonspecific answers, open to sharing data, open to contributions from nonusability experts, and open to the significant and changing impact that environment has on users and products.

Reader Take-Aways

- The small-scale usability test must be repurposed. Iteration, leading to the assumption that a finite number of problems will be discovered in each round of testing until all are eventually found by the completion of iterative testing, should be seen as less important than testing that is meant to offer an ongoing dynamic awareness of the product's status and the user's changing relationship with it.

- Borrowing from ecology, usability practitioners should create and maintain an ecosystem-like map of the user-centered system that is being evaluated. Such a mapping should be more cognizant of all the factors in the environment, not just the user–system interface, that influence the users' use of the system and also, through self-production, constantly serve to generate changes to the nature of the usability ecosystem that encompasses the users and the products they use.

- Consider employing more site visits to gain a better awareness of the user environment, and also experiment with paper prototyping and modeling to enable the gathering of new kinds of data, from different perspectives, that allow for a better appreciation of the depth and complexity of the system. Modeling, for example, promotes longitudinal evaluation and also aids designers because it allows them to introduce variables into testing. Much as the ecologist simulates the impact of environmental changes, the usability practitioner working within an ecological framework can proactively stay one step ahead of changes through modeling.

References

Adami, C. (2002). What is complexity? *BioEssays, 24*(12), 1085–1094.

Albers, M. J. (2003). *Communication of complex information: User goals and information needs for dynamic web information.* Mahwah, NJ: Lawrence Erlbaum.

Alstad, D. (2007). Populus [computer software]. Minneapolis: University of Minnesota.

Bangor, A., Kortum, P., & Miller, J. (2009). Determining what individual SUS scores mean: Adding an adjective rating scale. *Journal of Usability Studies, 4*(3), 114–123.

Becker, L. (2004, June 16). 90% of all usability testing is useless. *Adaptive Path.* Retrieved from http://www.adaptivepath.com/ideas/essays/archives/000328.php

Begon, M., Townsend, C. R., & Harper, J. L. (2006). *Ecology: From individuals to ecosystems* (4th ed.). Malden, MA: Blackwell.

du Gay, P., Hall, S., Janes, L., Mackay, H., & Negus, K. (1997). *Doing cultural studies: The story of the Sony Walkman* (Culture, Media and Identities series, Vol. 1). Thousand Oaks, CA: Sage.

Genov, A. (2005). Iterative usability testing as continuous feedback: A control systems perspective. *Journal of Usability Studies, 1*(1), 18–27.

Howard, T. (2008). Unexpected complexity in a traditional usability study. *Journal of Usability Studies, 3*(4), 189–205.

Kreitzberg, C. (2006). Can collaboration help redefine usability? *Journal of Usability Studies*, 3(1), 109–111.

Molich, R., Ede, M., Kaasgaard, K., & Karyukin, B. (2004). Comparative usability evaluation. *Behaviour & Information Technology*, 23(1), 65–74.

Morgan, M. (2004). 360 degrees of usability. In *Proceedings from CHI 2004: Conferences on human factors in computing systems*. Vienna. New York: Association for Computing Machinery.

Radcliffe, E. (2007). Introduction to popular ecology. In E. B. Radcliffe, W. D. Hutchison, & R. E. Cancelado (Eds.), *Radcliffe's IPM world textbook*. Retrieved from http://ipmworld.umn.edu

Redish, J. (2007). Expanding usability testing to evaluate complex systems. *Journal of Usability Studies*, 2(3), 102–111.

Snyder, C. (2003). *Paper prototyping: The fast and easy way to design and refine user interfaces* (Morgan Kaufmann Series in Interactive Technologies). San Francisco: Morgan Kaufmann.

Spool, J. (2007). Surviving our success: Three radical recommendations. *Journal of Usability Studies*, 2(4), 155–161.

Still, B., & Morris, J. (In press). The blank page technique: Reinvigorating paper prototyping in usability testing. *IEEE Transactions on Professional Communication*.

Worm, B., & Duffy, J. E. (2003). Biodiversity, productivity and stability in real food webs. *Trends in Ecology and Evolution*, 18(12), 628–632.

6

Usability and Information Relationships: Considering Content Relationships and Contextual Awareness When Testing Complex Information

Michael J. Albers

East Carolina University

CONTENTS

Introduction 110
    Importance of Relationships and Contextual Awareness 112
    Usability Tests for Relationships and Contextual Awareness 117
Building a Test Plan for Contextual Awareness Usability 120
    Factors to Consider When Building the Test Plan 121
        Mental Models 122
        Information Salience 123
        Information Relationships 123
        Contextual Awareness 124
Collecting Data 124
    Tests While a Person Is Still Developing Contextual Awareness 125
    Tests of the Quality of a Person's Contextual Awareness 125
    Data Analysis 126
Conclusion 126
Reader Take-Aways 128
References 129

Abstract

Modern communication situations are no longer highly structured; instead, they have shifted to complex situations revolving around information seeking, problem solving, and decision making. In response, most of the information that writers produce and most of the information that people want are complex, not simple. From a usability-testing standpoint, for a complex situation the focus must be on how people integrate multiple information elements. This integration depends not on the content of a text element itself but on forming relationships between those elements. A reader needs to figure out what information is relevant and how it connects to the current problem. Without proper information relationships, the reader does not gain an integrated understanding of information. As a result, he or she fails to have a clear understanding of the situation and its future evolution, and is unable to effectively make decisions that influence it. This chapter presents an argument for why focusing usability tests on how people understand relationships is important, and presents basic guidance.

Introduction

Readers often deal with clearly written technical information that is technically accurate, yet utterly fails to communicate. Usability tests run during development showed people found answers promptly with minimal errors, and so the design was signed off as usable. If asked, however, the readers even say it's clearly written, but they can't use it. Effective technical communication provides information that conforms to human behavior in complex situations and fits people's information needs. As such, technical communication operates within a highly complex and dynamic world (Albers, 2003; Ash, Berg, & Coiera, 2004; Mirel, 1998, 2003), and usability tests must confirm that the information communicates within that world.

Typically, the failure of these technical documents comes not from a lack of information; the text probably contains an excess of information. Post hoc studies of communication failures find many sources to blame: poor information architecture, poor organization, wrong grade level or writing style, or poor presentation. But instead of seeing these problems as a root cause, let's consider them as symptoms of a more fundamental problem: a problem stemming from the underlying complexity of the situational context and a failure of the information presentation to match that complexity. A significant problem is that documents fail to distinguish between simple and complex information and contain a presentation based on an assumption that all information is simple. Schroeder (2008) looked at the difficulties that people have with combining the tools available in a typical software program. He criticized standard manuals that explain each tool option without connecting it to other tools since most software operations require using multiple tools, and concluded that effective software interaction requires understanding the relationships between tools to accomplish a task. Likewise, Howard and Greer (this volume, Chapter 4) found similar issues with using writing handbooks. The individual pieces of information seem obvious, but when a person needs to use one to address a real-world question, it often proves problematical.

Most of the information that writers produce and most of the information that people want are complex, not simple (Albers, 2004). Complex information and complex information needs are multidimensional. There is no simple or single answer to a particular set of information needs, and, to add to the complexity, the situation's information contains a dynamic set of relationships that change with time and in response to situational changes. Discussing methods of communicating complex information, Mirel (1998) follows the same line of reasoning as Conklin's (2003) wicked problems when she points out that analyzing complex tasks requires seeing more than a single path:

> This broader view is necessary to capture the following traits of complex tasks: paths of action that are unpredictable, paths that are never completely visible from any one vantage point, and nuance judgments and interpretations that involve multiple factors and that yield many solutions. (p. 14)

Content needs have moved beyond providing information on single objects or procedures; much of the Web's current and future content involves communication of experiences and strategies that require much more complex information than the simple procedures of past technical documentation. A traditional usability test may have focused on finding single pieces of information or following linear procedural steps. Interestingly, even the adequateness of that linear sequence can be called into question, with results such as Howard's (2008) work with handbooks and Hailey's (this volume, Chapter 2) examination of understanding website genres. But beyond linear task sequences, in many situations, rather than looking up single elements, a person needs to find and integrate several pieces of information. Morrison, Pirolli, and Card (2001) found that only 25 percent of the people they studied searched for something specific, looking for a clearly defined X. Rather than needing single pieces of information, they found that 71 percent searched for multiple pieces of information. And the user then had to integrate those pieces of information into a coherent, useful answer.

From a design and usability-testing standpoint, for a complex situation the focus must shift from supporting *finding* (although, obviously, it is important) to supporting *integrating* multiple information elements. This brings up the question of how people integrate information and how to support it.

Information integration lies, not in a text element itself, but in the relationships between those elements. A reader needs to figure out what information is relevant and how to connect it to the current problem. Without proper information relationships, the reader does not gain an integrated understanding of information, but instead gains a collection of facts. Without relationships, information exists as a bunch of interesting factoids that do not help a person form an adequate mental picture of the situation. Collections of facts are less than useful for understanding and working with the open-ended problems that people encounter in complex situations (Mirel, 1998, 2003). Without the relationships, a person learns about X and Y, but not how X and Y relate to each other or to Z in terms of then current problem or situation (Woods, Patterson, & Roth, 2002). The text fails to communicate because the reader can't form the necessary information relationships.

Most people will accept the previous sentence without argument, but from a usability standpoint, we face an additional issue. Information relationships can change with different situations or as a situation evolves. How do we determine if the communication problem is a failure to form the information relationships, and how do we test for them? One possible answer lies in defining the person's contextual awareness (Albers, 2009a). Gaining a contextual awareness of the situation progresses from simply perceiving the raw data to having a comprehensive understanding that enables making accurate predictions of future developments. Kain, de Jong, and Smith's (this volume, Chapter 14) examination of hurricane risks and warnings reflects similar ideas of how people interpret information and make decisions about how they will react. It also leads directly to verifiable methods that can be used during prototyping and usability testing to ensure that contextual awareness goals have been met (Redish, 2007).

Contextual awareness is about a person's state of knowledge about the current situation and not the process used to obtain that knowledge. It deals with how a person has integrated the available information into something relevant for his or her current situation.

- Do they have an understanding of the relevant information?
- Do they know how the information applies to the situation?
- Do they know how different pieces of information relate to each other?

Design and content decisions and usability testing for measuring a person's contextual awareness need to focus on measuring information comprehension and use, and not primarily on how a person obtained it (e.g., measuring search times or paths). With complex situations, usability tests shift to a results-based approach of ensuring the person has contextual awareness, rather than a procedural-based approach of how he or she obtained it. Of course, making the overall process as simple as possible is also important, but with the understanding that the process (the path to the information and understanding it) can vary between people.

Importance of Relationships and Contextual Awareness

The importance of relationships has long been understood in learning and education. I frequently encounter students who know the information. If asked probing questions about the separate ideas, they have no problem giving a correct answer. But if asked to connect them, they draw a blank. Of course, that's one of the main purposes of education, to help students form those relationships. Those who can quote the book but not apply it have not formed mental relationships between different information elements.

> Learning in a subject area, such as science, involves understanding the rich set of relationships among important concepts, which may form a web or a network. Revisiting the same material at different times, in rearranged contexts, for different purposes, and from different conceptual perspectives is essential for attaining the goals of advanced knowledge acquisition. (Spiro, Feltovich, Jacobson, & Coulson, 1991, p. 25)

Likewise, the dynamic nature of complex information structuring and transforming was the focus of O'Malley's (1986) discussion when he described the users' information needs as follows: "[U]sers should be able to structure or restructure the information to suit their own unique purposes" (p. 396;). In the course of interaction with information, the person needs these:

- *Access to integrated complex information.* The multidimensional nature of information in complex situations requires that it be presented with some level of integration that clearly connects all the dimensions. A person often lacks the knowledge to or chooses not to perform extensive mental integration.
- *Adjustment of the presentation to fit the current goals and information needs.* Goals and information needs vary between people and with the same person at different times.
- *Views of the problem from multiple viewpoints.* The situation often contains various viewpoints. Cognitive tunnel vision can prevent a person from seeing an issue in more than one way and, consequentially, can hinder gaining a clear understanding.
- *Understanding of the relationships between information elements.* The important changes in a situation often only exist as measured by the change in the relationships, not as a change in the individual tasks (Cilliers, 1998). The design goal behind showing the information relationships "is to help people solve problems, rather than directly to solve problems posed to them" (Belkin, 1980, p. 134).

One significant difference between most of the information produced by technical communicators and information for learning is the situationalness of the information. Ignoring training material, the readers of technical information typically understand the basics but need to know specific information about the current situation (as opposed to the general situation) in order to make decisions. Understanding of a complex situation and the contextual awareness needed to make informed decisions comes when people can distinguish the information structure, grasp the relationships within it (Thuring, Hannemann, & Haake, 1995), and make inferences on the future evolution of the situation. Performing that mental assembly requires a person to form relationships between multiple information elements.

For example, a business executive knows how to analyze her monthly reports (the general situation), but needs information to help her understand *this month's* performance (the specific situation). The problem with these reports is not a lack of information (they usually provide an overabundance), but they fail to allow for easy integration and comprehension. The reports are not designed for integration; they are designed to present their own information. To use multiple reports requires retrieving and integrating the relevant information. However, people have a hard time integrating information and relating various data points to each other. Also, they have a hard time remembering or considering subtle cause-and-effect relationships that exist between the information being viewed and other information. Cognitive tunnel vision can cause them to ignore and forget to consider other information. Years of practice with paper reports have allowed an analyst to develop personal methods of compensating for the integration problems. The movement of this information to a computer screen has seriously impacted the ability of people to use their learned compensation methods by introducing another level of navigation and view limitations. When moving these reports into a computer system, the design team needs to acknowledge the complex nature of the interaction and that the purpose is not to present information, but to allow people to develop their contextual awareness.

As shown in Figure 6.1, the understanding of the situation arises from and exists within the arrows connecting the boxes, not solely from the information in the boxes themselves. This means the design team needs to worry about usability at two levels: interacting with the content (the traditional view of usability) and the formation of relationships (a paramount concern with complex information). With simple situations, content interaction overwhelms any relationship concerns. Tests of looking up information and how quickly it can be found or manipulated are sufficient. However, with complex information, simply having access to the information is no longer sufficient to ensure quality communication. Information comprehension and the ability to translate it into terms applicable to the current situation become the major factors in developing contextual awareness and making inferences on how the situation will evolve. A person who can quote facts and figures, but not connect them to the situation or make inferences, does not have contextual awareness. Use of complex information provides the ability to take those facts and figures, build the relationships between them in a way that makes sense within the current situation, and then use that information to make inferences and decisions. With this requirement, the design team has a much more difficult task of ensuring its information communicates at all of these levels.

In many problem-solving situations, a person has typically encountered a nonnormal or unexpected event and wants to make decisions about it. Decision-making research has found that there are different cognitive processes followed depending on whether the event is routine or is nonroutine

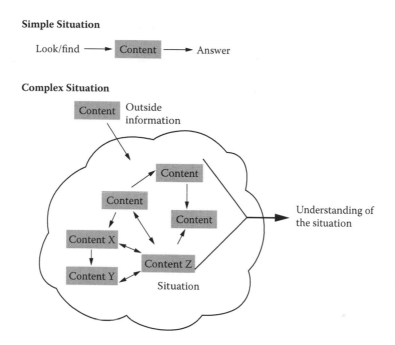

FIGURE 6.1
Relationships and understanding a situation: The reader needs to form relationships to connect the various content sources. Understanding arises from comprehending both the content and the relationships between the content.

or unexpected; the information that receives attention and the extent of information gathering change. Part of the change that is problematical for usability tests is that the reader shifts from rule-based thinking to knowledge-based thinking, which changes the amount and type of information required and the effective presentation format. In addition, Klein's (1993) recognition-primed model of decision making posits that experienced people do little problem solving, especially in routine situations. Instead, once they recognize a situation, they form a possible intention very rapidly, mentally evaluate it, and, if no major problems are evident, take action on it. Alternatives are not considered. In other words, they tend to use their prior knowledge and initial assessment to immediately pick a solution, and if it seems workable, then that is considered the answer. (Concerns about whether the solution is optimal are not part of the mental evaluation.) As a result, the quality of the response depends directly on how they initially perceive the situation, which depends directly on the quality of information they receive. However, when people confront a nonroutine situation, they have to shift to a knowledge-based approach. It's when they encounter a nonroutine situation that people turn to an information system, and the usability tests must ensure they can use a knowledge-based approach to understanding the situation.

From a design team's view, the issue can often be characterized as picking the low-hanging fruit. The information needs for a routine situation are easier to define and usually consistent, and so the decision is made to design a system to supply that information. The nonroutine situation issues are deferred until later. But once the system is rolled out, it fails to provide any great return, since it is not addressing the information issues where the people really need help. Golightly et al. (this volume, Chapter 13) consider essentially the same issues from a viewpoint of technical and contextual issues and how one often gets privileged over the other.

The flow of Figure 6.2 shows the transition of people's understanding from the initial activation of a mental model through developing contextual awareness. After developing contextual awareness, people can effectively consider decision choices and make extrapolations or mental simulations about the situation's future. They start with their mental model and bring in information from the situation to build up their contextual awareness based on the relationships within the available information. Some factors that affect the development of building relationships and a clear contextual awareness are as follows:

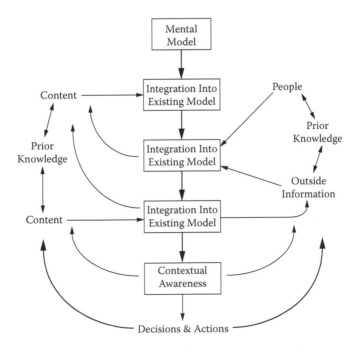

FIGURE 6.2
Moving from a mental model to contextual awareness. Relationships are formed via integrating new information with prior knowledge in a recursive manner to fill in the mental model and provide high-quality contextual awareness.

- The wrong mental model means information needs and relationships are misinterpreted and can lead to a major misunderstanding of the situation. Many accident reports contain elements where people seemed to ignore pertinent information that is obvious in hindsight. Working from the wrong mental model, the people incorrectly interpreted the information, ignored information that was irrelevant in their mental model but relevant to the actual situation, and made poor decisions.
- Highly salient, but irrelevant, information may cause a person to make the wrong choice since he or she wants to mentally integrate it into the other information (Hsee, 1996).
- Problem-solving involves constantly adjusting goals and subgoals to allow for the dynamic nature and quality of the information available (Klein, 1999; Orasanu & Connolly, 1993). This results in the cycles shown in Figure 6.2 as people build the mental information and relationship structure in a recursive manner, not in a single rise to full completeness (Mirel, 1998, 2003; Redish, 2007).
- As the amount of information increases, people increasingly ignore information incompatible with the decision they want to make (Ganzach & Schul, 1995). They skip information that might conflict with building the relationships they want to see. Information salience plays a major element here to essentially force the person to consider all of the pertinent information. One-size-fits-all solutions tend to have very poor information salience; providing more customized presentations helps to both reduce the overall information presentation and avoid the reader ignoring important text.
- Information that a person does not see, whether blocked by a window or on a previously visited page, is considered at a level of reduced importance compared to the current information, which exerts a profound effect on how people form information relationships. Thus, order of presentation, which a design team may have difficulty controlling, plays a substantial part of how a person integrates information.

Usability Tests for Relationships and Contextual Awareness

A traditional usability test is set up to observe how well people can perform tasks with a given tool, and most of the time it's assumed to be a task they intend to do, which assumes they understand, never mind recognize, they can do it. Perhaps the first few pages of a website could be tested for how well the product benefits come across. But complex systems require a completely different view of testing, and even the ability of the traditional test has been called into question (Schroeder, 2008). A complex system usability test needs to consider more than "Can a person look at the site and list 5 of the 8 product benefits?" Instead, it needs to consider if people comprehend the product benefits information and can apply it to their situation. Before people can effectively proceed within a situation, they must gain an understanding of the situation within the current context; in other words, they must develop their contextual awareness. A substantial factor in developing contextual awareness is being able to integrate many pieces of information (the cyclic nature of Figure 6.2), which requires that a person knows that the information exists, what it means, and how it is interrelated to other pieces of information. The need for this integration separates simple from complex situations.

The last part, "how it is interrelated," proves to be the most difficult for effective communication and for effective testing. It is easy to provide information and relatively easy to figure out what the audience already knows or provide explanations, but ensuring that the reader can take that information and relate it to other information elements is where many information sources fail. Thus, although we know complex and simple situations are different, we have a hard time defining what makes them different in a manner that can be operationalized into something that can be written about or developed into a usability test. Too often, a design team simply ignores the issue and leaves it for the reader to assemble the pieces. Unfortunately, a large body of research has found that people perform poorly when asked to perform this mental integration.

The usability test needs to take the early analysis, which should have built an instantiated version of Figure 6.3, and verify that when people interact with the information, the progressive integration and comprehension of the situation proceed fluently. In general, usability testing strives to address these issues by collecting one of two types of measurements. Each has its place in the usability toolbox, but they are not interchangeable.

Performance-based measures. These measure the outcome with factors such as time, correct answer, and so on. Since these are external measures that do not capture the person's internal thoughts and cognitive processes, they are better for tasks requiring procedural knowledge. Although essential within the scope of testing the entire system, they play a small role in testing for building relationships and developing contextual awareness.

Knowledge-based measures. Measures of internal thought processes that work by asking the person to describe the current situation. They work to capture a reader's mental models, internal thoughts, and developing view of a situation. They also give insight into how the person is developing contextual awareness and where the design is failing to contribute to supporting contextual awareness.

Usability tests of complex systems that focus on performance-based measures, such as click count or correct answers, are highly problematical. They violate Mirel's (1998) definition that complex tasks have indeterminacy of both task goals and criteria for task completion. Yet, performance-based measures

Global contextual awareness

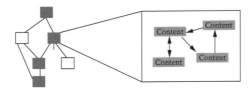

Reality for the overall situation   Relevant for the reader
for the specific situation

Local contextual awareness

FIGURE 6.3
Global and local views of a situation. Reality of the relationships for the information space. What the person needs the information for the specific situation (gray boxes are relevant content). With local awareness, only the content of the individual boxes is viewed; how they connect is ignored.

tend to require at least one, if not both, to have determined end points or correct answers to allow for meaningful interpretation of the measured values. However, collecting those numbers in any sort of meaningful way almost always requires simplifying the complex interaction into a simple interaction to avoid confounds. As a result, the process of interacting with the information is lost, and any resulting numbers do not necessarily reflect the quality of communicating information or the overall contextual awareness. Worse, it may indicate the system is highly usable for simple interactions. Desktop testing of the failed San Jose Police Department laptop-based system for computer-aided dispatch in their patrol cars (Santa Clara Grand Jury, 2005) probably showed good interaction and good performance-based measures. But once placed into the complex multitasking environment of a police car driving down the street, usability plummeted and the system became dangerous to use.

On the other hand, knowledge-based measures can explain the effectiveness of the design's communication ability and whether it will prove too brittle for real-world use. Rather than ensuring that user performance conforms to predefined paths, knowledge-based measures allow supporting user-defined paths toward understanding a situation. Unfortunately, a significant problem for testing complex systems is that many managers, system engineers, and software designers want numbers, which can only come from performance-based measures, but which cause them to overestimate the actual usability or, worse, misdirect the usability focus to low-level button-pushing issues.

Knowledge-based measures also help to keep the focus on global-level contextual awareness. Contextual awareness exists at both the local and global levels (Figure 6.3). Global contextual awareness spans the entire situation and puts all of the individual pieces into context and relative perspective. Local contextual awareness, on the other hand, narrows the understanding to a single element. The danger is that the understanding and potential decisions that make sense for a single piece are not appropriate for the overall situation. Many information architectures are described in terms of an outline or otherwise assume the information presentation fits a hierarchy. However, most information that people need to understand a situation lacks a hierarchical structure; instead, it forms a highly interconnected, weblike structure. The relationships a person forms result from reworking the information into a mental structure that fits the current situation. Thus, the hierarchy needs to exist to ensure all the information is present, but the presentation needs to be more than just the hierarchy.

The local and global contextual awareness can be addressed at different phases in system development. Early testing often focuses on local contextual awareness. Later tests with a more developed system must assess the global contextual awareness. On the other hand, care must be taken to not allow local contextual awareness issues to drive design decisions such that the system provides good local contextual awareness, but fails to support global contextual awareness (the problem of optimizing individual pieces without considering how they fit together). People often are guilty of falling victim to cognitive tunnel vision or giving current local variables excessive salience. They lock in on what they want to see or downplay the relative importance of information that is not directly before them.

Building a Test Plan for Contextual Awareness Usability

Usability tests need to ensure that communication is occurring and that the person has formed contextual awareness based on the information and relationships needed to make decisions. Providing a quality product with high usability requires a design and testing cycle focused on how the information will be used, how it fits within the situation, and why a person needs it.

Contextual awareness strongly depends on understanding the relationships within a mental model, finding those relationships within the real-world information, and using them to build a coherent story from available information. If a person is using a wrong mental model, he or she will ignore relevant information and relationships, and thus develop inadequate contextual awareness (Van Dijk & Kintch, 1983). The work that goes into designing a usability test plan needs to draw on the audience analysis that captured the mental models that the readers use and the relationships they form to connect the information to the situation. Creating and testing products that support building the relationships shown in Figure 6.1 and allowing the person to develop contextual awareness require a solid understanding of human–information interaction (HII) to increase the success of creating information that communicates effectively to all readers (Albers, 2008).

A confound for developing the test plan is that, in a complex situation, a person's goals tend to be ill-defined, as Redish (2007) explains with some specific examples:

> Furthermore, these initial, high-level goals may be vague, such as, "What in this patient's records will help me understand how to interpret this patient's current complaint and relate that to the patient's overall health?" or "What are we overstocked on and would putting that on sale be good? Is there a trend in this data that I should make my boss aware of?" (p. 106)

Besides being ill-defined, different groups of people come to understand a situation differently using different mental routes and assigning different priorities to information elements. Kain, de Jong, and Smith (this volume, Chapter 14) reveal strongly different feelings across the people they interviewed (local residents) and how they felt different rules and priorities apply to them rather than to tourists when a hurricane warning is issued. The audience analysis should have uncovered these different views and mental routes so the test plan can ensure the system supports them to maximize the information communication for all audience groups. The multidimensional processing strategies used by people (Mirel, 1998) limit the effectiveness of any test plan that tries to verify people follow a single "best way" of reaching contextual awareness. For simple situations, a "best way" exists, but for the complex information situations that form the overall subject of this book, a single way does not exist. Quite simply, if the situation can be broken down and completely defined by a task hierarchy, then relationships and contextual awareness are probably not highly relevant to that situation. However, if it is impossible to create the task hierarchy or it can only be created in a very forced manner, or the hierarchy forms an interconnected web, then testing a person's path to building contextual awareness becomes an essential element of the test plan.

Factors to Consider When Building the Test Plan

The first step in designing an effective system should not be to define the data needs, but to define the communication situation in terms of the necessary information processes and mental models of the users (Rasmussen, 1986).

A design that supports complex situations must help the readers identify the important elements of the situation and the relationship between the elements. As such, it needs to support a reader's goals and not be an acontextual approach to communication. One-size-fits-all or a "design for the novice that the experts will understand too" are both examples of acontextual design. Likewise, there are no average users. Each group forms a fuzzy ball within the overall audience space with an individual somewhere within that ball. The outer edges of that fuzzy ball as well as the center point need to be captured (Albers, 2003; and see Figure 1.1, this volume).

A person's mental model combined with the information salience of the system work together to allow the formation of relevant information relationships and a high-quality contextual awareness. The mental model, or salient information, requires tight integration with the presentation, as readers often ignore any information they think they don't need (Hallgren, 1997).

Mental Models

- *Define the information comfort zone.* People will stop collecting information and make decisions when they feel they have sufficient information. Unfortunately, there is no fixed end point for complex information as there is for procedures. The test must ensure that the feeling of having sufficient information corresponds to actually having sufficient information.

- *Consider both routine and nonroutine situations.* The focus of helping to build contextual awareness should fall within the nonroutine events, where contextual awareness is essential and difficult to efficiently obtain. When faced with nonroutine events, a person's decision making changes to knowledge based, making it the point when the information is most needed (Woods & Hollnagel, 2006) and contextual awareness is the hardest to obtain.

- *Define the different levels of relevant information and relationships for different audience groups.* Part of the design and test process must be to understand the information needs and to ensure that each audience group can obtain those needs.

- *Acknowledge that experience contributes to how a person interacts with and understands the information.* A person with a high knowledge level needs a different presentation than a person with a low knowledge level. It is highly problematical to assume an expert will get full understanding from a design aimed at a low-knowledge reader (McNamara, 2001). People with different knowledge levels need the same information, but they need to receive it differently. It is not a simple matter of how much information to provide; as Klein (1999) points out, "[E]xperts don't have more data. They have the ability to identify which cues to use, whereas novices tend to place too much emphasis on cues that are less diagnostic" (p. 69). The test plans need to consider the cues which direct the reader to relevant information and uncover places where cues are misleading.

Information Salience

- By basing decisions on a good (and correct) mental model, an experienced person focuses his or her attention on relevant information and does not risk being overloaded, unlike an inexperienced person who attempts to monitor too much information. The situation analysis has determined the relevant information and relationships; now the usability test can verify that information gets perceived with a higher salience. More than just helping the experienced person, it will also assist the less experienced person in focusing on the proper information.

- Does the highly salient information help define the environmental or contextual cues that bring into focus a problem's existence and what is needed to resolve uncertainty between potentially competing problems and solutions? Besides simply verifying that cues are perceived, the order of information acquisition and ensuring it is seen when it is relevant to a situation's development need to be captured.

- The order in which people view information strongly influences the salience of later information. Klein (1999) found that people evaluate based on the order in which they receive information, accentuating the importance of presenting information in an order relevant to the situation and for information salience to help drive the order of processing. The information may be hard to integrate and relate to the current situation; relevant and irrelevant information is mixed together, and the information has improper salience (Albers, 2007).

Information Relationships

- The highly salient information and their interrelationships can change as the person progressively understands the situation or as the situation develops (Albers, 2004, 2007). A person has many goals and subgoals active at once, although their relative priority is constantly shifting. As such, the information relevant to the person changes as he or she shifts between goals. Also, the relative information salience of the same information element shifts with respect to the current top-priority goal. Although a person may be shifting between goals quite rapidly, he or she can focus on only one or two (i.e., give them top priority) at once.

- Verify people's expectations for cueing of information relationships between information elements. A fundamental problem with many existing approaches to providing information is the lack of support for building information relationships. The system contains the information but lacks clear user-relevant and visibly apparent connections between information elements.

Contextual Awareness

- Earlier analysis should have defined a person's goals, information needs, and what constitutes good contextual awareness for each major goal. Now the test needs to verify that those goals and information needs are met in a manner that makes sense with respect to the situation. An important element to test is the person's comprehension of the overall situation and the information presented. With the strong cognitive basis of situations that demand high contextual awareness, the tests must focus on that global comprehension.

- Build tests of both routine and nonroutine situations. The focus of helping to build contextual awareness should fall within the nonroutine events, where contextual awareness is essential and difficult to efficiently obtain. A person's decision making changes to knowledge based when he or she is faced with nonroutine events, making it the point when the information is most needed (Woods & Hollnagel, 2006).

Collecting Data

Contextual awareness builds over time to a useful level for developing an intention and making a decision. Thus, as part of the test the person needs time to build a knowledge base. Since contextual awareness tends to be developed over time, any user test scenarios must be realistic and give time to develop the contextual awareness. Typically, time is not a major factor; in most situations relevant to contextual awareness, the person has the time to search for information and review previous information. On the other hand, a good design can minimize the amount of time a person spends searching and minimize the time needed to obtain adequate information. Poor design can mislead the person with irrelevant information or lead him or her down dead-end information paths.

User testing directed toward determining how well a system supports contextual awareness needs to focus on information perception or comprehension rather than information availability. Only through iterative user testing can the designers ensure that the people perceive and comprehend the information correctly.

Tests While a Person Is Still Developing Contextual Awareness

Test information salience. Ask directed questions to find out if the person's expectations of what should be the most salient information map onto the situation's reality and the presentation. Holes in the mapping need to be explored to determine whether they comprise a local or global problem and what factors are causing the holes.

Test information relationships. Ask the person how the information relates to other information. Since a significant part of developing contextual awareness exists in understanding information relationships, the ways a person comes to make those connections and what impedes them must be uncovered. It is important that the people show deep understanding by stating the relationships in their own words and not verbatim repeating text from the presentation.

Tests of the Quality of a Person's Contextual Awareness

Test the comprehension of the information. Pause and ask the person to give a summary of what he or she knows. During the early stages of interacting with the information, a person may have only poorly developed contextual awareness, but as he or she continues to interact, the contextual awareness should improve. Test deep knowledge, rather than surface knowledge. The person can be asked to describe the situation. The person should give a description of the situation in his or her words, without parroting the system content. A person who does not understand a situation will often be able to repeat the pertinent words and phrases of what he or she read, but not have an understanding of how they fit into the situation.

Ask for predictions of the future behavior or development of the situation. Also, ask what information led to those predictions.

Ask the person to make a decision and explain why. The focus on the analysis is on the explanation of why and not on the correctness of the decision. With poor contextual awareness, people can know something is occurring or that a particular piece of information exists, but they cannot easily find relevant, related information or do not understand how it relates to the overall situation. Good contextual awareness does not guarantee a good decision, nor does poor contextual awareness mean the decision will be incorrect. But by understanding the underlying logic for the decision, the test can uncover if people grasp the situation or what part of the design misled them.

Ask about cause-and-effect relationships within the situation. Does the person know what caused events to occur and how that cause-and-effect ripples through the entire situation? A person with low contextual awareness may view each event as independent or, at least, not see how they relate.

Data Analysis

A significant element of the post-test analysis should be to identify information that was poorly or incorrectly perceived with respect to the test scenario. Poor design can hide information (a problem in any usability test) or give highly relevant information low salience.

The analysis needs to reconstruct the mental model building and situational thought processes while it builds a picture of how the person becomes fully aware of the relationships within the overall situation and uncovers the roadblocks to contextual awareness. And it must accept that different audience groups build different mental models, have different situational thought processes, and need different subsets of the information with different presentations.

Avoid analysis for the "average" user. There are no average users. At best, the user's abilities form a (skewed) bell curve that will often have multiple peaks. Personas could be viewed as sitting at the top of each peak.

Points to consider as part of the analysis:

- Follow the click stream. Review the test logs to see how the person interacted with the system. In a complex situation there is no single correct path, but an analysis can uncover if the person was following a logical path or randomly interacting with the information.

- Uncover the environmental or contextual cues that brought into focus the problem's existence and what is needed to resolve uncertainty between potentially competing problems and solutions.

- Uncover common information errors or misunderstandings, and why those errors exist (Albers, 2009b). Later design work can work to improve the information flow to minimize these problems.

Conclusion

Traditionally, technical communicators developed texts for highly structured situations, with the basic goal of efficiently completing a task. However, a collection of information elements alone, no matter how they are arranged or organized, does not address a person's information needs beyond simple look-ups. Someone can find that a value equals 5, but for more complex information needs, the person needs to form relationships between the information elements. It's not knowing 5 that is important, but knowing what 5 means with respect to the other information in the situation. Likewise, an underlying mantra of technical communication seems to be "Make the information as simple as possible," which is almost always an oversimplification of the communication needs. Modern communication situations are no longer highly structured; instead, they have shifted to complex situations revolving around information seeking, problem solving, and decision making, which call for the reader to develop a high level of contextual awareness. Design based on contextual awareness gives us a basis for determining what "simple as possible, but no simpler" means. At the other end of the spectrum, it gives a basis for breaking up the often convoluted text obtained from the subject matter exerts (SMEs), knowing how to reformat it, and establishing what to include and what really can be considered information that "everyone who does this already knows."

A strong focus of current research agendas deals with how people search for and find information within an information system. Obviously, finding relevant information is the first step in acquiring an understanding of a situation. However, finding the information is but a first step; after it is found, the person needs to interpret it and decide if and how it applies. Even if a person gets presented with information that is accurate and reliable, that information is essentially useless to the person unless he or she is able to interpret it and apply it to his or her current situation (Albers, 2007). After getting the person's attention, the ultimate test of whether information communicates is what actions a person performs after reading it. Contextual awareness gives a foundation from which to perform the analysis, design, and testing to ensure a person can understand and relate information to the current situation. Usability testing focused on all four HII phases (Albers, 2008) helps to ensure that the underlying system communicates information, rather than simply provides access to information.

Interact: A person needs to be able to manipulate and arrange the information to fit into a presentation that makes sense for his or her situation and information needs.

Find: A person needs to be able to find the information in the right order, and the presentation needs the proper salience to ensure that relevant information is not overshadowed by irrelevant or less important information, which may be easier to present.

Interpret: A person's understanding of a problem or situation comes from mentally forming relationships, not from simply knowing the information. Different people have different approaches to interpreting information, and many factors can influence the interpretation.

Use: A person ultimately needs to shape the information that has been assembled into a coherent view of the current situation and make a decision. Unfortunately, the correctness of decisions cannot be a substantial measurement of complex system usability since people can have good information and understanding and still make bad decisions (Endsley, 1995).

For complex information needs, a document suits a person's needs when it provides a structure fitting the problem and containing specific information relevant to the current situation. Such a structure supports building information relationships and, in turn, developing a deeper understanding resulting in good contextual awareness. Usability testing of complex information must focus on how effectively and how efficiently the information supports developing those relationships and contextual awareness. Granted, a design team cannot predetermine a person's activities or how he or she will interpret information, but, at the same time, the team must realize the influence the design has on how a person interprets information. The reader should not be forced to shoulder the entire responsibility for using and understanding the information (Chalmers, 2004). How the design team decides to structure the information plays a major role in the resulting clarity of the content, how easily a person can form relationships within that content, how well he or she understands it, and, ultimately, how well it communicates.

How to effectively structure usability tests in complex situations and how to perform the analysis, design, and testing remain open questions. In particular, we need better metrics to capture levels of contextual awareness rather than surface knowledge and usability. For that matter, we need to develop clear metrics of what it means to have good contextual awareness. Development of metrics to support contextual awareness will provide a strong value-add for the technical communicator developing the system's information content.

Reader Take-Aways

- The goal of a usability test in a complex situation is to ensure that the information is being efficiently communicated and that the person is building information relationships.
- Developing an understanding of a complex situation requires understanding the relationships between information elements. A usability test needs to examine how people know the information exists, what it means, how it is interrelated to other pieces of information, and how they build those relationships.

- Contextual awareness provides a means of discussing how people understand both a situation and their ability to make predictions about its future development.

- Complex systems require completely different views of testing and need to consider more than "Can people look at the site and list five of the eight product benefits?" Instead, they need to consider if people comprehend the product benefits information and can apply it to their situation.

- Usability testing in complex situations needs to focus on knowledge-based measures that measure the internal thought processes. Typically this involves asking the person to describe the current situation at various points and asking what information led to those conclusions. Performance-based measures (task time, click counts, etc.) provide little useful information.

References

Albers, M. (2003). Multidimensional audience analysis for dynamic information. *Journal of Technical Writing and Communication, 33*(3), 263–279.

Albers, M. (2004). *Design for complex situations: Analysis and creation of dynamic Web information.* Mahwah, NJ: Lawrence Erlbaum.

Albers, M. (2007, October 22–24). Information salience and interpreting information. Paper presented at the 27th Annual International Conference on Computer Documentation, El Paso, TX.

Albers, M. (2008, October). Human-information interaction. In *paper presented at the 28th Annual International Conference on Computer Documentation,* September 22–24, 2008, Lisbon Portugal.

Albers, M. (2009a). Design aspects that inhibit effective development of user intentions in complex informational interactions. *Journal of Technical Writing and Communication, 39*(2), 177–194.

Albers, M. (2009b, October 4–7). Information relationships: The source of useful and usable content. Paper presented at the 29th Annual International Conference on Computer Documentation, Indianapolis, IN.

Ash, J., Berg, M., & Coiera, E. (2004). Some unintended consequences of information technology in health care: The nature of patient care information system-related errors. *Journal of the American Medical Informatics Association, 11*(2), 104–112.

Belkin, N. (1980). Anomalous states of knowledge as a basis for information retrieval. *Canadian Journal of Information Science, 5,* 133–143.

Chalmers, M. (2004). A historical view of context. *Computer Supported Cooperative Work, 13,* 223–247.

Cilliers, P. (1998). *Complexity and postmodernism: Understanding complex systems.* New York: Routledge.

Conklin, J. (2003). *Wicked problems and fragmentation.* Retrieved from http://www.cognexus.org/id26.htm

Endsley, M. (1995). Toward a theory of situation awareness in dynamic systems. *Human Factors, 37*(1), 32–64.

Ganzach, Y., & Schul, Y. (1995). The influence of quantity of information and goal framing on decisions. *Acta Psychologia, 89,* 23–36.

Hallgren, C. (1997). Using a problem focus to quickly aid users in trouble. In *Proceedings of the 1997 STC Annual Conference.* Washington, DC: Society for Technical Communication.

Howard, T. (2008). Unexpected complexity in a traditional usability study. *Journal of Usability Studies, 2*(3), 189–205.

Hsee, C. (1996). The evaluability hypothesis: An explanation of preference reversals between joint and separate evaluations of alternatives. *Organizational Behavior and Human Decision Processes, 46,* 247–257.

Klein, G. (1993). A recognition-primed decision (RPD) model of rapid decision making. In G. Klein, J. Orasanu, R. Calderwood, & C. Zsambok (Eds.), *Decision making in action: Models and methods* (pp. 138–147). Norwood, NJ: Ablex.

Klein, G. (1999). *Sources of power: How people make decisions.* Cambridge, MA: MIT Press.

McNamara, D. S. (2001). Reading both high and low coherence texts: Effects of text sequence and prior knowledge. *Canadian Journal of Experimental Psychology, 55,* 51–62.

Mirel, B. (1998). Applied constructivism for user documentation. *Journal of Business and Technical Communication, 12*(1), 7–49.

Mirel, B. (2003). Dynamic usability: Designing usefulness into systems for complex tasks. In M. Albers & B. Mazur (Eds.), *Content and complexity: Information design in software development and documentation* (pp. 233–261). Mahwah, NJ: Lawrence Erlbaum.

Morrison, J. B., Pirolli, P., & Card, S. K. (2001, March 31–April 5). A taxonomic analysis of what World Wide Web activities significantly impact people's decisions and actions. Interactive poster, presented at the Association for Computing Machinery's Conference on Human Factors in Computing Systems, Seattle, WA.

O'Malley, C. (1986). Helping users help themselves. In D. Norman & S. Draper (Eds.), *User-centered system design: New perspectives on human-computer interaction.* Hillsdale, NJ: Lawrence Erlbaum.

Orasanu, J., & Connolly, T. (1993). The reinvention of decision making. In G. Klein, J. Orasanu, R. Calderwood, & C. Zsambok (Eds.), *Decision making in action: Models and methods* (pp. 3–20). Norwood, NJ: Ablex.

Rasmussen, J. (1986). *Information processing and human-machine interaction: An approach to cognitive engineering.* New York: North-Holland.

Redish, J. (2007). Expanding usability testing to evaluate complex systems. *Journal of Usability Studies. 2*(3), 102–111.

Santa Clara Grand Jury. (2005). *Problems implementing the San Jose police computer aided dispatch system.* Retrieved from http://www.sccsuperiorcourt.org/jury/GJreports/2005/SJPoliceComputerAidedDispatch.pdf

Schroeder, W. (2008). Switching between tools in complex applications. *Journal of Usability Studies, 3*(4), 173–188.

Spiro, R., Feltovich, P., Jacobson, M., & Coulson, R. (1991). Cognitive flexibility, constructivism, and hypertext: Random access instruction for advanced knowledge acquisition in ill-structured domains. *Educational Technology, 5,* 24–33.

Thuring, M., Hannemann, J., & Haake, J. (1995). Hypermedia and cognition: Designing for comprehension. *Communications of the ACM, 38*, 57–66.

Van Dijk, T., & Kintch, W. (1983). *Strategies of discourse comprehension*. New York: Academic Press.

Woods, D., Patterson, E., & Roth, E. (2002). Can we ever escape from data overload? A cognitive systems diagnosis. *Cognition, Technology, & Work, 4*, 22–36.

Woods, D. & Hollnagel, E. (2006). *Joint cognitive systems: Patterns in system engineering*. Boca Raton, FL: CRC Press.

7

Continuous Usability Evaluation of Increasingly Complex Systems

Vladimir Stantchev

Public Services and SOA Research Group, Berlin Institute of Technology

Fachhochschule für Oekonomie und Management

CONTENTS

Introduction 134
Overview of Usability Evaluation Methodology 135
    Process Evaluation of Domain Service 136
    Evaluation of the Existing Information Systems 136
    Identification of Decision Paths and Actions That Can Benefit
        from Increasingly Complex Systems 137
    Implementation 137
Usability Evaluation of Increasingly Complex Systems in Health Care 138
    Usability Evaluation Techniques 138
    Design of the System 138
    ASUR Model of the System 140
    Overview of Evaluation Results 140
Usability Evaluation in Increasingly Complex Systems for Project
    Portfolio Management 142
    Tasks and Activities in Project Portfolio Management 144
    Increasingly Complex Systems for Project Portfolio Management 145
    Continuous Evaluation of PPM Systems 146
    Experimental Usability Evaluation 149
    Assumptions 150
    Domain Services of an Increasingly Complex PPM System 151
    Service Coverage 151
Conclusion 153
Reader Take-Aways 153
References 154

Abstract

Current trends in usability evaluation research focus on providing the right mix of empirical and technical methods. This chapter presents an approach for continuous evaluation of increasingly complex systems based on such a mix. The approach is iterative and covers usability aspects continuously throughout the following phases: (1) evaluation of existing tasks and processes within the specific domain service; (2) evaluation of the information systems that currently support the service; (3) identification of decision paths and actions (together with interaction patterns) that can benefit from a more complex system; as well as (4) design and implementation of the complex system, with focus on usability and integration. The verification of the approach was conducted in two complex application domains—health care, more specifically clinical environments, and industrial engineering, more specifically project portfolio management.

Introduction

Recent research on usability studies has questioned established methods and their suitability for testing more complex systems (Redish, 2007). There are also calls for a better balance in usability studies between empirical observation and rhetoric, and technical communicators are regarded as best suited to deal with the challenges of such a new approach (Johnson, Salvo, & Zoetewey, 2007). Following these works, we propose a new methodology for evaluating complex systems that is based on a combination of empirical observation and data gathering, on the one side, and technical communication (expert interviews and workshops) between domain experts and system analysts, on the other.

Two complex domains serve as verification areas for this approach and present sometimes contrasting ways of perceiving complexity. The first domain is health care, where our particular focus is on increasingly complex systems that support operating rooms for minimally invasive surgery. The second domain is industrial engineering. There we evaluate increasingly complex systems for project portfolio management (PPM) in the context of product life cycle management (PLM) and innovation management.

In these areas, experts are very focused on their domains and notorious for their reluctance to adhere to recommendations from external analysts. Furthermore, they have strongly differing objectives in the areas of usability and computer-supported work. While in the health care domain we are aiming to provide a context-aware, process-driven "ambient intelligence" system (Stantchev, 2008, 2009a, 2009b) for optimized workflows in a life-critical application, in the PPM domain we deal with a heavily data-driven and nevertheless subjective approach. Here a decision to start or terminate a product development project is made by an expert, based on a wide range of objective information (e.g., project costs and requirements) and subjective categories (e.g., strategic alignment or degree of innovation) (Stantchev & Franke, 2009; Stantchev, Franke, & Discher, 2009).

In this chapter, we first outline the methodology with its specific steps for the continuous evaluation of user requirements regarding functionality and usability of an increasingly complex system. Then we proceed to discuss the verification of the approach in the two expert domains. A list of key reader take-aways concludes the chapter.

Overview of Usability Evaluation Methodology

Our methodology focuses on continuous usability evaluation of increasingly complex systems. To cope with the growing complexity we use an iterative approach (see Figure 7.1) that covers usability aspects continuously throughout the following phases:

1. Evaluation of existing tasks and processes within the specific domain service
2. Evaluation of the information systems that currently support the service

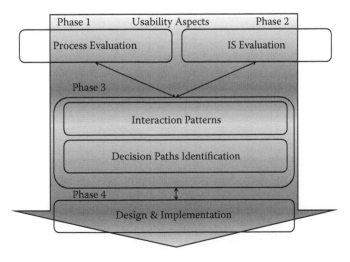

FIGURE 7.1
Overview of the usability evaluation methodology.

3. Identification of decision paths and actions (together with interaction patterns) that can benefit from a more complex system (e.g., a mixed reality system)

4. Design and implementation of the complex system, with focus on usability and integration

The process is iterative, so that we go through these phases when new requirements need to be reflected by the system. The following subsections will describe each of these phases in more detail.

Process Evaluation of Domain Service

In order to capture the current state of the domain service, we use a standard process-oriented approach—system analysis (Krallmann, Schoenherr, & Trier, 2007). It consists of several steps as depicted in Figure 7.2, and is designed as a general blueprint for process optimization projects (Krallmann et al., 2007).

For the process evaluation of the domain service, we focus on the situation analysis part of the approach. Here we use notations such as event-driven process chains (EPCs) and business process modeling notation (BPMN). We use usability evaluation techniques such as cognitive walkthrough, action analysis, field observation, and questionnaires (Holzinger, 2005). Our user groups are professionals from the problem domains (e.g., clinicians from hospitals with large surgical departments and specialized clinics, or project portfolio managers from industrial companies).

Evaluation of the Existing Information Systems

In this phase, we focus on the evaluation of the information systems (IS) that currently support the service. This activity is the second main aspect

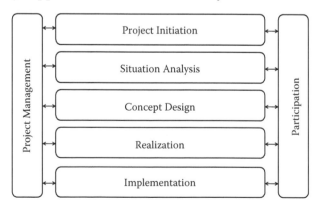

FIGURE 7.2
Process model of system analysis.

of the situation analysis (Krallmann et al., 2007). During this phase, we enhance the process models from the previous step with details about the information systems used at every step of the process. This gives us a complete set of the current process activities, decision paths, as well as information systems currently in use for the provision of the domain service.

Identification of Decision Paths and Actions That Can Benefit from Increasingly Complex Systems

The starting point for this phase is the complete set of activities, decision paths, and information systems that we compile in the previous step. The methodology here is often domain specific. For example, in the health care domain we use a mixed-reality system as an increasingly complex system and the ASUR (adapters, systems, users, and real objects; Dubois, Gray, & Nigay, 2002) design approach. There we start with a classification that maps these artifacts to the ASUR design approach. We then discuss with clinicians who fall in the user (Component U) category which decision paths and actions are currently underperforming and would benefit from a mixed-reality system. Here we also identify relationships among the ASUR components ($\Rightarrow$, $\rightarrow$, and $=$). The result is a list of decision paths and actions that will be addressed by the mixed-reality system in the current iteration of the design and implementation process. More details about the application of the ASUR design approach in health care can be found in Stantchev (2009a).

In the domain of industrial engineering, we focus on interaction patterns between users and information systems at every process step. This typically entails a more in-depth analysis about needed information for the decision paths, as well as data quality aspects. Examples include project templates, project data (e.g., costs, duration, and resources), as well as project categories and priority drivers (e.g., business or technical). More details about service coverage of information systems in this domain are provided in Stantchev et al. (2009), while specific knowledge and learning aspects are discussed in Stantchev and Franke (2009).

Implementation

During this phase, we address usability aspects twofold. On the one side, we account for usability as a key criterion for technology selection. On the other side, acceptance tests and system transition deal explicitly with usability aspects and nonfunctional properties (NFPs). We use service orientation as an architectural model, and the assurance of NFPs (Stantchev & Malek, 2009; Stantchev & Schröpfer, 2009) is a key usability aspect in such distributed, loosely coupled environments.

Usability Evaluation of Increasingly Complex Systems in Health Care

Our application scenario focuses on the surgical sector—one of the largest cost factors in health care, and at the same time a place where high creation of value takes place. For the effective utilization of the surgical sector, we need well-performing pre- and postoperative processes. The surgery (equipments and specialists) is a fixed (and expensive) resource. So pre-and postoperative processes need to be aligned and provide for an optimized utilization of this resource. In this section we show exemplarily how we conduct continuous usability evaluation, generate scenario descriptions as text, then convert them to the EPC notation and use this EPC notation to create the ASUR model of the system.

Usability Evaluation Techniques

"One of the basic lessons we have learned in human-computer interaction (HCI) is that usability must be considered before prototyping takes place" (Holzinger, 2005). This applies particularly to the health care domain. Nevertheless, usability studies are still not considered an obligatory part of design in this domain. A comprehensive overview of usability evaluation techniques is presented in Holzinger. It differentiates between inspection methods (heuristic evaluation, cognitive walkthrough, and action analysis) and test methods (thinking aloud, field observation, and questionnaires). These techniques are categorized according to their applicability in different phases of the system development process; to their time requirements and the number of users and evaluators (and the complexity of equipment) needed for the evaluation; as well as to their intrusiveness. A historic overview and recent developments in usability research of augmented and mixed-reality systems in the health care domain are presented in Behringer, Christian, Holzinger, and Wilkinson (2007).

Our approach considers intrusiveness as a particularly important aspect in health care; therefore, we apply cognitive walkthrough and action analysis as inspection methods. These methods can also be applied in different phases of our iterative methodology and are therefore well suited for continuous evaluation. They also require high expertise from the evaluators, who are either clinicians themselves or design specialists with extensive domain knowledge. As test methods, we use questionnaires and expert interviews.

Design of the System

Approaches to redesign and reorganize perioperative patient flow and work processes in operating rooms often bring changes in operating room architecture (Sandberg et al., 2005). Figure 7.3 shows an overview of perioperative and postoperative processes in our application scenario. These were evaluated using our approach as already described and the already introduced usability evaluation techniques.

The perioperative processes start with a notification from an operating room nurse or an anesthesia nurse that the clinical staff should transport the next patient to the operating room. This action takes place in the ward (location area 1). Then a transport service or a nurse moves the patient from the ward to the operating room area (location area 2). In the main registration

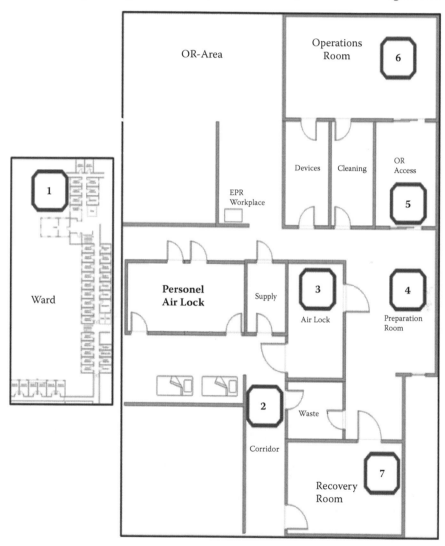

FIGURE 7.3
A health care scenario where increasingly complex systems can be considered.

area, the clinicians transfer the patient from the ward bed to an operating room table (location area 3). Next, the staff moves the operating room table to the preparation room (location area 4), where the anesthesia nurse or the anesthetist prepares the patient for the operation. The stop that follows is in the induction area, where the patient is anesthetized (location area 5). Finally, the staff moves the patient to the operating room, where the preparation for the operation starts, for example, operation-specific bedding, sterile coverage, and so on (location area 6). After the operation ends, the patient is taken out of the operating room and back via the induction area (location area 5) and the preparation room (location area 4) to the recovery room (location area 7). The process ends with the transport of the patient back to the ward (location area 1).

The corresponding EPC model is shown in Figure 7.4. It describes the sequence of the actions, together with possible decision paths. The location areas for every action are depicted with the same numbers as in Figure 7.3.

An important prerequisite for a mixed-reality system that supports this process is object and person location sensing. Such localization can be done by increasingly complex systems via different technologies, for example, Ultra Wide Band (UWB), Bluetooth (BT), or wireless local area network (wireless LAN, or WLAN) location applications. We have described in detail the position-sensing techniques we use in Stantchev, Hoang, Schulz, and Ratchinski (2008).

ASUR Model of the System

As already stated, we derive process models such as EPCs and BPMN from the early steps of our usability evaluation and then model their artifacts in the ASUR description (Dubois et al., 2002). The aim of our system is to allow a person to assess the overall process by blending the real world of a clinic with virtual information in a mixed-reality system. Therefore, the adapters need to allow the person to control actual movement of the real objects by mapping it to the mixed-reality system (see Figure 7.5 and Stantchev, 2009a, for more details about the different ASUR components—adapters, systems, users, and real objects—and their relationships).

Overview of Evaluation Results

The continued evaluation and assurance of NFPs of our system as a specific usability aspect are evaluated in several works, for example, Stantchev and Malek (2009) and Stantchev and Schröpfer (2009). The questionnaires and expert interviews we used as usability evaluation test methods were addressed toward clinicians who use our system. An overview of the surveyed group is given in Table 7.1, and a summary of results is presented in Table 7.2. Overall, there is a substantial ($\Delta > 50\%$) increase in usability throughout one iteration of our methodology.

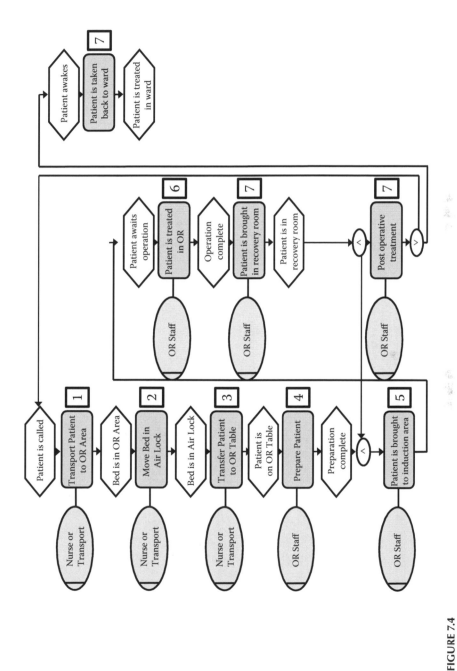

FIGURE 7.4
A sample event-driven process chain within the application scenario.

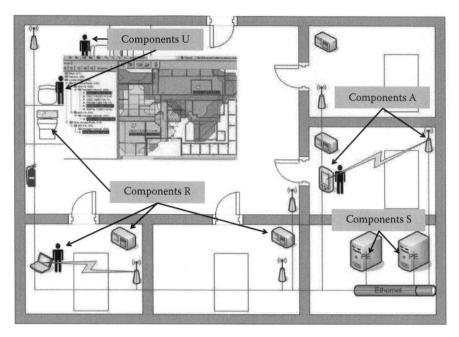

FIGURE 7.5
Overview of the ASUR components in the health care scenario.

TABLE 7.1

Group Profile

Group Characteristics	Number of Participants	%
Nurses	10	33.33
Surgeons	6	20
Anesthetists	3	10
Management	1	3.33
Other	10	33.33
Total	30	100

Usability Evaluation in Increasingly Complex Systems for Project Portfolio Management

Portfolio management is one of the key activities within innovation management. Authors such as Wheelwright (1984); Clarke, de Silva, and Sapra (2004); Cooper and Edgett (2007); and Calantone, Di Benedetto, and Schmidt (1999) have written extensively on this subject. There is also a variety of methods these authors have proposed for project selection and portfolio optimization. Quality criteria for project selection and priority such as value

TABLE 7.2

Summary of Usability Evaluation Results

Dependent Variable	Before	After First Iteration
Average patient preparation time (min)	31.20	16.30
Average additional preparation tasks needed (number)	12	7
Average number of process errors (perioperative)	6	2
Average number of process errors (postoperative)	5	2
Clinicians: user satisfaction with the increasingly complex system (%)	46	83
Patients: user satisfaction with the increasingly complex system (%)	52	89

creation and strategic fit are becoming even more important (Rajegopal, Waller, & McGuin, 2007). Such criteria—more specifically, value enhancement, strategic conformity, and risk balancing—are the deciding factors for resource distribution to projects (Cooper & Edgett). We regard project portfolio management (Rajegopal et al., 2007) and the increasingly complex systems that support it as promising areas to apply our continuous evaluation methodology.

The application of portfolio management further to the management of complex project landscapes is a transfer from its origin in financial theory. The portfolio management process is implemented within the innovation and product development functions, the central components of product life cycle management. PLM itself is a complex life cycle approach that serves as an integrating point for all the information produced throughout all phases of a product's life. This information (and thus PLM itself) can be relevant to everyone in an enterprise, both managerial and technical positions (Sudarsan, Fenves, Sriram, & Wang, 2005). It relates directly to a wide range of professional communication interactions among engineers, developers, managers, as well as key suppliers and customers.

For our usability evaluation, we regard PPM in its context of project, program, multiproject, and portfolio management (see Figure 7.6). Projects and programs are singular tasks. They are explicitly separated from one another and are the responsibility of their project managers.

With the operation of many project initiatives, conflicts occur within the whole project landscape in terms of overlapping knowledge areas, intraproject coordination, resource and competence staffing, as well as budget allocation. A portfolio serves as a precisely defined collection of associated

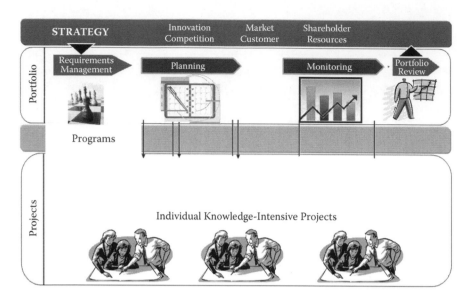

FIGURE 7.6
Project landscapes and project portfolio management (PPM).

projects and programs. Therefore, the organization and administration of a portfolio plus its overall implementation to achieve the company's strategic and business goals are closely related and interdependent. The components of a portfolio—namely, projects and programs—are generally quantifiable and measurable, so they can be evaluated and prioritized against each other (Project Management Institute, 2006). An optimized global allocation of resources needs a complete view across the whole corporate project landscape where all existing portfolios should be considered. Overall, we can regard portfolio management as the central management of one or several portfolios in terms of identification, prioritization, authorization, organization, realization, and the controlling of its associated projects and programs (Stantchev et al., 2009).

Tasks and Activities in Project Portfolio Management

There are three general views of portfolio management in project landscapes—organizational, process oriented, and chart oriented. All of them are supported by increasingly complex systems. The first is an organizational view with regard to its hierarchical administration of projects and the central collection and consolidation of project information such as project goal, budget cost, timeline, resource demand, and risk class in superior project portfolios (Rajegopal et al., 2007). The second process-related view describes the iterative and cyclic process of selecting and prioritizing new requests and active projects (Cooper & Edgett, 2007). A third understanding of the word *portfolio* is the multi-axis portfolio chart, which is used as a graphical management decision tool (Pepels, 2006). As its own organizational unit, the project and portfolio management office (PMO) governs as a neutral instance and is authorized to audit the accuracy of the process compliance.

The following steps typically compose a product development process: project proposal acceptance; requirement specification and preliminary organizational activities; conceptual work on product and process design (e.g., *failure modes and effects analysis* [FMEA], *quality function deployment* [QFD], and feasibility studies); the actual product development, including quality management and prototyping; market launch and distribution marketing, and the start of series manufacturing (in the case of mass products). A so-called stage gate process was developed to provide a generic pattern to formalize, structure, and thus improve the efficiency of these innovation management and product development tasks (Cooper, Edgett, & Kleinschmidt, 2001). This process is subdivided into clearly defined phases (stages) together with gates between them. They act as thresholds to measure and guarantee a certain level of quality. To pass a gate, specific partial results, measures, and values as the gate's input value have to be achieved, for instance a high-quality test performance, the completeness of the project request's business case, or a minimal fulfillment of evaluation criteria values. The output result of a gate states the decision whether the project shall be conducted and thus proceed to the next stage. There it is benchmarked against higher thresholds regarding the requirements to pass a further gate. Thereby, the resource supply must be also committed, and the financial budget allocated. Phases and steps of the associated PPM process are shown in Figure 7.7 and described in detail in Stantchev et al. (2009) and Stantchev and Franke (2009).

Increasingly Complex Systems for Project Portfolio Management

Currently there are more than two hundred different systems available in the area of software for project management support. These include systems for management of singular projects (10 percent), multiproject management systems (70 percent), and enterprise project management systems (EPM; 20 percent). Systems for management of singular projects support scheduling and critical path estimations. They also allow for visualization of results and resource planning. Multiproject management systems offer extended functionality for management and coordination of multiple projects with respect to resources, time frames, and cost forecasting. The EPM systems cover planning and management of all projects within an enterprise throughout the life of a project. They also serve as tools for project portfolio planning and forecasting, aggregating business and technology parameters to support more informed portfolio decisions.

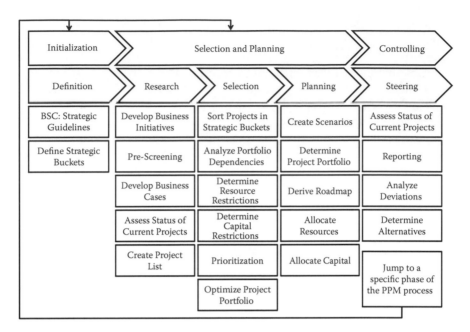

FIGURE 7.7
Activities within project portfolio management (PPM).

Continuous Evaluation of PPM Systems

As part of our continuous usability evaluation in this area, we conducted a survey with a focus group from companies in several industrial categories. Our survey covers Phases 1–3 from our methodology and uses the approach presented in Banker, Bardhan, and Asdemir (2006). We are conducting the verification for every dimension of our maturity models—management, governance, process, system, resource management, and social aspects. Thereby, we assess quantitative and qualitative parameters during the survey. This allows us to more clearly establish correlations between PPM maturity and project success. Table 7.3 shows the demography of our focus group. The representation of industrial categories in the focus group was derived from a much larger (greater than five thousand) data set of central European companies. The response rate was 79.6 percent, and we accounted for nonresponse bias by comparing demography variables (e.g., the number of employees and revenue) with those of nonparticipating companies. We conducted the survey as an interview that was supported by a twelve-page questionnaire.

Figure 7.8 shows the relevance of PPM as one aspect of our evaluation—more than 90 percent of the respondents regard the recommendations of the PPM process as highly relevant to the actual portfolio decision. Figure 7.9 shows our results concerning the hierarchy level of PPM application—more than 75 percent of the respondents apply PPM at the level of a business unit, or companywide. Finally, Figure 7.10 shows the decision criteria for project

TABLE 7.3

Group Profile of Enterprises for the Continuous Evaluation of Increasingly Complex PPM Systems

Industrial Category	Number of Participating Enterprises	%
Automotive	9	22.5
Industrial products	15	37.5
High-technology	4	10
Consumer	10	25
Transportation	2	5
Total	**40**	**100**

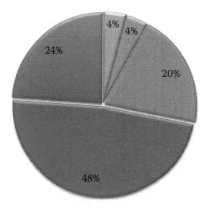

Consequences of Portfolio Recommendations

- Not regarded
- Regarded as information
- Regarded as decision-relevant
- Mostly accepted as decisions
- Unconditionally accepted

FIGURE 7.8

Acceptance of results of the project portfolio management (PPM) process in the portfolio decision.

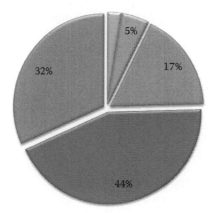

Hierarchy of Portfolio Management

The application of Project Portfolio Management is...

- Individual
- Department-specific
- Includes multiple product line
- Within a business unit
- Companywide

FIGURE 7.9

Hierarchy of project portfolio management (PPM).

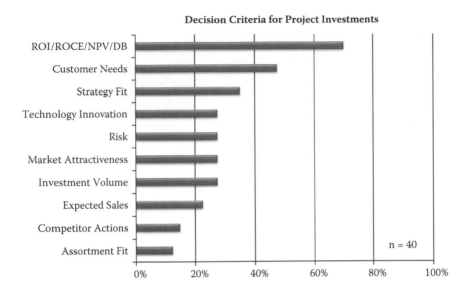

FIGURE 7.10
Decision criteria for project investments.

investments that our respondents most commonly considered. It shows that almost all considered criteria are increasingly complex. Examples are customer needs, strategy fit, innovation and technology, risk, market attractiveness, as well as assortment fit. While return on investment (ROI) was most commonly used as a decision criterion, it also exhibits an increasing complexity, as it typically denotes the expected ROI from a given project investment.

Another aspect of our evaluation regarded the usage of increasingly complex systems to support PPM. Figure 7.11 shows an overview of results. Our results show that missing integration and applications that do not cover the process completely are perceived as major usability problems for increasingly complex systems in the context of PPM.

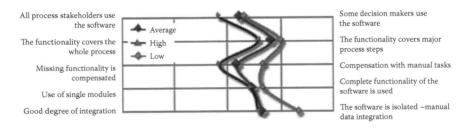

FIGURE 7.11
Major usability problems of increasingly complex project portfolio management (PPM) systems in the view of our respondents.

One of our main objectives in the overall project is the definition and evaluation of a holistic maturity model for PPM in enterprise project landscapes. Figure 7.12 shows the aggregated view of maturity levels of all respondents in the six different dimensions—management, governance, process, systems, resource management, and social aspects. Of particular relevance are the comparatively low levels in the dimensions of systems and resource management. They reflect a significant gap concerning increasingly complex systems within PPM—the high requirements of the areas of management, governance, and process are not met from existing software systems. Furthermore, resource management seldom accounts for competence and knowledge profiles of the personnel resources.

Experimental Usability Evaluation

Our experimental usability evaluation focuses on the configuration of an increasingly complex PPM system. The system is a standard EPM system which is then used as an experimental setup for usability evaluation. We apply a standard software selection process with application-specific criteria, similar to the one used in Stantchev et al. (2008). We derived the criteria from Stantchev and Franke (2009). Furthermore, we used studies from Gartner and Forrester in the area of PPM software to determine a standard software solution for our experimental evaluation; see also Stantchev et al. (2009).

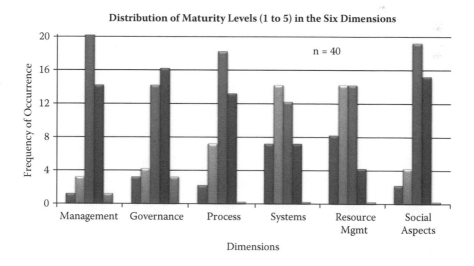

FIGURE 7.12
Maturity levels of the respondents: Bars in every dimension denote the occurrence of a certain maturity level (e.g., twenty-two respondents have a maturity level of 3 in the dimension *management*).

In order to evaluate service coverage, we considered a project landscape that was close to reality, but with a reduced granularity (see Table 7.4). The risk assessment and the strategic drivers were the ones used in the real world.

Assumptions

Our assumptions for the experimental setup were as follows:

- Project duration is aggregated at the level of a year (e.g., duration can be one year, two years, and so on).
- Projects have constant resource needs throughout their duration.
- We regard five permanent resource-skill types; they all use the same cost center.
- We do not regard temporary or external resources.
- Only the following project cost centers exist: internal labor, administrative overhead, and capital.
- Project costs have a uniform distribution throughout the project duration.
- Although we can categorize projects in four classes, we do not differentiate strategic buckets with regard to governance workflow, budget, and resource availability.

TABLE 7.4

Data Matrix of the Experimental Setup of an Increasingly Complex PPM System as Compared to a Real Project Landscape

Project Data	Experimental Setup	Real Setup
Costs	9	**2,700**
Cost types	3	15
Cost departments	1	5
Granularity	(Annually) 3	(Monthly) 36
Resources	15	720
Skill types	5	20
Granularity	(Annually) 3	(Monthly) 36
Financial benefits	9	180
Types	3	5
Granularity	(Annually) 3	(Monthly) 36
Strategic drivers	12	12
Risk	15	15
Data per project	75	**3,647**
Number of projects	12	100
Data set (Data per Project * Number of Projects)	900	364,700

- Our selection process is based on the initial portfolio decision (i.e., we do not account for already-running projects at the beginning).

Comparisons with existing portfolios in industry show that they exhibit a more granular structure (resulting in increasing complexity) in the areas of costs, resources, and financial benefits. Risk factors and strategic drivers are generally the same.

Domain Services of an Increasingly Complex PPM System

The Microsoft Office Project Portfolio Server (MOPPS) consists of three modules: (1) Builder, (2) Optimizer, and (3) Portfolio Dashboard. The Builder structures the input of the business case, the gathering of all proposed projects in a central database for reviews, and *go* or *stop* decisions. Potentially authorized projects are kept until the deadline for the portfolio process. This allows us to conduct the selection and prioritization actions using the Optimizer. It includes weighting strategic factors; prioritizing according to our multidimensional evaluation framework; considering restrictions, dependencies, and must-do status; as well as detailed portfolio analysis using scenario-based techniques, efficient frontier, and charts. The objective is the selection of the most valuable projects within the available budget. During the plan phase, we export the data of the selected projects to the Microsoft Project Server, where we define the specific project road map and assign concrete workers and machines instead of the abstract resource types that we used during the portfolio decision. The advancement of the project (as represented in the Microsoft Project Server) is continuously monitored by the Portfolio Dashboard during the manage phase. Here we can generate reports, require resource or cost adaptations, and monitor early warning indicators. Furthermore, every running project is continuously reevaluated concerning its benefits relative to other running or newly proposed projects. Figure 7.13 shows the result of the portfolio process conducted by MOPPS. Selected projects are automatically transferred to the life cycle step entitled Project Selected, and declined projects to Suspended. A suspended project can also be ultimately canceled.

Service Coverage

Overall, the evaluated system offers a fair coverage of the PPM domain services (see Table 7.5). The further integration with fine-grained resource data (e.g., competence and experience profiles of personnel resources) leads to an increasing complexity and requires a new iteration of our usability evaluation methodology starting from Step 1.

Project Name	Application Decision	Current Workflow Status	Next Workflow Status
Customizable automatic lathe	✓	Portfolio selection	Project selected
Xenon dipped-beam headlight	✓	Portfolio selection	Project selected
Super glue applications	✓	Portfolio selection	Project selected
High-tempature superconductor	X	Portfolio selection	Suspended
VW PQ 35 Plattform	✓	Portfolio selection	Project selected
Laser Remote Welding	✓	Portfolio selection	Project selected
Improved unit injector	X	Portfolio selection	Suspended
Software for electronic drives	✓	Portfolio selection	Project selected
Lotus-surface lacquer	X	Portfolio selection	Suspended
Modular plant system X1	X	Portfolio selection	Suspended
Small Integrated temperature sensor	X	Portfolio selection	Suspended
Integrated safety application	✓	Portfolio selection	Project selected

FIGURE 7.13
Result of the portfolio process in the Microsoft Office Project Portfolio Server (MOPPS) as an example for an increasingly complex project portfolio management (PPM) system.

TABLE 7.5

Coverage of the Domain Services as Usability Aspects of the PPM System (Excerpt)

Domain Service	Coverage	Note
Stage gate	Yes	
Scenario analysis	Yes	
Driver costs calculation	Yes	
Effect analysis for restrictive factors	Yes	
Portfolio monitoring with individual indicators	Yes	
Variance calculation of planned, current, and expected costs	Yes	
Resource demand estimation	Yes	
Custom criteria types	Yes	
Integration of proposal data	No	Manually
Templates for project assessment	No	
Project activity management	Partially	Little visualization

Conclusion

This chapter presented an approach for continuous evaluation of increasingly complex systems based on such a mix of empirical and technical methods. The iterative and holistic essence of the approach allows us to better address usability aspects in increasingly complex systems. The verification of the approach was conducted in two complex application domains—health care, more specifically clinical environments, and industrial engineering, more specifically project portfolio management.

Future work is focused primarily in the further development of the methodology with respect to professional communication and its application in additional problem domains.

Reader Take-Aways

Usability evaluation of increasingly complex systems requires novel approaches. They should adhere to several requirements.

- Iterative approaches are better suited to evaluate increasingly complex systems. They allow for a better consideration of different complexity levels.
- Usability evaluation considers usability aspects during the different phases of the engineering of the system.

- In our specific case, we apply a combination of empirical observation and data gathering, on the one side, and technical communication (expert interviews and workshops) between domain experts and system analysts, on the other.

We presented the evaluation of the approach in two complex application domains—health care and project portfolio management in industrial engineering.

- The health care domain is characterized by complex applications and activities. Furthermore, these complex systems are typically used in life-critical situations.
- The area of project portfolio management is one of the most complex activities in industrial engineering. Project portfolio management systems evaluate technical information about new products, financial data, and project fulfillment indicators in an increasingly complex way and provide recommendations about optimized project portfolios.
- In both fields, we were able to identify relevant usability problems and address them during the next upgrade (or migration) of the increasingly complex system.

References

Banker, R. D., Bardhan, I., & Asdemir, O. (2006). Understanding the impact of collaboration software on product design and development. *Information Systems Research, 17*(4), 352–373.

Behringer, R., Christian, J., Holzinger, A., & Wilkinson, S. (2007). Some usability issues of augmented and mixed reality for e-health applications in the medical domain. *HCI and Usability for Medicine and Health Care, 4799*, 255–266.

Calantone, R., Di Benedetto, C., & Schmidt, J. (1999). Using the analytic hierarchy process in new product screening. *Journal of Product Innovation Management, 16*(1), 65–76.

Clarke, R., de Silva, H., & Sapra, S. (2004). Toward more information-efficient portfolios. *Journal of Portfolio Management, 31*(1), 54–63.

Cooper, R., Edgett, S., & Kleinschmidt, E. (2001). *Portfolio management for new products.* Cambridge, MA: Perseus.

Cooper, R. G., & Edgett, S. J. (2007). *Ten ways to make better portfolio and project selection decisions* (technical report). Retrieved from http://www.planview.com/docs/Planview-Stage-Gate-10-Ways.pdf

Dubois, E., Gray, P. D., & Nigay, L. (2002). ASUR++: A design notation for mobile mixed systems. In *Mobile HCI '02: Proceedings of the 4th international symposium on mobile human-computer interaction* (pp. 123–139). London: Springer.

Holzinger, A. (2005). Usability engineering methods for software developers. *Commununications of the ACM, 48*(1), 71–74.

Johnson, R., Salvo, M., & Zoetewey, M. (2007, December). User-centered technology in participatory culture: Two decades "beyond a narrow conception of usability testing." *IEEE Transactions on Professional Communication, 50*(4), 320–332.

Krallmann, H., Schoenherr, M., & Trier, M. (2007). *Systemanalyse im Unternehmen: Prozessorientierte Methoden der Wirtschaftsinformatik.* Munich: Oldenbourg Wissenschaftsverlag.

Pepels, W. (2006). *Produktmanagement: Produktinnovation, Markenpolitik, Programmplanung, Prozessorganisation.* Munich: Oldenbourg Wissenschaftsverlag.

Project Management Institute. (2006). *Standard for portfolio management.* Newton Square, PA: Project Management Institute.

Rajegopal, S., Waller, J., & McGuin, P. (2007). *Project portfolio management: Leading the corporate vision.* London: Palgrave Macmillan.

Redish, G. (2007). Expanding usability testing to evaluate complex systems. *Journal of Usability Studies, 2*(3), 102–111.

Sandberg, W., Daily, B., Egan, M., Stahl, J., Goldman, J., Wiklund, R., et al. (2005). Deliberate perioperative systems design improves operating room throughput. *Anesthesiology, 103*(2), 406–418.

Stantchev, V. (2008). Service-based systems in clinical environments. In *Proceedings of the second information service design symposium.* Berkeley: University of California, School of Information.

Stantchev, V. (2009a). *Enhancing health care services with mixed reality systems.* Berlin: Springer.

Stantchev, V. (2009b). Intelligent systems for optimized operating rooms. *New Directions in Intelligent Interactive Multimedia Systems and Services, 2,* 443–453.

Stantchev, V., & Franke, M. R. (2009). Managing project landscapes in knowledge-based enterprises. In *Best practices for the knowledge society: Knowledge, learning, development and technology for all: Proceedings of the second world summit on the knowledge society—wsks 2009* (Vol. 49, pp. 208–215). Berlin: Springer.

Stantchev, V., Franke, M. R., & Discher, A. (2009, May). Project portfolio management systems: Business services and web services. In *ICIW '09: Proceedings of the 2009 fourth international conference on Internet and Web applications and services* (pp. 171–176). Los Alamitos, CA: IEEE Computer Society.

Stantchev, V., Hoang, T. D., Schulz, T., & Ratchinski, I. (2008). Optimizing clinical processes with position-sensing. *IT Professional, 10*(2), 31–37.

Stantchev, V., & Malek, M. (2009). Translucent replication for service level assurance. In *High assurance services computing* (pp. 1–18). Berlin: Springer.

Stantchev, V., & Schröpfer, C. (2009). Service level enforcement in Web-services based systems. *International Journal on Web and Grid Services, 5*(2), 1741–1106.

Sudarsan, R., Fenves, S., Sriram, R., & Wang, F. (2005). A product information modeling framework for product lifecycle management. *Computer-Aided Design, 37*(13), 1399–1411.

Wheelwright, S. C. (1984). Strategy, management, and strategic planning approaches. *Interfaces, 14*(1), 19–33.

8

Design Considerations for Usability Testing Complex Electronic Commerce Websites: A Perspective from the Literature

Julie Fisher

Monash University

CONTENTS

Introduction 158
Usability and Complex E-Commerce Websites 158
    Website Usability Testing.. 159
Reliability and Validity in Usability Testing Research................................. 160
Research Approach 160
Designing a Reliable and Valid Usability Test 162
    Selecting Websites.. 162
    Design of the Scenario and/or Task.. 164
    Identifying and Recruiting Participants 165
        Number of Participants... 165
        Selecting Participants 166
    Research Instrument... 171
    Conducting the Usability Test and the Setting 171
Conclusion 172
Reader Take-Aways 174
References... 175

Abstract

Electronic commerce (e-commerce) for transacting business has become increasingly important for small and large organizations alike. With this comes a rise in the complexity of e-commerce systems enabling everything from consumers buying goods or services online to the management of internal business transactions. The need to ensure such systems are usable is without question a priority. Those undertaking usability testing in industry and research contribute to our understanding of how to build effective, usable systems. The literature reports a wide variety of approaches to website usability testing and raises issues on designing reliable and valid usability tests. This chapter explores usability testing of complex e-commerce websites. A review of current research involving laboratory-based testing identified the key design considerations important for ensuring reliable and valid results. The chapter provides guidelines for researchers and technical communicators for the design and conduct of usability testing, particularly in the context of complex systems.

Introduction

Introducing an e-commerce system into an organization is a challenging task. E-commerce systems are complex, requiring the development of not just the front-end website but also back-end support systems including stock management, delivery, and transaction-processing systems. Successful e-commerce needs to "provide a seamless process of placing and receiving an order, making appropriate logistical arrangements, and making and receiving payments electronically, and externally, among all players in strategic business networks" (McKay & Marshall, 2004, p. 30).

For any business contemplating moving to e-commerce the stakes are high, and the usability and effectiveness of the website must be of prime importance (Schneider, 2007, p. 542). E-commerce website design has become a key research area as our understanding of the importance of design and its impact on users increases. Many of the usability and design issues relate to any system; however, for complex systems such as e-commerce there are other considerations, in particular designing for a broad, often unknown audience. The research discussed in this chapter explores how usability testing conducted in a laboratory setting is used and reported in e-commerce research and what we can learn from this about usability testing complex systems. The research specifically addresses the following questions:

- How has usability testing been conducted in research relating to e-commerce websites?
- What are the design considerations for usability testing complex e-commerce systems to ensure reliable and valid results?

Usability and Complex E-Commerce Websites

Designers of complex websites involving e-commerce present significant usability challenges and frequently fail to meet business and user expectations (Stylianou & Jackson, 2007). Schneider (2007, p. 150) notes that up to 70 percent of e-commerce website customers leave without buying anything. He suggests even the best sites can lose up to 50 percent of their customers because the site is confusing or difficult to use, and argues that usability testing would help increase customer satisfaction and sales. Complex e-commerce systems may not have a clearly defined audience compared to most other software, and the users' tasks are more varied as are the reasons for use (Agarwal & Venkatesh, 2002). Shneiderman (2000) argues, "Designing for experienced frequent users is difficult enough, but designing for a broad audience of unskilled users is a far greater challenge" (p. 85).

E-commerce adds further to design problems because as websites become more complex, so do the usability issues. Research by Nadkarni and Gupta (2007) identified links between perceived website complexity and user satisfaction, finding that complexity inhibits user satisfaction. They recommend that designers provide different levels of complexity to cater for both goal-directed and experiential users. Stylianou and Jackson (2007) explored users' experiences of using the Internet and using a complex e-commerce system. They concluded that because e-commerce is a more complex activity, self-efficacy is important in determining use. With the high costs of establishing and maintaining complex e-commerce systems and the subsequent expectations of business owners for success, usability is a key consideration.

Website Usability Testing

Researchers have made significant contributions to our knowledge of usability and website design. Usability testing has become an important empirical data-gathering technique for research relating to the design of e-commerce websites (Palmer, 2002). The literature describes a wide variety of approaches and instruments (Treiblmaier, 2007). However, issues have been raised around the different ways that usability tests are conducted. For example, Lewis (2001) asks if we really understand the difference between methods and how these can be compared. Abdinnour-Helm, Chapparro, and Farmer (2005) note that usability-testing instruments are often inadequately tested. Treiblmaier examined Web analysis studies and identified many different approaches, concluding that some of the scales used have been "developed over and over again" and suggesting that "unambiguous scales" are lacking. Gray and Salzman (1998) examined five usability evaluation experiments and questioned how well results can be compared. Many of the issues raised apply to all forms of usability testing; however, when designing complex e-commerce systems, how usability testing of these systems is managed is more critical because the costs are higher and the chances of failure greater. The purpose of this chapter is not to review the different methods and approaches to usability testing but rather to reflect on the key considerations that researchers must manage when designing usability tests, particularly for complex systems.

Reliability and Validity in Usability Testing Research

Reliability is one important criterion for judging the efficacy of research outcomes. Reliability is "Stability—over time; Representative—across subgroups; Equivalence—across indicators" (Neuman, 2003, p. 183). Reliability is about consistency both internal (the scales are internally consistent) and external (the scales are consistent over time) (Bryman & Cramer, 1992, p. 70). Reliability is also the extent to which the measurement instrument, if applied again, would yield the same results (Leedy, 1997, p. 35; Neuman, p. 179).

Validity is how well the constructs used measure the things to be measured (Neuman, 2003). Internal validity establishes if there are differences between groups (Gray & Salzman, 1998). Gray and Salzman describe two issues relating to construct validity: whether the researchers were manipulating the variables they thought they were and whether they were measuring what they thought they were. External validity relates to the generalizability of the results: how well the experiment, if repeated, would yield the same results (Leedy, 1997, p. 34).

To ensure reliability and validity in usability-testing research, the design of the test should take into account how the websites were selected, the variability of the participants, how the instrument was designed, how the test was conducted, and any changes that might occur over time (Abdinnour-Helm et al., 2005; Aladwani, 2002; Leedy, 1997, p. 35). As stated by Gray and Salzman (1998), "A well conducted, valid experiment permits us to make strong inferences regarding two important issues: (A) cause-and-effect and (B) generality" (p. 208). Irrespective of whether the system tested is a small website or a complex e-commerce system, the reliability and validity of results are important.

Research Approach

Recently and relevant to this research, Treiblmaier (2007) examined Web analysis studies, suggesting that such studies provide insights into current approaches to Web analysis and are useful for those planning similar research. Research reported in the literature, describing usability testing of e-commerce websites, was the focus of this study.

For this research, only papers where it was clear the research approach could be reasonably considered to be a usability test were examined, or, as Hartson, Terence, and Williges (2001) suggest, "any method or technique used to perform formative usability evaluation (ie, usability evaluation or testing to improve usability)" (p. 376). Only usability tests conducted in laboratory conditions involving e-commerce websites were included. It is acknowledged that good usability tests have been conducted using qualitative methods and not in laboratory conditions; however, for this research the focus was on quantitative studies only.

Papers were excluded if the research did not cover some form of usability, the websites did not have e-commerce functionality, the research involved respondents answering a survey or interview outside of a laboratory setting, or the websites were specifically designed for the study. This, as far as possible, ensured the research conditions were similar and to some extent comparable.

A wide range of publications were examined to identify relevant papers. The leading information systems journals and other key journals publishing e-commerce research were included. A limited search was made of the conference literature as conference papers generally are shorter and provide less opportunity for authors to describe their research process in depth. However, three papers from leading information systems conferences were included. Nineteen papers published in the last ten years were found to meet the criteria.

Each of the papers that met the criteria was analyzed, and the following identified:

- The topic under investigation
- Outcomes or findings
- How the author(s) described the research approach, including how the usability test was conducted, the task or scenario, and the data collection instrument
- The number of participants and their demographic details

It should be noted that the search was limited and it is likely other papers reporting testing of e-commerce websites were missed; however, for the purposes of this research the papers identified do illustrate the variety of approaches and reporting of usability-testing research involving e-commerce websites. A further aim was to identify some of the key considerations important to ensure reliable and valid results when testing complex systems, considerations that are important for both technical communicators and researchers.

Designing a Reliable and Valid Usability Test

Hallahan (2001) argues that a usability test, in the context of website testing, involves typical users performing tasks in a laboratory. The facilities for usability tests are similar to those of an experiment: They take place in a controlled setting and all participants undertake the same tasks at the same time. Many companies also use testing laboratories (Shneiderman & Plaisant, 2010, p. 156).

An examination of the literature found a variety of approaches to laboratory-based usability testing and a variety of ways the testing was reported. The reliability and validity of the research design will impact on the quality and generalizability of the results. In an industry context, the quality of the outcomes will impact ultimately on the usability of the system, and for complex e-commerce systems this could mean the success or failure of the business. This chapter identifies five important activities requiring particular consideration when designing usability tests and in particular testing complex e-commerce websites. These are the selection of the website(s) or parts of the website, the design of the task and/or scenario, the selection of participants, the design of the research instrument, and the conduct of the test, including the test environment. Each of these, if not carefully designed, will impact on the reliability and validity of the results. Next I discuss each of these five activities, and what is important to consider and why with respect to reliability and validity. Examples from the literature are provided to help illustrate the issues.

Selecting Websites

When an organization is undertaking usability testing, the system and website will be known factors. Complex e-commerce systems are composed of many subsystems; therefore, what must be determined is how and when the different subsystems will be tested. Which parts of the website are targets for testing will influence participant selection; this is discussed later. In a research context, the researcher might consider website selection based on complexity, the type of website, the level of interactivity, the level of personalization, the products or services sold, and the appeal of a website to a particular audience (Schneider, 2007, p. 148; Shneiderman & Plaisant, 2010, p. 471). These might also be considerations when testing in industry.

Table 8.1 summarizes the websites selected and the task(s) the participants undertook in the papers this research examined. The number of websites that researchers investigated, as described in Table 8.1, varied from one to twenty; in some cases, the number was not mentioned. Gray and Salzman (1998) discuss the problem of construct validity where three different system

TABLE 8.1

Websites and Tasks Used in Usability Testing

Author	Website(s) Selected	Task
1. Abdinnour-Helm et al. (2005)	A clothing store	Participants used the interactive component of the website to define and view outfits.
2. Aladwani (2002)	A bookstore	Participants searched for and purchased a specified book.
3. Chua and Tan (2007)	A bookstore	Participants purchased books and gift cards.
4. Ethier and Hadaya (2004)	Four stores selling CDs	Participants shopped for one specified and one unspecified CD.
5. Featherman, Valacich, and Wells (2006)	E-service websites	Participants investigated bill-paying-service websites and payment processing.
6. Fiore and Jin (2003)	A clothing website	Participants explored mix-and-match functionality.
7. Green and Pearson (2006)	A department store	Participants searched for and bought an item.
8. Geissler, Zinkhan, and Watson (2001)	Commercial websites	Participants explored homepage complexity.
9. Guo and Poole (2009)	A bookshop	Participants selected, searched for, and bought books.
10. Kuan, Bock, and Vathanophas (2008)	Two online travel websites	Participants explored and bought travel packages.
11. Kumar, Smith, and Bannerjee (2004)	Two bookshop websites	Participants created a new account and interacted with shopping carts.
12. Massey, Khatri, and Ramesh (2005)	Two Web services for mobiles and PCs	Participants explored different Web services on mobile phones.
13. Nel, van Niekerk, Berthon, and Davies (1999)	Twenty different e-commerce websites	Participants evaluated the websites.
14. Rosen, Purinton, and Lloyd (2004)	E-commerce websites	Participants searched for information.
15. Roy, Dewit, and Aubert (2001)	A bookstore	Participants found and bought a book.
16. Singh, Dalal, and Spears (2005)	Twenty small-business websites	Participants explored the Web pages.
17. Teo, Oh, Liu, and Wei (2003)	Three online stores	Participants shopped for a new computer system.
18. Vijayasarathy and Jones (2000)	A variety of stores with online catalogs	Participants bought a product of choice.
19. Zhang, Keeling, and Pavur (2000)	Ten Fortune 500 companies	Participants explored the homepages.

applications were used in a usability test. They argue that information on system differences was needed.

There are many approaches for selecting websites for research. Guo and Poole (2009) undertook a pre-analysis before selecting the target sites. They looked particularly at the size and complexity of the sites and if the participants were likely to be familiar with the website. Another study (Singh, Dalal, & Spears, 2005) prescreened many websites to ensure relevancy to the audience. Roy, Dewit, and Aubert (2001) selected their websites because they were accessible and understandable to the participants and because they offered similar products and transactions.

For complex systems with various subsystems, there will be different content types, different media for displaying information, different levels of complexity, different tasks performed, different interactions, and different download speeds. Technical communicators performing usability testing of subsystems will need to be mindful of the fact that a test designed for one subsystem may not be appropriate for another.

Design of the Scenario and/or Task

Designing the scenario or task the participants will undertake during a usability test requires careful consideration irrespective of whether the usability test is for research or an industry-based test. Scenarios are "typically tasks that are representative of what a typical user may do with the system" (Abdinnour-Helm et al., 2005). Scenarios and tasks may be complex or short depending on the purpose of the testing. The wording should reflect the user's own words and not direct the user on how to complete the task (Dumas & Redish, 1994, p. 174). A study by Clemmensen, Hertzum, Hornbaek, Shi, and Yammiyavar (2008) found that East Asian users need more contextual information in scenarios.

The tasks must be manageable and suitable for a laboratory setting. Cordes (2001) notes that the tasks should be something the user can perform. Complex e-commerce websites will enable a wide range of tasks. These can be quite challenging tasks such as ordering, paying, or managing stock levels, or something quite simple such as searching. Schneider (2007, p. 147) lists reasons why users might visit an e-commerce website; these include learning about products, buying products, getting financial information, finding out general information about a company, and obtaining information such as service policies. The task should take into account what tasks are possible. The tasks should be interesting and suitable for the participants involved (Cordes).

Booking a family holiday overseas may be an unrealistic task for a group of 18–25-year-old students who are less likely to have children. The design of the scenario or task impacts on representative reliability; in other words, would the outcomes be the same if the same test and task were applied to a different demographic group? Enough detail also needs to be provided to ensure representative reliability. If the task is vaguely described or little detail is provided or not described at all, then it may be difficult to repeat and replicate the results.

As illustrated in Table 8.1, participants were set different tasks depending on the study. In some cases such as buying a book, the task was quite simple; other tasks were more complex such as bill paying and browsing and buying travel packages. The tasks reported varied considerably in the level of detail provided. In research by Abdinnour-Helm et al. (2005), the tasks were written by usability experts and pretested. Teo, Oh, Liu, and Wei (2003) tried to increase the realism of the tasks by asking participants to pretend to be home computer users looking for a new system. Most studies paraphrased the task used; some provided very little detail. In three studies, no task or scenario was provided. In one study, the task was described as follows: "The students were asked to evaluate the Websites while thinking of the site as a whole" (Rosen, Purinton, & Lloyd, 2004, p. 21).

For those designing a usability test for a complex system, the findings from the research highlight the importance of carefully selecting the tasks to be used. The tasks must be "doable" and should provide a good level of detail. It is advisable that the tasks or scenarios be pretested, and they should be "real" to the users, ensuring more reliable results. Further, the demographics of the participants and their role need to be considered in task design.

Identifying and Recruiting Participants

One of the most important elements of a usability test, particularly for complex e-commerce systems, is participant selection. Complex systems by their nature are likely to be used by a wide variety of users, both internal and external to the organization, who will use the system for a range of different tasks. Poor participant selection will impact on both the reliability and validity of the results.

Number of Participants

The number of participants is important where statistical testing is involved and for "statistical conclusion validity" (Gray & Salzman, 1998). For statistical testing, thirty is the generally accepted minimum number of participants required (Caulton, 2001) depending on the statistical tests to be conducted. If the sample size is low, then the heterogeneity of the participants might be an issue and threaten the validity of the conclusions (Gray & Salzman; Hartson et al., 2001).

Participants should be selected to ensure that there is sufficient coverage of different user characteristics to produce good results (Dumas & Redish, 1994). This point is made in the context of practice and where complex e-commerce systems are involved and range of participants is likely to be broad, ensuring that coverage of user characteristics will be important. If participants are not homogeneous—that is, participants' use and understanding of what is being tested are distinctive—then testing should include subgroups (Caulton, 2001). Selection of participants is a threat to internal validity where there are different kinds of participants in different experimental groups (Gray & Salzman, 1998). Given the nature of complex e-commerce systems that are potentially accessible to everyone, identifying and selecting participants on the basis of subgroups are paramount.

Selecting Participants

The literature argues for the use of "real users" (Gould & Lewis, 1985; Hallahan, 2001). Real users are generally those who are most likely to use the final system (Gould & Lewis). Users of complex e-commerce systems are likely to include those within the organization and those outside. Tests can also be conducted with *representative users*, users who represent, but are not, real users; however, the results may not be as accurate or reliable (Whitefield, Wilson, & Dowell, 1991). The difficulty for usability testing of complex e-commerce websites is that real users can theoretically be anyone with Internet access. In reality, *real users* could be defined as participants who have a high probability of interacting with the target website(s) and who are likely to undertake tasks similar to those selected. The challenge for technical communicators is in identifying participants who are real users and those who are representative.

Selecting participants who are not interested in the target website(s) will impact on the validity of the results (Maguire, 2001). Participants who are not likely to transact business with a particular e-commerce website are also inappropriate. Developing a user profile is widely recommended. The profile would contain information about the target audience. It should include all the characteristics the users should have and the characteristics that will be different. Table 8.2 summarizes the information provided in the papers on the number of participants and participant demographics. The paper number corresponds to the authors detailed in Table 8.1.

As detailed in Table 8.2, in all but one study students were used. Dennis and Valacich (2001) ask if it is possible to have credible results when using undergraduate students undertaking "pretend tasks" they may not be interested in. Abdinnour-Helm et al. (2005) argue such users may be appropriate providing they are not dissimilar to Web users generally and are likely to perform the tasks being investigated; however, some justification is needed. Limited justification was provided for using student participants in most of the reported research. A number of the studies noted the relevance of the task to their student participants because, for example, students regularly purchase online.

For complex e-commerce systems, recruitment of participants should ensure the selected participants are interested in the website, are likely to purchase, and have similar characteristics to the audience for which the

TABLE 8.2

Participant Details

Paper No.	Participants
1	76 undergraduate (UG) psychology and e-commerce students. 47% were female. Average age: 22.7 years. Most had some Internet experience. 47% had purchased from the Web.
2	387 UG business students. 69% were female. Average age: 19.7 years. Most had Internet experience.
3	186 college students. Participants were paid.
4	215 business administration students. Participants were paid.
5	526 business students. 34% were female. A small percentage had made online payments.
6	103 UG students from different courses. 72% were female. Age: 18–25 years. Experienced with the Internet. 32% had previously purchased from the Web. Students were given an incentive.
7	375 UG business students.
8	169 UG business students. 47% were female. The majority was experienced Internet users. Students were offered an incentive.
9	354 students from a variety of majors. 59% were female. Average age: 21.1 years.
10	101 students. Participants were paid.
11	123 MBA students in Hong Kong. All held senior management positions. Range of ages: mid-20s to early 40s.
12	76 UG students in a range of courses, postgraduate students, and university employees. 37% were female. Participants were paid. Range of ages: 18–40 years.
13	36 students. 44% were female. Ages: 20–21 years. Ten had not used the Internet before.
14	211 UG and graduate students. 50% were female. Range of ages: 18–25 years. The majority had used the Internet. 47% had never purchased from the Web.
15	35 UG students and 31 university personnel. 48% were female. Range of ages: 20 to over 45 years. Education, occupation, and income were described. All had used the Internet. 31% had purchased from the Web.
16	540 UG students. 45% were female. Participants were given course credits.
17	54 UG information systems students.
18	201 upper-level business students. 40% were female. Average age: 23 years.
19	40 junior- and senior-level business students.

system is designed. Other considerations for selecting participants include factors such as gender, cultural background and language proficiency, age, and Internet experience. Issues relating to gender, culture, and age are discussed next.

Gender

The research highlights that for those conducting usability testing on complex systems, consideration must be given to the gender of the participants.

A number of studies have found significant differences between male and female responses to website design (Cutmore, Hine, Maberly, Langford, & Hawgood, 2000; Cyr, 2009; Simon, 2001). Stenstrom, Stenstrom, Saad, and Cheikhrouhou (2008) examined the difference that gender makes to the way people respond; they found that Web navigation structure that was deep as distinct from wide was better suited to males. Cyr and Bonanni (2005) found differences in the way that men and women respond to website information design, color, and animation. Further research by Cyr identified differences between men and women when it came to trust, satisfaction, and website design. When designing usability tests for complex e-commerce websites, gender has to be a serious consideration. Where usability testing is conducted, for example, on a website selling a variety of goods, consideration has to be given to the task to ensure it is of interest to both male and female participants.

As detailed in Table 8.2, 37 percent of studies did not report the gender breakdown of their participants. In four cases more than 60 percent of the participants were of one gender. One study by Featherman, Valacich, and Wells (2006) examined risk; however, only 34 percent of the participants were female. As noted by Cyr (2009), there are differences in relation to trust between men and women; the gender breakdown needed to be more even for the findings to be valid. In research by Fiore and Jin (2003), 72 percent of the participants were female. They acknowledge this may have impacted on their results as they were investigating Web marketing and user responses; however, there was no attempt to allow for this in the study's design. Kumar, Smith, and Bannerjee (2004) found that gender influenced perceptions of ease-of-use factors. Given research by Cyr and Bonanni (2005) indicating significant differences in the way males and females respond to aesthetics and website design, this makes the results problematic. Zhang, Keeling, and Pavur (2000), without providing details of the gender of the participants, drew conclusions regarding the presentation, navigation, and quality of the homepages of a number of Fortune 500 companies. The companies selected included those in the general merchandise, computer software, electronics, and forest products industries. Presuming women participated in the usability test, we must ask if female participants would be as interested and engaged as male participants with the website of a company selling motor vehicle parts, for example.

Cultural Background and Language Proficiency

In practice, when a company is testing a complex system, those designing the test will need an understanding of who their audience is from the perspective of culture and language proficiency. Research by Sears, Jacko, and Dubach (2000) concluded that cultural background, among other factors, has an influence on how users rate websites. Findings of a study comparing Chinese and German users by Rau, Gao, and Liu (2007) found that Chinese users were more satisfied with a Web portal they believed was useful than German users. Research by Clemmensen et al. (2008) concluded that for valid results in cross-cultural usability tests, understanding the differences in cognition of those from different cultural backgrounds is required. Where the usability test might also involve observations, Clemmensen et al. (2008) note that Western users, because they are more likely to express their emotions, are easier to read than users from East Asia.

Age

Anyone between the ages of eighteen and seventy is a potential participant for a usability test of a complex e-commerce website. This raises issues; Canzer (2006, p. 181) argues that older users will use different sociocultural references when looking at and interpreting images on a website. Different age groups will derive different meanings, triggering different responses from different images. Often participants' ages are not provided, although what different age groups are studying can help us draw some conclusions about age. Many studies, however, mention using "college" students; outside of the United States, this is of little help in determining ages. For example, in Australia a university student can be seventeen or sixty-plus years old. It is therefore important to consider and appropriately note the age range of participants as student ages can vary dramatically.

Of the studies examined, eight studies provided no information on participants' ages, four provided an average, and seven reported the age range. Featherman et al. (2006) studied online bill payment; participant ages were not provided, but given that the participants were described as undergraduate students it could be assumed that most were between the ages of eighteen and twenty-five years old. No justification was provided as to why this group might be suitable. The authors note that a very small number had paid a bill online, suggesting that this demographic may not have been suitable. Kuan, Bock, and Vathanophas (2008) explored users buying travel packages; the participants were business students, so presumably the majority was under the age of twenty-three. The authors explained that international websites were chosen to avoid websites students may have seen before, but no reasons were given as to why these websites and why these tasks would be of interest to this group. In some cases, the age of the participants may present a challenge to the validity of the results. Another study examining homepage quality did not report participant ages. A number of the company homepages used were likely to have been designed with an older audience in mind.

The research provides pointers to selecting participants for usability testing of complex Web-based systems. Gender must be considered given that women do respond differently. If a complex e-commerce website has broad appeal, then test participants must include women; otherwise, the results are likely to be skewed. Research also tells us that the percentage of female participants must be considered; having just one woman, for example, may not be sufficient for meaningful results. The cultural and language backgrounds of potential users need to be considered also in selecting participants. People from different backgrounds bring different perspectives. The age of participants should be considered, and decisions made with reference to which aspect or subsystem is being tested. Users' ages may differ depending on the tasks being performed.

TABLE 8.3

Usability Test Instruments

Paper No.	Research Instrument	Validation
1	Used a revised version of the End User Computing Satisfaction instrument.	Factor analysis.
2	Constructs based on other studies.	Factor analysis.
3	Own instrument based on another modified instrument.	Factor analysis.
4	Constructs based on questions and statements from other studies.	Factor analysis.
5	Items drawn from the literature.	Factor analysis.
6	Used a modified scale and included their own items.	Instrument pilot tested; tested for face validity.
7	Constructs based on another study.	Factor analysis.
8	Based on other studies but modified.	Pretested the instrument.
9	Adapted other instruments.	Reliability analysis and confirmatory factor analysis conducted.
10	Items adapted from other studies.	Pretested the instrument; reliability analysis conducted; factor analysis.
11	Used existing literature and other instruments to develop their instrument.	Regression analysis.
12	Constructs based on other instruments.	No details of instrument validation provided.
13	Adapted from a previously used instrument.	No details instrument validation provides.
14	Instrument adapted from other studies.	Pretested the instrument; factor analysis.
15	Questionnaire developed based on other studies and questions.	Reliability analysis.
16	Questionnaire based on input from participants similar to those used in the study.	Instrument pilot tested; used colleague to test face validity; factor analysis.
17	Adapted other instruments.	Assessed content validity using other experts; pilot tested instrument; factor analysis.
18	Adapted an instrument from another study.	Factor analysis.
19	Developed own instrument; constructs obtained from the literature and interviews with Web developers.	Factor analysis.

Research Instrument

Although it would not be expected that usability testing in practice would require a valid, reliable instrument, what we learn from the research is that pretesting of instruments will help ensure that survey items are not ambiguous and that they test what they are supposed to test.

Abdinnour-Helm et al. (2005) and Palmer (2002) include detailed discussions of research instruments. It is not the intention, therefore, to provide a detailed discussion of usability test instruments; what is important is the extent to which reported research details the instruments used and how they were validated. Table 8.3 describes the instruments used in the literature and any instrument testing conducted.

Ensuring that the instrument is useful for the wider research community, applying the instrument across a range of subjects and conditions is important (Palmer, 2002). Often, the usability test instruments used have not been adequately tested for validity or reliability (Abdinnour-Helm et al., 2005). Where an instrument has not been previously tested, this can be done through pretesting or pilot testing previously validated instruments (Palmer) and/or undertaking a factor or reliability analysis (Aladwani, 2002). A factor analysis was the most common approach reported in the literature for determining the reliability and validity of research instruments.

Many studies use the same or similar testing approaches such as the use of Likert scales drawing on statements and questions used in previous studies. This is a reasonable approach to take, particularly where previously used statements and questions may have been tested. Given that researchers are not always replicating previous studies, questions and statements selected are often modified in line with the research focus. Zhang et al. (2000) argue their study was exploratory, and with no statistically valid instrument available a factor analysis was used to assess the validity of the instrument. Aladwani (2002) undertook a literature review to identify sixty-four items. Ambiguous or duplicate items were eliminated, then two other researchers looked at the items and suggested changes. Singh et al. (2005) generated a list of items through focus groups with students. These were then refined.

The literature on the design of research instruments is useful for technical communicators designing complex system testing. The literature can help identify how items for the instrument can be generated and how participants might respond. Further, where previous items have been used and tested, these will add to the robustness of the results.

Conducting the Usability Test and the Setting

Shneiderman and Plaisant (2010, pp. 157–158) suggest that in an industry environment, a test plan should be developed and the procedures to be undertaken pilot tested. They also note that sometimes it is useful in an industry setting to combine the more rigid controls used in research with the usual testing process used in industry. The research literature can therefore provide further guidance for those designing testing of complex Web-based systems. For research, ensuring the usability tests are conducted under the same conditions in the same location assists with internal validity; and for causal construct validity, how the test was conducted and the method used should be described (Gray & Salzman, 1998). For complex system testing, however, a variety of testing conditions may be more appropriate.

A test plan will include the procedures to be followed, including instructions to participants, the time spent on an activity, and how the data were gathered. One threat to causal construct validity occurs when the way in which the experiment is undertaken is different from the way someone else might think to run it (Gray & Salzman, 1998).

Some studies provided the time taken to complete the task, or there was a time limit (Abdinnour-Helm et al., 2005; Massey, Khatri, & Ramesh, 2005). Aladwani (2002), at the start of his experiment, had participants explore the website for ten minutes to eliminate the possibility of some participants being more familiar with the site than others, which could have resulted in confounding effects.

Many of the studies examined provided only very brief details of how the usability test was conducted. Most simply described the task that participants were to perform. Few described how the task or scenario was presented to participants, such as whether in written or oral form.

Usability testing of complex e-commerce systems requires careful planning and management because of the number of subsystems likely to be involved. It could be argued today that the usability test environments are not likely to vary a great deal in quality given the state of current technology and bandwidth, and may not be an issue in an industry setting. If, however, a complex e-commerce website being tested was slow to download, then it might be necessary to explain how the testing conditions dealt with this.

Conclusion

This chapter has explored the range of approaches, as described in the literature, taken for conducting and reporting usability testing in research, and has used this as a basis for guidance for designing usability tests for complex e-commerce systems. The outcomes of this study demonstrate that we are still some way off from understanding what is needed to provide reliable and valid results from usability testing. This research has sought to address the issue describing where some of the more serious problems arise, and it proposes solutions.

In summary, designing usability testing for complex e-commerce websites presents challenges for the tester whether the testing is conducted in an industry or research environment. E-commerce systems are complex from the perspective of both the back-end systems, which are required to communicate with numerous other systems, in an organization, and the front-end website designed to meet the changing needs of a diverse audience whose reasons for visiting the site will be unknown. This chapter has identified a number of those design challenges. Figure 8.1 describes the five usability test design considerations; it details each consideration and its characteristics, and proposes what actions those designing a usability test for a complex e-commerce system need to take to ensure the outcomes are useful.

As we continue to see poorly designed complex e-commerce systems, the need for reliable and valid usability tests continues to be an imperative. Schneider (2007, p. 146) argues that website design is different, and that "failure to understand how the Web is different from other presence-building media is one reason that so many businesses do not achieve their Web objectives." From a research perspective we can, as researchers using well-

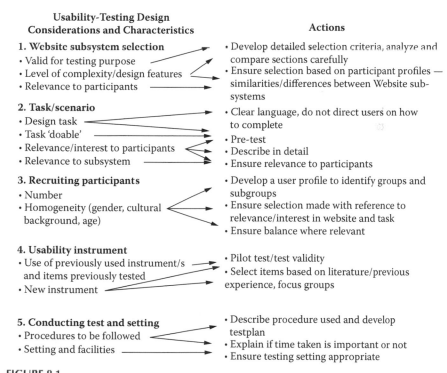

FIGURE 8.1.
Usability-testing design considerations and actions for complex systems.

designed usability tests, add to the body of knowledge for both researchers and industry of how to develop effective, usable websites that will ultimately help businesses achieve their Web objectives.

Johnson, Salvo, and Zoetewey (2007) argue that broadly for the field of usability to be healthy and productive, we need scientific outcomes that are replicable. It is suggested in a research context that it is difficult to reliably compare the results of different usability test approaches because of a lack of standard comparable criteria and a lack of standard definitions and measures (Hartson et al., 2001; Palmer, 2002). It is important therefore that in usability testing research and in industry, we remain ever vigilant to how we design tests to ensure that the outcomes are valid and replicable, particularly when dealing with complex systems.

There is no question that e-commerce will continue to grow and may in the future, if not already, dominate the marketplace. In order to build our knowledge of what improves the usability of complex e-commerce websites, we need good research. More work is needed to better understand how we can improve the usability-testing design processes for complex systems. This chapter makes a start.

Reader Take-Aways

- Complex e-commerce systems will be composed of numerous subsystems, all of which may require testing. The first consideration, therefore, is how and when each of the subsystems will be tested. The subsystem selected will impact on participant selection.
- The task or scenario must take into account, apart from the tasks supported by that subsystem, the participants and their backgrounds. Participants should be able to relate to the task, and the task must be relevant to the subsystem being tested. Pretesting of the task or scenario is recommended.
- Participant recruitment is often poorly managed. If the audience is unknown, selection must be done to carefully ensure a wide range of user characteristics are considered, including age, gender, and cultural background. Developing a profile is important.
- Reference to the literature can assist with developing the usability test instrument(s), and pilot testing is recommended. Care needs to be taken in planning the testing procedures, including decisions regarding the facilities to be used.

References

Abdinnour-Helm, S., Chapparro, B., & Farmer, S. (2005). Using the end-user computing satisfaction instrument to measure satisfaction with a Web site. *Decision Sciences, 36*(2), 341–364.

Agarwal, R., & Venkatesh, V. (2002). Assessing a firm's Web presence: A heuristic evaluation procedure for the measurement of usability. *Information Systems Research, 13*(2), 168–186.

Aladwani, A. (2002). The development of two tools for measuring the easiness and usefulness of transactional websites. *European Journal of Information Systems, 11*, 223–234.

Bryman, A., & Cramer, D. (1992). *Quantitative data analysis for social scientists* (Vol. 1). London: Routledge.

Canzer, B. (2006). *E business: Strategic thinking and practice*. Boston: Houghton Mifflin.

Caulton, D. A. (2001). Relaxing the homogeneity assumption in usability testing. *Behaviour & Information Technology, 20*(1), 1–7.

Chua, W., & Tan, B. (2007, December). Effects of website interactivity on consumer involvement and purchase intention. Paper presented at the International Conference on Information Systems, Montreal.

Clemmensen, T., Hertzum, M., Hornbaek, K., Shi, Q., & Yammiyavar, P. (2008, December). Cultural cognition in the thinking-aloud method for usability evaluation. Paper presented at the International Conference on Information Systems, Paris.

Cordes, R. (2001). Task-selection bias: A case for user-defined tasks. *International Journal of Human–Computer Interaction, 13*(4), 411–419.

Cutmore, T., Hine, T., Maberly, K., Langford, N., & Hawgood, G. (2000). Cognitive and gender factors influencing navigation in a virtual environment. *International Journal of Human-Computer Studies, 53*(2), 223–249.

Cyr, D. (2009, June). Gender and website design across cultures. Paper presented at the European Conference on Information Systems, Verona, Italy.

Cyr, D., & Bonanni, C. (2005). Gender and website design in e-business. *International Journal of Electronic Business, 3*(6), 565–582.

Dennis, A., & Valacich, J. (2001). Conducting experimental research in information systems. *Communications of the Association for Information Systems, 7*, 1–40.

Dumas, J., & Redish, J. (1994). *A practical guide to usability testing*. Norwood, NJ: Ablex.

Ethier, J., & Hadaya, P. (2004, December). Business-to-consumer Web site quality and Web shoppers' emotions: Exploring a research model. Paper presented at the International Conference on Information Systems, Washington, DC.

Featherman, M., Valacich, J., & Wells, J. (2006). Is that authentic or artificial? Understanding consumer perceptions of risk in e-service encounters. *Information Systems Journal, 16*(2), 107–134.

Fiore, M., & Jin, H. (2003). Influence of image interactivity on approach responses towards an online retailer. *Internet Research, 13*(1), 38–48.

Geissler, G., Zinkhan, G., & Watson, R. (2001). Web home page complexity and communication effectiveness. *Journal of the Association for Information Systems, 2*, 1–48.

Gould, J., & Lewis, C. (1985). Designing for usability: Key principles and what designers think. *Communications of the ACM, 28*(3), 300–311.

Gray, W., & Salzman, M. (1998). Damaged merchandise? A review of experiments that compare usability evaluation methods. *Human-Computer Interaction, 13,* 203–261.

Green, D., & Pearson, M. (2006). Development of a website usability instrument based on ISO 9241-11. *Journal of Computer Information Systems, 47*(1), 66–72.

Guo, M., & Poole, M. (2009). Antecedents of flow in online shopping: A test of alternative models. *Information Systems Journal, 19*(4), 369–390.

Hallahan, K. (2001). Improving public relations websites through usability research. *Public Relations Review, 27,* 223–239.

Hartson, R., Terence, A., & Williges, R. (2001). Criteria for evaluating usability evaluation methods. *International Journal of Human–Computer Interaction, 13*(4), 373–410.

Johnson, R., Salvo, M., & Zoetewey, M. (2007). User-centered technology in participatory culture: Two decades beyond a narrow conception of usability testing. *IEEE Transactions on Professional Communication, 50*(4), 320–332.

Kuan, H. H., Bock, G. W., & Vathanophas, V. (2008). Comparing the effects of website quality on customer initial purchase and continued purchase at e-commerce websites. *Behaviour & Information Technology, 27*(1), 3–16.

Kumar, R., Smith, M., & Bannerjee, S. (2004). User interface features influencing overall ease of use and personalization. *Information & Management, 41,* 289–302.

Leedy, P. (1997). *Practical research planning and design* (6th ed.). Upper Saddle River, NJ: Merrill Prentice Hall.

Lewis, J. (2001). Introduction: Current Issues in Usability Evaluation. *International Journal of Human–Computer Interaction 13*(4), 343–349.

Maguire, M. (2001). Context of use within usability activities. *International Journal of Human–Computer Studies, 55,* 453–483.

Massey, A., Khatri, V., & Ramesh, V. (2005, January). From the Web to the wireless web: Technology readiness and usability. Paper presented at the 38th Hawaii International Conference on System Sciences, Big Island, HI.

McKay, J., & Marshall, P. (2004). *Strategic management of e-business.* Milton, QLD, Australia: Wiley.

Nadkarni, S., & Gupta, R. (2007). A task-based model of perceived website complexity. *MIS Quarterly, 31*(3), 501–524.

Nel, D., van Niekerk, R., Berthon, J-P., & Davies, T. (1999). Going with the flow: Web sites and customer involvement. *Internet Research: Electronic Networking Applications and Policy, 9*(2), 109–116.

Neuman, W. (2003). *Social research methods.* Boston: Allyn & Bacon.

Palmer, J. (2002). Web site usability, design and performance metrics. *Information Systems Research, 13*(2), 152–168.

Rau, P-L. P., Gao, Q., & Liu, J. (2007). The effect of rich Web portal design and floating animations on visual search. *International Journal of Human-Computer Interaction, 22*(3), 195–216.

Rosen, D., Purinton, E., & Lloyd, S. (2004). Web site design: Building a cognitive framework. *Journal of Electronic Commerce in Organizations, 2*(1), 15–28.

Roy, M., Dewit, O., & Aubert, B. (2001). The impact of interface usability on trust in Web retailers. *Internet Research: Electronic Networking Applications and Policy, 11*(5), 388–398.

Schneider, G. (2007). *Electronic commerce* (7th ed.). Toronto: Thomson.

Sears, A., Jacko, J. A., & Dubach, E. M. (2000). International aspects of World Wide Web usability and the role of high-end graphical enhancements. *International Journal of Human–Computer Interaction, 12*(2), 241–261.

Shneiderman, B. (2000). Universal usability. *Communications of the ACM, 43*(5), 85–91.

Shneiderman, B., & Plaisant, C. (2010). *Designing the user interface: Strategies for effective human–computer interaction.* Boston: Pearson.

Simon, S. (2001). The impact of culture and gender on Web sites: An empirical study. *Data Base for Advances in Information Systems, 32*(1), 18–37.

Singh, S., Dalal, N., & Spears, N. (2005). Understanding Web home page perception. *European Journal of Information Systems, 14*(3), 288–302.

Stenstrom, E., Stenstrom, P., Saad, G., & Cheikhrouhou, S. (2008). Online hunting and gathering: An evolutionary perspective on sex differences in website preferences and navigation. *IEEE Transactions on Professional Communication, 51*(2), 155–168.

Stylianou, A., & Jackson, P. (2007). A comparative examination of individual differences and beliefs on technology usage: Gauging the role of IT. *Journal of Computer Information Systems, 47*(4), 11–18.

Teo, H., Oh, L., Liu, C., & Wei, K. (2003). An empirical study of the effects of interactivity on Web user attitude. *International Journal of Human–Computer Studies, 58*, 281–305.

Treiblmaier, H. (2007). Website analysis: A review and assessment of previous research. *Communications of the Association for Information Systems, 19*, 806–843.

Vijayasarathy, L., & Jones, J. (2000). Print and internet catalog shopping: Assessing attitudes and intentions. *Internet Research, 10*(3), 191–202.

Whitefield, A., Wilson, F., & Dowell, J. (1991). A framework for human factors evaluation. *Behaviour & Information Technology, 10*(1), 65–79.

Zhang, Z., Keeling, K., & Pavur, R. (2000, November). Information quality of commercial Web site homepages: An explorative analysis. Paper presented at the International Conference on Information Systems, Brisbane, Australia.

Section III

Designing for Complexity: Methods of Conceptualizing Design Needs of Complex Systems

We are at a critical point in usability testing, especially related to complex systems, where we need tested models that we can evaluate if not employ for our own use. The authors in this section of the book, each focused on different subject areas, provide how-to methods for conceptualizing complex user-centered systems.

Heather Shearer draws on activity theory, and especially its ability to operationalize context, to show how something like the USDA's 1992 Food Guide Pyramid can be more effectively evaluated so that issues of complexity, such as contradictions in a system or the internalization or externalization of system elements, are not overlooked. Stan Dicks provides a case sudy of a genomics research software project as a means of demonstrating that complexity can be managed if researchers collaborate, a system's state can be saved at any point in the process, and, among other things, customization is built into both the development process and software so that a wide range of current and future methodologies can be accommodated. Finally, Bob Watson looks at usability aspects unique to application programming interfaces (APIs). Specifically, he shows how technical writers can work with API design teams to integrate usability into the software development process.

9

An Activity-Theoretical Approach to the Usability Testing of Information Products Meant to Support Complex Use

Heather Shearer

Montana Tech of the University of Montana

CONTENTS

Introduction 182
Activity Theory: Operationalizing Context... 185
    Overview of Activity Theory (e.g., Rubinstein, 1957)and
        Cultural-Historical Activity Theory 187
The USDA Food Guide Pyramids: Overview and Re-View 193
    Overview... 193
    Re-View: Activity-Theoretical Approach to the Usability Testing
        of MyPyramid 195
        Understand and Accommodate the Object of
            Users' Activity 196
        Choose Methods That Capture Data about All Levels of
            Activity in the System 199
        Commit to a Time Frame That Is Long Enough to Observe
            Changes in the Activity System 201
Reader Take-Aways 203
Notes 203
References... 204

Abstract

This chapter discusses the benefits of using activity theory to study and assess complex use situations outside of the workplace, such as those associated with the use of health information products. The author first explains what activity theory is by focusing on the unique benefits—most notably, its ability to operationalize context—that the framework provides to usability researchers. Then, the author applies activity to a usability-testing example: the revision of the USDA's 1992 Food Guide

Pyramid. The chapter concludes by suggesting that researchers using an activity-theoretical approach to usability testing should (1) determine how different levels of activity reinforce each other in a given activity system, (2) identify where contradictions occur in that activity system, and (3) observe how internalization and externalization affect artifact use in that system.

Introduction

On April 26, 2005, the United States Department of Agriculture (USDA) released its revision of the 1992 Food Guide Pyramid. The USDA called this revision the MyPyramid Food Guidance System. The public face of this new system is the revised pyramid graphic, which contains a URL that directs users to the MyPyramid website, MyPyramid.gov. The USDA and the American Dietetic Association (ADA) cited two reasons for the revision. First, scientific advances in nutrition necessitated a change to the content of the 1992 Food Guide Pyramid. Second, too few consumers used the 1992 Food Guide Pyramid to direct their eating habits (USDA, 2005, p. 4).

As can be seen from a comparison of Figure 9.1 and Figure 9.2, the MyPyramid graphic is a radical departure from its information-dense

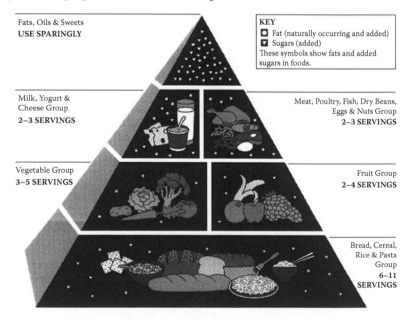

FIGURE 9.1
1992 Food Guide Pyramid.

FIGURE 9.2.
Unofficial (labeled) MyPyramid graphic.

predecessor. Colored stripes—which correspond to different food groups—divide MyPyramid. These stripes are intended to represent the relative proportions of each food group that a user should consumer daily.[1] The concept of "[m]oderation is represented by the narrowing of each food group from bottom to top. The wider base stands for foods with little or no solid fats, added sugars" ("Rebuilding the Pyramid," 2005, p. 1). A stick figure climbs the side of the pyramid, a reminder to users to be physically active. The slogan, "Steps to a Healthier You," is accompanied by a Web address that directs users to a website where they can find nutritional information tailored to their needs.

Far from being a success, the MyPyramid Food Guidance System—which cost $2.4 million and took five years to develop—was met with criticism by users, health professionals, and usability professionals. Some of the criticism naturally centered on the particular advice dispensed by the new guidelines. However, a good deal of the criticism also focused on the design of the MyPyramid graphic. One reviewer described it as "a low point in ergonomic design" (Anderson, 2005). Reviewers reacted negatively to the use of color coding to convey food group information, and MyPyramid was compared to the Homeland Security Advisory System, another unpopular government-designed information product that relies on color to convey information. From a usability perspective, the most critical complaint voiced was that the revised graphic, MyPyramid, could not stand on its own; in effect, the designers forced users to read the manual (i.e., the MyPyramid.gov website) in order to understand the specific habits promoted by the graphic.[2]

What went wrong? And could this design failure have been prevented? In what follows, I demonstrate that the design failures associated with MyPyramid were the result of the researchers' view of usability and their subsequent approach to usability testing. Specifically, the researchers viewed usability in terms of simply "delivering" information to users, when they should have viewed usability in terms of activity.

A number of researchers in technical communication argue that approaches to usability testing need to be modified so that they can more accurately assess usability situations involving texts meant to support complex activity—activity, that is, that cannot be satisfactorily broken down into discrete tasks for analysis (Albers, 2003; Mirel, 2002; Redish, 2007; Spinuzzi, 2003). For summary purposes here, this research can be characterized as holding at least one tenet in common: the belief that a focus on *use* rather than *the user* in usability testing is what is needed.

Barbara Mirel has been one of the most vocal voices in the collective call for reform. In "Advancing a Vision of Usability" (Mirel, 2002), she explains what a new perspective on "use" entails:

> Usability leaders must distinguish between ease of use and usefulness. Ease of use involves being able to work the program effectively and easily; usefulness involves being able to do one's work-in-context effectively and meaningfully. Stripping usefulness from ease of use and focusing primarily on the latter is an incomplete recipe for usability or user-centeredness. (p. 169)

Approaching the usability of complex tasks from the goal of ensuring *only* ease of use is an incomplete approach because "[c]omplex tasks ... cannot be formalized into fixed procedures and rules" (Mirel, 2002, p. 170). Rather, ways of achieving success are contextual and change from person to person, and from context to context. Therefore, in order to accommodate complex work, usability specialists must ensure that "the right sets and structures of interactivity will be in place for the right users to perform the right possible actions for the right situations" (Mirel, 2002, p. 170). In other words, designs must tolerate flexibility, and usability testing must evolve to assess that flexibility.

Existing research on complex user interaction often deals with situations germane to workplace settings (e.g., Hayes & Akhavi, 2008; Redish, 2007). Yet, many important information products, such as those associated with health communication, are designed for use outside of the workplace (e.g., Howard, 2008; Schroeder, 2008). In addition, many health information products, especially those designed by government or nonprofit agencies, are often intended to support complex activity. Consequently, to the end of expanding the field's research on complex user interaction, this chapter examines the redesign process of the 1992 Food Guide Pyramid—the process that resulted in the MyPyramid Food Guidance System.

Most relevant to this discussion is the complex nature of the tasks that food guides intend to support. If one considers all of the activity that goes into selecting, buying, and preparing foods according to the food pyramids' recommendations, one quickly realizes how complex the activity is, and how it might defy accurate and useful description via traditional routes such as task analysis. Nor can one ignore the emotion-laden dimensions of food-related activity. Anyone who has attempted to eat any kind of "special" diet can attest to this. Yet, standard usability testing is not on its own equipped to account for the design and testing of tools for complex activities like food selection, preparation, and consumption *or* to account for the mental and material problems that can arise as one tries to adhere to a certain type of diet.

Researchers working with information products similar to the food guide pyramids need a framework that allows them to examine goal-directed activity that takes place in messy and changing social settings. In addition, those involved with the usability testing of health information products need a research method that allows them to better understand the goals and motivations of the potential users of their products. One framework that offers promise in these regards is *activity theory*.

Activity Theory: Operationalizing Context

In technical communication, activity theory has been mainly used as an *analytical* framework put to the service of better understanding how texts regulate organizational activity. Researchers use activity theory for this purpose because it provides a macro-level view of how humans use artifacts (in this case, genres) to accomplish work in social settings. Also, because one version of activity theory, cultural-historical activity theory (CHAT), examines how conflict functions within and between activity systems, it has proven itself especially useful in analyzing workplace communication. It is my view, however, that activity theory can also be used to guide usability testing and, by extension, the design of many information products intended to support complex activity. Two characteristics of activity theory prompt me to recommend it:

- Activity theory permits a fine-grained (micro) and large-grained (macro) view of use. In this way, it negotiates the tension between task-based views of usability and more holistic views. What this means in practical terms is that activity theory offers both synchronic and diachronic views of use.
- Activity theory operationalizes context, thereby making an accurate assessment of complex systems and complex use possible. It accomplishes this through the delineation of its unit of analysis, the activity system.

The ability to operationalize context is, in fact, the primary benefit that activity theory offers to researchers. Reductive notions of context have been critiqued by many researchers, but perhaps Spinuzzi (2000) offers one of the most incisive critiques in his discussion of technical communication's reliance on the "holy trinity" of audience, purpose, and context. He claims that context is

> most frequently appealed to as a useful ellipses, one that allows us to focus on specific aspects of an analysis while ignoring the rest. We talk of political context, social context, cultural context, ethical context, situational context (to pile vagueness upon vagueness), etc. And these contexts allow us to bracket off what we want to study (the "text") from everything else in the universe ("the context"). (Spinuzzi, 2000, p. 216)

Even though Spinuzzi's critique is targeted at audience analysis, the implications of his critique bear directly on usability testing since the conception of context used by technical communicators in usability testing is tacitly tied to audience analysis. And the major problem identified by Spinuzzi should be one that is troubling to researchers investigating complex use. That is, if context is always a fuzzy, ill-defined adjunct to use, how can complex use— which *is* contextual in an essential way—be accounted for?

In comparison to the models of context critiqued by Spinuzzi, context in activity theory is thoroughly distributed throughout and within the activity system. Nardi (1997) explains,

> Activity theory ... proposes a very specific notion of context: the activity itself is the context. What takes place in an activity system composed of objects, action, and operations, *is* the context. Context is constituted through the enactment of activity involving people and artifacts. Context is not an outer container or shell inside of which people behave in certain ways. People consciously and deliberately generate contexts (activities) in part through their own objects; hence context is not just "out there." (p. 38)

To put it simply: Context is not separate from activity—it *is* the activity. It cannot be separated off from the use because activity and context are one and the same.

A number of researchers have discussed how such a redefinition of context might impact research on usability. For example, Mirel's work on computer documentation (1998, 2000, 2002) describes a model of design that privileges situated-use perspectives on the context–activity relationship. The main purpose of her project is to shed light on the usability-testing needs specific to creating successful documentation for complex tasks. In line with activity-oriented models of use, she describes knowledge as existing "in the connections" between "material, social, cultural, institutional, technological, and individual forces" (Mirel, 1998, p. 16, quoted in Spinuzzi, 2000, p. 217). Similarly, Grice and Hart-Davidson (2002) study ways to accommodate the needs of users who "'inhabit' information environments in which they work, learn, and socialize," suggesting that successful usability testing needs to focus on not the success of one tool used in isolation, but on the "success and usability of the entire system" (pp. 159, 166). For these researchers, as for others who view their objects of inquiry from the perspective of activity, context is action and is distributed among all aspects of a use situation.

In many ways, how an approach to usability treats context may be the measure of whether it is up to the challenge of accounting for complex use. Those approaches that treat context as a separate but distinguishing feature of any instance of use cannot adequately account for complex use. Those that offer a thorough integration of contextual elements *within* the system can.

What researchers have yet to suggest, however, is that the concept of context has outlived its usefulness for assessing complex use situations. To put it plainly, an activity-theoretical approach to usability testing eliminates the need for the concept of context because the co-constituting elements of the activity system "stand in" for context. In this way, activity theory operationalizes context by allowing activity-theoretical researchers to rely on the parts of the activity system to account for matters typically ascribed to context. And in many ways, this is what distinguishes an activity-theoretical approach to usability testing from other complex use models.

In the next section, I summarize the aspects of activity theory that researchers can use to describe and define complex use.

Overview of Activity Theory (e.g., Rubinstein, 1957) and Cultural-Historical Activity Theory

Activity theory is a framework developed by Leont'ev (1977), Luria, (1986), Engeström (1987), and others from Vygotsky's cultural-historical psychology. It views human activity as goal directed, social, historical, and mediated by the use of artifacts. Activity theory has been thoroughly described elsewhere (e.g., Engeström; Kaptelinin & Nardi, 2006; Russell, 1995; Wertsch, 1985). The overview offered here presents the features of activity theory that should be of interest to usability researchers who are tasked with assessing information products intended to support complex use.

The first benefit that activity theory offers to usability researchers is the particular view of consciousness that underlies the theory. As Nardi (1997) notes, "The object of activity theory is to understand the unity of consciousness and activity" (p. 4). This object is not necessarily a psychological one, nor has acting on this object been accomplished mainly in psychological terms, at least not as *psychology* is generally defined in the United States. This should be apparent after reviewing the work of Vygotsky, but Nardi (1997) offers insights into the point that make it clear:

> Unlike anthropology, which is also preoccupied with everyday activity, activity theory is concerned with the development and function of

individual consciousness. Activity theory was developed by psychologists, so this is not surprising, but it is a very different flavor of psychology from what the West has been accustomed to, as activity theory emphasizes naturalistic study, culture, and history. (p. 4)

In other words, activity theory is built to account for far more than what one might assign to the descriptor *psychological*. One only has to recall its foundation—in Vygotsky's *cultural-historical* psychology—to see how Vygotsky's notion of human psychology differs from popular Western accounts, especially those that typically inform theories of usability and usability-testing practices. Nardi (1997) summarizes activity theorists' view of psychology and consciousness, which has been extrapolated from Vygotsky's work:

Activity theory incorporates strong notions of intentionality, history, mediation, collaboration and development in constructing consciousness. ... Activity theorists argue that consciousness is not a set of discrete disembodied cognitive acts (decision making, classification, remembering), and certainly is not the brain; rather, consciousness is located in everyday practice: you are what you do. And what you do is firmly and inextricably embedded in the social matrix of which every person is an organic part. This social matrix is composed of people and artifacts. Artifacts may be physical tools or sign systems such as human language. Understanding the interpenetration of the individual, other people, and artifacts in everyday activity is the challenge that activity theory has set for itself. (p. 4)

For designers of information products intended to support complex use, this description of consciousness should offer hope. Namely, activity theory's model of consciousness suggests that it is *only* through the use of artifacts that humans have consciousness. An ethical challenge facing usability researchers, then, is how to design for use when "use" is inextricably tied to human consciousness.

An important part of activity theory's account of activity is *internalization* and *externalization*. Together, these two concepts explain how psychological development vis-à-vis tool use develops over time, and they offer usability researchers a way to account for that development. Internalization "is the transformation of external activities into internal ones. ... [N]ot only do mental representations get placed in someone's head, but the holistic activity, including motor activity and the use of artifacts, is crucial for internalization" (Kaptelinin, Nardi, & Maccaulay, 1999, p. 30). Internalization is desirable because it can help users to "identify optimal actions before actually performing an action externally. In some cases, internalization can make an action more efficient" (Kaptelinin et al., p. 30). In other words, internalization can equip users with a mental "practice run" of the conscious activities they take in the pursuit of ultimately achieving some objective.

Externalization is the transformation of "internal activities into external ones" (Kaptelinin et al., 1999, p. 30). Given that internalization is, by and large, considered to be beneficial, one might wonder why usability researchers would bother with externalization at all. In fact, externalization is a necessary process for refining, correcting, eliminating, or changing unconscious behavior. In addition, externalization makes collaborative work possible:

> Externalization is often necessary when an internalized action needs to be repaired, or scaled, such as when a calculation is not coming out right when done mentally or is too large to perform without pencil and paper or calculator (or some external artifact). Externalization is also important when collaboration between several people requires their activities to be performed externally in order to be coordinated. (Kaptelinin et al., p. 31)

This dual focus on internalization and externalization allows usability researchers to examine different levels of tool use in relationship to efficiency and learning, and this distinguishes an activity-theoretical approach to studying usability from those that do not focus on development.

The second benefit that activity theory offers to usability researchers is a set of terms and concepts that can be used to describe complex use situations by operationalizing context and, by extension, complex use. In this way, activity theory permits a simultaneous view of use at the micro (task) level and the macro (system) level. Figure 9.3 depicts the most basic view of an activity system. Table 9.1 defines the parts of the basic system in addition to defining Leont'ev's (1977) key additions of *actions, goals,* and *operations.*

Even though activity theory organizes levels of activity into a hierarchy (see Figure 9.4), it is important to note that each level of activity can transform into a different level, given the right circumstances. For example,

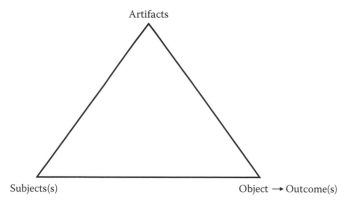

FIGURE 9.3
The activity system.

TABLE 9.1

Leont'ev's Basic Activity Theory Vocabulary

Activity (System)	Activity (or an "activity system") is the primary unit of analysis in activity theory, and it is "the minimal meaningful context for understanding individual actions" (Kuutti, 1996, p. 28). Activity occupies the top position in the hierarchy of activity in an activity system (Kaptelinin & Nardi, 2006, p. 64).
Subject	The subject is the person or group of persons "whose agency is chosen as the point of view in the analysis" (Center for Activity Theory and Developmental Work Research, n.d.).
Artifacts	Artifacts are the "instruments, signs, procedures, machines, methods, laws, [and] forms of work organization" that mediate the interactions among the elements of an activity system (Kuutti, 1996, p. 26). Artifacts are often referred to as *tools* or *instruments*.
Object(ive)	The object is the "'raw material' or 'problem space' at which the activity is directed and which is molded or transformed into *outcomes* with the help of "artifacts (Hasu & Engeström, 2000, p. 63). In other words, the object is the thing or concept that motivates a person to act; it is "perceived as something that can meet a need of the subject.... The object motivates the subject—it is a motive" (Kaptelinin & Nardi, 2006, p. 59). In fact, this object-oriented nature of activity is its "basic, constituent feature," so much so that "the very concept of activity implies the concept of the object of activity." The only thing that distinguishes one activity system from another, Leont'ev explains, is their differing objects: "It is the object of the activity that endows it with a certain orientation. The expression 'objectless activity' has no meaning at all" (Leont'ev, 1977). Importantly for research on usability, what constitutes an object is flexible enough to include many types of products and processes. According to Leont'ev, the object of an activity "may be both material and ideal; it may be given in perception or it may exist only in imagination, in the mind" (1977).
Actions	Actions are the "basic 'components' of separate human activities" (Leont'ev, 1977), and they occupy the middle level on the hierarchy of activity. The object of any activity is realized through *actions* on the part of the subject. Actions are oriented toward the object or motive of the activity system and are carried out by an individual or group of individuals. Actions are short-lived and have clearly demarcated beginnings and endings (Center for Activity Theory and Developmental Work Research, n.d.). In addition, actions are conscious and take goals as their objects (Leont'ev, 1977).
Goals	Goals are the objects of actions.
Operations	Operations are the lowest level of work in an activity system. They are "functional subunits of actions, which are carried out automatically," and they are oriented toward the conditions of an activity system (Kaptelinin & Nardi, 2006, p. 63; Kaptelinin et al., 1999, p. 29). Unlike activities and actions, operations do not have their own aims. Instead, they "adjust" to help subjects to achieve the goals of actions.

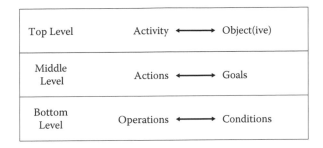

FIGURE 9.4
Levels of activity in the activity system.

operations often start out as actions since "actions transform into operations when they become routinized and unconscious with practice. ... Conversely, an operation can become an action when 'conditions impede an action's execution through previously formed operation'" (Kaptelinin et al., 1999, p. 29; Leont'ev, 1977).

To explain how an action might become an operation and vice versa, one could use the example of knitting. For someone learning to knit a scarf (an activity with an object), making a knit stitch is an action with a goal that must be consciously attended to. As one progresses in skill, however, completing the knit stitch becomes an operation because it no longer requires the conscious attention of the knitter. But completing a knit stitch can once again require the seasoned knitter's attention when something impedes the operation of making a knit stitch. For instance, if the knitter "drops" or twists a stitch while knitting, he or she must consciously attend to fixing that stitch before moving on; the creation of that one stitch requires conscious attention. In this way, the hierarchy of interactions between humans and their material world is dynamic, which differentiates activity theory from other systems models like GOMS (goals, operators, methods, and selection rules) that do not allow pieces of the hierarchy to shift roles or locations (Kaptelinin et al., 1999, p. 29; Nardi, 1996, p. 38).[3]

To Leont'ev's (1977) model of activity theory, cultural-historical activity theory adds the concepts of *community, division of labor, rules,* and *contradictions* (see Figure 9.5 and Table 9.2). The addition of these concepts to the general theory of activity deepens and refines activity theory's ability to account for complex use by delineating the influence of "the social" within use situations.

All of these concepts are defined in terms of their role in constituting (an) activity (system). It helps to keep in mind that activity is at root a *process* and is *dynamic.* Leont'ev (1977) writes that "activity is a highly dynamic system, which is characterized by constantly occurring transformations." When one studies activity systems, one must guard against breaking them apart into discrete units, even if vocabulary like *actions, operations, goals,* and so on makes

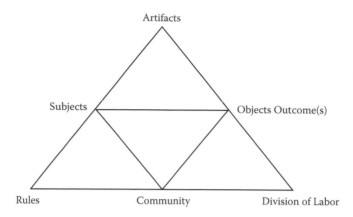

FIGURE 9.5
Cultural-historical activity theory (CHAT) triangle.

TABLE 9.2

Cultural-Historical Activity Theory Vocabulary

Community	*Community* refers to a collection of "multiple individuals and groups who share the same general object" (Hasu & Engeström, 2000, p. 63). This group may or may not overlap with the subject of the study; they are a group by virtue of having an object in common, not necessarily *the* object of the subject.
Division of Labor	*Division of labor* refers to two aspects of the activity system: "the horizontal division of tasks between members of the community and the vertical division of power and status" (Hasu & Engeström, 2000, p. 63).
Rules	Rules are the "explicit and implicit regulations, norms and conventions that constrain actions and interactions within the activity system" (Hasu & Engeström, 2000, p. 63).
Contradictions	Contradictions are the driving force behind the development of new activity systems, and as such, they are a process. Specifically, they are a multileveled set of interactions (first, between two separate activity systems, and, second, among the elements of an individual activity system) that cause imbalance in a system that leads to movement and change (Artemeva & Freedman, 2001; Center for Activity Theory and Developmental Work Research, n.d.; Ilyenkov, 1977, cited in Engeström, 1987). The concept of contradictions allows researchers to account for change in an activity system.

that kind of analysis tempting. In other words, we must keep the parts and the sum of the parts in mind at the same time. Leont'ev reminds us that

> activity is not an additive process. Hence actions are not separate things that are included in activity. Human activity exists as action or a chain of actions. If we were to mentally subtract from activity the actions which realise it there would be nothing left of activity. This can be expressed another way. When we consider the unfolding of a specific process— internal or external—from the angle of the motive [object], it appears as human activity, but when considered as a goal-oriented process, it appears as an action or a system, a chain of actions.

The challenge, then, is keeping both views of the activity in mind at once: as object oriented (which relates to the overall aim of the activity) and as goal oriented (which relates to the aims of individual actions).

When one takes these concepts in toto, one has activity theory's unity of analysis, the activity system. This expansive focus results in a rather large (and, some would say, unwieldy) unit of analysis. Analyzing activity requires researchers to maintain the integrity of the dynamic activity system and—at the same time—to discover how each part of the system relates to the others.

Because of its ability to account for micro and macro aspects of activity and because of its flexibility, activity theory is an elegant and robust framework for analyzing complex activities. In the following sections, I demonstrate how an activity-theoretical approach to usability testing has the potential to reform researchers' view of what use is and how it should be accounted for.

The USDA Food Guide Pyramids: Overview and Re-View

Before I demonstrate what an activity-theoretical approach to the usability testing of the MyPyramid Food Guidance System might look like, I will offer a brief overview of the actual research and design processes that resulted in the 2005 MyPyramid Food Guidance System. I follow the overview with an application of activity theory to the revision of the 1992 Food Guide Pyramid.

Overview

The redesign of the 1992 Food Guide Pyramid began with an assessment of whether the pyramid communicated its "concepts and message," whether "consumers use[d] the Pyramid to make food choices," and if barriers to usage of the pyramid existed (USDA, 2005, p. 4). In order to gather information related to these assessment issues, research conducted eighteen focus groups ($n = 178$) with users. The results of this research indicated that users could identify general Pyramid messages, but not the finer details that its designers wanted to communicate; users were confused about what constituted a serving of food; and users wanted more individualized information. (USDA, p. 4)

However, it must be pointed out that even when trying to identify barriers to usage that existed, the researchers' view of usability amounted to one of ensuring the receipt of information. Notably, they did not, for example, assess how "contextual" issues (e.g., lack of funds, problems with transportation, physical format of pyramid tool) might have caused problems for users. That is, rather than inquiring about the ways that a tool like the pyramid *could* (or did or did not) fit into users' lives, the researchers chose to frame the usage problems in terms of the sender–receiver model of communication, with the users being on the receiving end of the message.

Table 9.3 describes the research conducted at the different levels of the MyPyramid research and design process. Usability testing was conducted at two points: in finalizing the MyPyramid graphic (Stage IV) and in testing the MyPyramid website, MyPyramid.gov (Stage VI).

Stage IV usability testing was conducted through a WebTV survey. Researchers used the survey to gather preference data about the final four graphics and final five slogans that survived the focus groups in Stage II and Stage III. The fact that researchers did not collect performance data about the usability of the MyPyramid graphic is rather shocking given the performance problems that researchers identified in Stage I. For example, users had trouble

TABLE 9.3

Research and Design Stages of MyPyramid

Stage Description	Method and Population
Stage I (May 2000–June 2000): Determine weaknesses in Food Guide Pyramid.	Focus groups ($n = 178$)
Stage II (February 2004): Assess potential new food guidance system messages.	Focus groups ($n = 75$)
Stage III (October 2004): Get feedback on potential graphics and slogans.	Focus groups ($n = 77$)
Stage IV (December 2004): Determine which of the final four graphic concepts should become the "new national symbol for healthy eating and physical activity" (USDA, 2005, p. 27).	Remote WebTV test for preference data ($n = 200$)
Stage V (February 2005): Perform final graphics testing.	Remote WebTV test for preference data ($n = 200$)
Stage VI (February 2005): Determine MyPyramid.gov usability.	Formative testing and one-on-one interviews ($n = 18$)

recalling the position of the food groups on the 1992 Food Guide Pyramid, yet researchers did not collect data on whether users could match up the different colors on the MyPyramid graphic to the corresponding food groups.

To test the MyPyramid.gov website (Stage VI), researchers presented users with a prototype of the website and, through one-on-one interviews, used a "formative design" to assess the site's ease of use (USDA, 2005, p. 36). This passage from the research report offers a glimpse into what the researchers tested for and how they defined *ease of use*:

> Even though testing revealed that MyPyramid.gov did not suffer from any Category I usability issues, there were some areas identified for potential improvement. ... Some respondents ... exhibited some frustration searching for more detailed nutrition information about topics on the Web site. Specifically, participants were interested in learning the different nutrients included in different types of vegetables (e.g., dark green versus orange vegetables) or in more details about how beans, nuts, and seeds could be appropriate substitutes for meat. Some of the information was included on the Web site, but respondents were unable to locate it. (USDA, pp. 36–37)[4]

There is no doubt that preference data are helpful in designing effective and affective tools. And formative testing is useful in finding out whether or not a website can be navigated. But neither of these approaches to usability testing addresses the complex activities that the website and the graphic ostensibly support (i.e., the locating, buying, preparing, and eating of certain foods). To put it another way, what was left out of the usability testing of the MyPyramid Food Guidance System was an appreciation for and concern with, to rely on Mirel's term, the usefulness of the information product(s).

As discussed earlier, activity theory responds to the shortcomings evidenced in the redesign of the 1992 Food Guide Pyramid. The challenge for usability researchers, however, is in translating activity theory into testing protocols appropriate for information products like MyPyramid—products that are used to support complex, mundane activities of everyday life. In what follows, I respond to that challenge by answering the following question: What would an activity-theoretical approach to redesigning the 1992 Food Guide Pyramid look like? In answering this question, I take initial steps toward translating activity theory into an applied usability-testing process that accommodates the usability demands of information products intended to support complex use.

Re-View: Activity-Theoretical Approach to the Usability Testing of MyPyramid

Adopting an activity-theoretical research and design process requires usability researchers to

- understand and accommodate the object of users' activity;
- choose a set of methods that capture data about all levels of activity (object, actions, and operations) in a system; and
- commit to a time frame that is long enough to observe "changes in objects over time and their relation to the objects of others in the setting being studied" (Nardi, 1996, p. 47).

That is, researchers need to observe the activity system long enough and in such as fashion as to determine how elements in the system are *co-constituted* (Spinuzzi, 2003, p. 36).[5]

Understand and Accommodate the Object of Users' Activity

An activity system is constructed from the perspective of the system's subject. Because it is the object of a system that calls the system into being, activity-theoretical researchers are required to discover the object of their subjects' activity. This means that researchers should first understand users' objects and then design an artifact in response to those objects. The USDA researchers went about this backward in that they designed an artifact (MyPyramid) and established objects without researching what subjects themselves might posit as objects. In other words, they created something that some people might not ever find a use for or, worse, expected users to adopt the object set out for them by the researchers. Since researchers designed the tool according to their own object (i.e., translate the dietary guidelines for Americans into a graphics food guide), they lacked insight into what an appropriate artifact for aiding in the achievement of the subjects' own objects might look like.

In revising the 1992 Food Guide Pyramid, the initial task that researchers would undertake is setting up the macro boundaries of relevant activity systems by identifying appropriate subjects and those subjects' objects. Researchers would need to ask the following questions:

- What motives or needs do users or potential users have with respect to the kinds of behaviors addressed by a food guide?
- What are users' attitudes toward the 1992 Food Guide Pyramid? Do they feel that they would use it (or a revision)? Do they want to use it?

The answers to these questions will be unique, at least in some way, to each subject in each activity system. Activity theory rejects idealized representations of artifact use. Consequently, an activity-theoretical approach to usability testing requires not that researchers identify multiple audiences (e.g., "types" of people), but that they identify multiple objectives and corresponding activity systems. Figure 9.6 offers a variation on the theme of *use* that researchers might encounter.

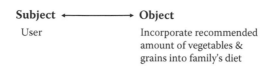

FIGURE 9.6
Subject–object example.

Activity-theoretical researchers would also pose questions that relate to the remaining aspects of the activity system. The object calls an activity system into being, but the other aspects of the system—subject, actions, operations, artifacts, rules, division of labor, and contradictions—co-constitute and motivate the system. A series of questions, formulated to populate the activity systems, should guide this. For example, in considering the redesign of the 1992 Food Guide Pyramid, researchers might ask,

- What artifacts other than the pyramid are available to users, and what is the relationship among these artifacts? Are the artifacts shared among users in the system? For instance, what food storage and preparation tools are available? What about access to transportation?

- Are competing artifacts in evidence? For instance, does a user employ another food guidance system—say, the Weight Watchers Momentum Plan®—to make food choices?

- What is the division of labor in the system with respect to the object? For instance, who is responsible for purchasing food? For meal planning? For preparing food? For consuming food?

- What rules shape the interactions between artifacts and users in the activity system? For instance, what regulations about food purchases are placed on subjects (e.g., some subjects might receive WIC money or supplemental nutritional assistance)? Are there cultural assumptions about food consumption (e.g., teenage boys need animal protein because "they're growing") that need to be addressed?

- How is work toward the objective distributed across time and space? Where does the labor take place? In homes? In farmers' markets? In Wal-Mart? In grocery stores? In schools? In restaurants? How is the artifact used in these settings? Are there barriers to usage in these settings?

- How do people gather knowledge about how to use the artifact? From the pyramid itself? From the media? In school?

- How much "time and effort" are necessary to be able to use the tool to achieve goals? What level of education is necessary before users are comfortable using the tool to guide their food intake habits?

Next, activity-theoretical researchers would try to determine what changes to the 1992 Food Guide Pyramid need to be made so that users are equipped to reach their objectives. To accomplish this, researchers would no doubt address what users already do to get around roadblocks that they encounter as they attempt to achieve their goals. As Spinuzzi's research (2003) demonstrates, worker innovations in dealing with usability problems can be instructive to researchers and designers. Perhaps more importantly (or at least as importantly), researchers would need to identify the roadblocks that users were not able to circumvent on their own. The health information tool being designed would ideally support the circumnavigation of these roadblocks.

As researchers begin to fill out the dimensions of the activity system, they will eventually arrive at a point where the system can be represented in basic diagram form. Figure 9.7 depicts what *one* such diagram might look like. This diagram can be enhanced with additional information, including diagrams of neighboring activity systems. The benefit of this diagram is that it depicts use as a confluence of forces coming together to motivate the user toward an objective. This is, in other words, a picture of complex use.

In order to gather this intake information, researchers would not simply rely on focus groups (as the actual researchers did). According to Nardi (1996), activity theory implies "[t]he use of a varied set of data collection techniques … without undue reliance on any one method" (p. 47). Perhaps this is good advice in any case, but this advice is especially important to remember when dealing with research on food consumption. After all, self-reported accounts of food habits are often unreliable, so gathering information through many different channels would improve any research project dealing with nutrition

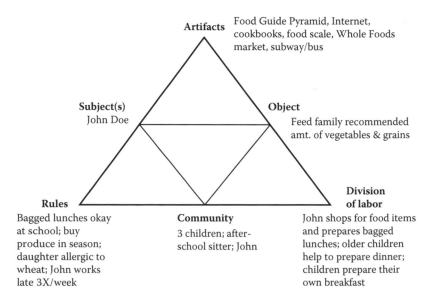

FIGURE 9.7
Activity system example.

information products (see Horner, 2002; Lara, Scott, & Lean, 2004; Mahabir et al., 2006; Rumpler, Rhodes, Moshfegh, Paul, & Kramer, 2007).[6] But more to the point, sketching out an activity system simply requires more information—and a wider variety of information—than is usually gathered in a traditional course of design and testing.

Choose Methods That Capture Data about All Levels of Activity in the System

The three levels of activity in the activity system—activity, actions, and operations—carry with them the social, historical, and material dimensions that make the activity system a rich and robust unit of analysis. Because the different levels of activity in a system mutually construct and inform each other, researchers need to gather data about all three levels so that they can understand the whole system. In order to accomplish this, researchers would need to use methods that allow them to collect relevant data, and their choices would be based on the system(s) being studied:

> Activity theory does not prescribe a single method of study. It only pre-
> scribes that a method be chosen based on the research question at hand. ...
> [A]ctivity theory starts from the problem and then moves to the selection
> of a method. (Kaptelinin & Nardi, 2006, p. 71)

For activity theorists, the goal is to construct a model of activity (represented by the activity system) that illustrates how the levels of activity co-constitute one other. For example, in order to determine how work toward the target goal is distributed across time and space (one question among many that researchers would pose in order to establish the boundaries of an activity system involving a food guide), researchers might use focus groups. But that would not be sufficient. Researchers would also have to go into the field to observe subjects working toward goals in the system itself or have a suitable surrogate for firsthand observation (such as video recordings). Indeed, I argue that when researching complex activities from an activity-theoretical perspective, there is really no way around the need for fieldwork because of the issue involving the misreporting of food intake. But because field research is not on its own sufficient to gather information about all levels of activity in a system, researchers would have to use other methods as well.

Table 9.4 presents possibilities for activity-theoretical usability research on the revision of the 1992 Food Guide Pyramid. The purpose of the table is to link research tools with specific aspects of the activity system to demonstrate how each tool can account, individually, for a part of the activity system with the goal of eventually gathering a critical mass of data that would sufficiently account for complex use in the activity system(s) under study.

The information in Table 9.2 demonstrates at least two points: (1) The activity system requires multiple approaches in order to capture data that are indicative of the different parts of the system, and (2) in many cases, already established research tools can be used to capture these data.

TABLE 9.4

Methods of Conducting Activity-Theoretical Research for the 1992 Food Guide
Pyramid Revision

Research Tool Deployed	Aspect of CHAT Activity System Addressed	Data Gathered
Subject sampling	Subject(s)	Current users of the pyramid who intend or desire to include the pyramid to achieve objectives.
Focus groups	Object; contradictions	Subject-identified objectives; problems associated with using the pyramid to mediate activity in the system.
Intake surveys	Object; artifacts; communities	Historical information, and demographic information about individual subjects in an activity system.
Field observations	Actions; operations; contradictions; rules	Video recordings of goal-directed activity (actions) in the system *and* video recordings of routinized, nonconscious activity (operations). Observations recorded (either electronically or otherwise) about how actions are distributed across time and space in the system (e.g., trips to supermarket, and time and procedures for preparing and consuming meals).
In situ interviews	Operations	Explanations from subjects about the purpose, role, and nature of unconscious operations.
Postobservation interviews	Object; actions; operations; contradictions; rules	Insight and clarification from subjects about observed activity in the system; information from subjects about constraints on activity (e.g., what regulations attend food stamp usage).
Artifact inventory	Artifacts	Lists of tools available for use in the system for achieving the object, for example, food preparation devices, transportation, and means of trade (e.g., currency and food stamps).
Subject relations map	Subjects; division of labor	Information about distribution of labor in system (e.g., Who procures food? Who produces the means to procure food? Who dictates which foods are consumed? Who consumes food?).
Directed talk-aloud protocol (video footage of subjects is reviewed with subjects; researchers inquire about specific actions or operations)	Actions; operations	Metalevel explanation about goal-directed and unconscious activity in the system. Most notably, explanations after the fact about unconscious operations and their relationship to actions.
Activity system diagrams	All aspects of activity system	Graphical representation of how all levels of activity relate to and co-constitute one another.

Commit to a Time Frame That Is Long Enough to Observe Changes in the Activity System

According to Kaptelinin and Nardi (2006), "[A]ctivity theory requires that human interaction with reality" be "analyzed in the context of development" (p. 71). This requirement stems from activity theory's beginnings in Vygotsky's cultural-historical psychology, and relates to both Vygotsky's zone of proximal development and the concept of externalization–internalization. Kaptelinin and Nardi (2006) explain the impact that this concern with development will have on methodology:

> Finally, activity theory requires that human interaction with reality should be analyzed in the context of development. ... Activity theory sees all practice as the result of certain historical developments under certain conditions. Development continuously reforms practice. That is why the basic research method in activity theory is not that of traditional laboratory experiments, but that of the formative experiment which combines active participation with monitoring of the developmental changes of the study participants. (pp. 71–72)

Researchers who are investigating complex use situations need to attend to the progression from externalization to internalization and vice versa. For example, researchers might observe that as a user becomes more familiar with the concepts of an artifact like the pyramid, he or she relies less on the physical artifact (whether that artifact be printed or Web-based materials) to prepare specific meals in line with his or her objective. In fact, as activity systems develop, researchers should *expect* some level of internalization to occur. If it does not occur, this might be a sign that the information product is not useful. Externalization would also be important to document since externalization can be a sign of usability problems. Externalization is not always symptomatic of problems, however. It can often serve an important role in collaboration, and if this occurs, usability researchers should be aware of its implications for design. For instance, a user with the objective of working with a dietitian on following dietary advice set out by a food guide might find that keeping a food diary—rather than relying on memory—is a useful and important part of that collaboration. In a case like this, researchers would want to ensure that the artifact being designed does indeed facilitate such collaboration.

Both the type of data sought and the purpose of collecting the data necessitate that researchers establish a sufficient time frame for research. In most cases, this probably means that researchers will need to spend at least as much time, but probably more, gathering data and observing users as they would with traditional approaches to usability testing. Since activity systems are not static—that is, they are not a "snapshot" of a consistent situation—one goal of researchers should be to see how the activity system being observed evolves over time. In particular, researchers should study

- how the different levels of activity mutually reinforce and build on one another;
- where contradictions (breakdowns) occur in the system; and
- how the processes of internalization and externalization affect artifact use.

The length of the study should be dictated by the system being studied. In the case of the 1992 Food Guide Pyramid revision and similar information products, I propose that observations lasting from as short as a week to as long as a month would not be unusual. What length of observation would be needed to capture the interaction among levels of activity in the "average" suburban family? In a one-parent family in an urban environment? In a single-person household? In a college student living with four other classmates (or on his or her own)? No doubt, each of these would demand unique observation parameters, and activity-theoretical researchers would need to accommodate them.

This commitment to observation has serious implications for usability testing, most notably in terms of cost. How feasible is it, really, to commit to an "as long as it takes" timetable when most likely (and as was the case for the 1992 Food Guide Pyramid revision) someone else is paying for the research? In his discussion of genre tracing, an activity-theoretical method for studying how genres mediate activity, Spinuzzi (2003) acknowledges these problems:

> Since it is labor intensive, genre tracing cannot be quickly deployed the way that usability testing and some forms of participatory design can. And since it involves careful, prolonged study of activities, unlike contextual design, it requires trained researchers. Also, since genre tracing requires significant time and resource commitments from researchers, participants, and organizations, it may not be suitable for all organizations. Finally, like many observational approaches, genre tracing can get bogged down in data. (p. 56)

Spinuzzi's (2003) response to these formidable drawbacks is that if "commitments can be marshaled," the results of the research would be sufficient to inform the design of usable information products (p. 56). For a government project like the revision of the 1992 Food Guide Pyramid, commitments were, in all likelihood, present.

What would the resulting revision of the 1992 Food Guide Pyramid look like? It's impossible to say without having conducted research with actual users. However, as this chapter demonstrates, usability testing from an activity-theoretical perspective is a vastly different endeavor than is usability testing as it is commonly practiced. In large part, this difference stems from the way that the activity theory operationalizes context and from its insistence on accounting for development—via artifact use—in the activity system.

It is encouraging to note that even though activity-theoretical usability testing presents challenges to researchers (most notably in terms of time and money), the research tools that researchers would use to conduct a usability study of a tool already exist and are in use. What activity-theoretical usability research presents, then, is the opportunity and means to put existing research methods to the service of a new usability framework that enables researchers to more appropriately account for complex use situations.

Reader Take-Aways

- Activity theory's unit of analysis, the activity system, allows researchers to operationalize context in order to assess complex use situations. The benefits of this are twofold. First, this allows researchers to move away from a container-contained concept of context. Second, this allows researchers to dispense with the concept of context altogether and use in its place the parts of the activity system.
- An activity-theoretical approach to assessing complex use situations requires researchers to understand and accommodate the object of users' activity. This means that usability researchers are, from the very beginning of the research and design process, responsible for understanding what users' objectives are and accommodating design elements to those objectives. An implication of this is that researchers should design for different users' objectives, not different users.
- Activity theory does not favor one method of data collection over another; the theory requires that researchers choose methods that evolve from the problem being examined *and* that researchers collect data at all levels in the activity system.
- Researchers using an activity-theoretical framework must commit to a time frame that is long enough to observe changes in the activity system. One key way in which activity theory accounts for complex use is in studying how use develops over time.

Notes

1. I have labeled the stripes on MyPyramid so that readers can mentally fill in the colors. Official black-and-white versions of MyPyramid distributed by the USDA are not labeled.
2. Many of the initial reviews were posted on blogs written by U.S. consumers, the target audience of MyPyramid.

3. GOMS is a usability framework common in the field of human–computer interaction and was created by Stuart Card, Thomas Moran, and Allen Newell (1983).
4. The USDA research report on the development of the MyPyramid Food Guidance System describes Category I issues as "severe usability issues that prevent successful operation of the site or completion of specific tasks that induce irrecoverable errors" (2005, p. 36).
5. Spinuzzi (2003) writes, "By coconstitution, I mean that even though we can analytically separate the [elements of an activity] and even though we need to use different methods and theoretical constructs to study each one, they are ultimately intertwined. They do not share a cause-and-effect relationship in which (for example) cultural-historical activity determines actions and operations, or actions are simply contextualized by activity, or operations simply make up actions and activities. Changes in an activity can be initiated at any level of scope" (p. 36).
6. Several reasons are cited in nutrition and related research for this unreliability. Most of the research attempts to account for underreporting caloric intake, though there are also studies that examine the opposite (overreporting caloric intake). Further, the literature can also be divided into those studies that examine unintentional misreporting and those that examine intentional misreporting. The former is more relevant to usability researchers because the reasons cited for unintentional misreporting often deal with the commonly used tools—questionnaires, diaries, and so on—used to measure food intake.

References

Albers, M. (2003). *Communication of complex information: User goals and information needs for dynamic web information.* Mahwah, NJ: Lawrence Erlbaum.

Anderson, J. (2005). *The new food pyramid: A low point in ergonomic design.* Retrieved from http://www.ergoweb.com/news/detail.cfm?id=1100

An Evaluation of Dietary Graphic Alternatives: The Evolution of the Eating Right Pyramid. (1992). *Nutrition Reviews, 50*(9), 275–282.

Card, S., Moran, T., & Newell, A. (1983). *The study of human-computer interaction.* New York: Routledge.

Center for Activity Theory and Developmental Work Research. (N.d.). *Activity system.* Retrieved from http://www.edu.helsinki.fi/activity/pages/chatanddwr/activitysystem

Engeström, Y. (1987). *Learning by expanding: An activity-theoretical approach to developmental research.* Retrieved from http://lchc.ucsd.edu/MCA/Paper/Engestrom/expanding/toc.htm

Grice, R., & Hart-Davidson, W. (2002, November). Mapping the expanding landscape of usability: The case of distributed education. *ACM Journal of Computer Documentation, 26*(4), 159–167.

Hasu, M., & Engeström, Y. (2000). Measurement in action: An activity-theoretical perspective in producer user interaction. *International Journal of Human-Computer Studies, 53*, 61–89.

Hayes, C., & Akhavi, F. (2008). Creating effective decision aids for complex tasks. *Journal of Usability Studies, 3*(4), 152–172.

Horner, N. K. (2002). Participant characteristics associated with errors in self-reported energy intake from the Women's Health Iniative Food-Frequency Questionnaire. *American Journal of Clinical Nutrition, 76*(4), 766–773.

Howard, T. (2008). Unexpected complexity in a traditional usability study. *Journal of Usability Studies, 3*(4), 189–205.

Ilyenkov, E. V. (1977). *Dialectical logic: Essays on its history and theory.* Moscow: Progress.

Kaptelinin, V., & Nardi, B. (2006). *Acting with technology: Activity theory and interaction design.* Cambridge, MA: MIT Press.

Kaptelinin, V., Nardi, B., & Maccaulay, C. (1999). The activity checklist: A tool for representing the "space" of context. *Interactions, 6*(4), 27–39.

Kuutti, K. (1996). Activity theory as a potential framework for human-computer interaction research. In B. Nardi (Ed.), *Context and consciousness: Activity theory and human computer interaction* (pp. 17–44). Cambridge, MA: MIT Press.

Lara, J., Scott, J., & Lean, M. (2004). Intentional mis-reporting of food consumption and its relationship with body mass index and psychological scores in women. *Journal of Human Nutrition Dietetics, 17*(3), 209–218.

Leont'ev, A. (1977). *Activity, consciousness, and personality.* Retrieved from http://www.marxists.org/archive/leontev/works/1977/leon1977.htm

Luria, A. R. (1976). *Cognitive development:* Its social and cultural foundations. Cambridge: Harvard University Press.

Mahabir, S., Baer, D., Giffen, C., Subar, A., Campbell, A., Hartman, T., et al. (2006). Calorie intake misreporting by diet record and food frequency questionnaire compared to doubly labeled water among postmenopausal women. *European Journal of Clinical Nutrition, 60*, 561–565.

Mirel, B. (1998). "Applied constructivism" for user documentation. *Journal of Business Communication, 12*(1), 7–49.

Mirel, B. (2000). Product, process, and profit: The politics of usability in a software venture. *Journal of Computer Documentation, 24*, 185–203.

Mirel, B. (2002). Advancing a vision of usability. In B. Mirel & R. Spilka (Eds.), *Reshaping technical communication: New directions and challenges for the 21st century* (pp. 165–187). Mahwah, NJ: Lawrence Erlbaum.

Nardi, B. (1996). Studying context: A comparison of activity theory, situated action models, and distributed cognition. In B. Nardi (Ed.), *Context and consciousness: Activity theory and human-computer interaction* (pp. 35–52). Cambridge, MA: MIT Press.

Nardi, B. (1997). Activity theory and human-computer interaction. In B. Nardi (Ed.), *Context and consciousness: Activity theory and human-computer interaction* (pp. 4–8). Cambridge, MA: MIT Press.

Rebuilding the Pyramid. (2005, June). *Tufts University Health & Nutrition Letter, 1*, 4–5.

Redish, J. (2007). Expanding usability testing to evaluate complex systems. *Journal of Usability Studies, 2*(3), 102–111.

Rubinshtein, S. L. (1957). *Existence and consciousness.* Moscow: Academy of Pedagogical Science.

Rumpler, W., Rhodes, D., Moshfegh, A., Paul, D., & Kramer, M. (2007). Identifying sources of reporting error using measured food intake. *European Journal of Clinical Nutrition, 62*(4), 544–552.

Russell, D. (1995). Activity theory and its implications for writing instruction. In J. Petraglia (Ed.), *Reconceiving writing, rethinking writing instruction* (pp. 51–78). Hillsdale, NJ: Lawrence Erlbaum.

Schroeder, W. (2008). Switching between tools in complex applications. *Journal of Usability Studies, 3*(4), 173–188.

Spinuzzi, C. (2000). How organizational politics takes us beyond rhetorical categories. Exploring the blind spot: Audience purpose and context in "Product, Process, and Profit." *ACM Journal of Computer Documentation, 24*(4), 213–219.

Spinuzzi, C. (2003). *Tracing genres through organizations: A sociocultural approach to information design.* Cambridge, MA: MIT Press.

U.S. Department of Agriculture & Center for Nutrition Policy and Promotion. (2005). *Research summary report for MyPyramid Food Guidance System.* Washington, DC: Author.

Wertsch, J. V. (1985). *Vygotsky and the social formation of mind.* Cambridge, MA: Harvard University Press.

10

Designing Usable and Useful Solutions for Complex Systems: A Case Study for Genomics Research

R. Stanley Dicks
North Carolina State University

CONTENTS

Introduction 208
Usability Issues in Designing Usability into the Genome Project 210
    The Necessity for a Single, Unified User Interface 211
    The Need for Domain Expertise and Field Testing and
        Observation.. 212
        Acquiring the Necessary Domain Expertise.............................. 213
        The Necessity for Field Studies and Observation 213
    Developmental Methodology 213
        Requirements for Design 214
        Usability Analysis through Field Studies and Contextual
            Inquiries.. 214
    Primary Design and Usability Challenges 215
        The Necessity for Supporting Collaborative,
            Complex Work.. 215
        The Necessity for Supporting the Paradox of Structure
            for Complex System Design 216
    Design and Usability Methodologies... 218
        Field Observation... 218
        User Involvement 218
        Extended Contextual Analysis... 219
Conclusion 220
Reader Take-Aways 221
References... 221

Abstract

This chapter discusses the difficulty of performing usability analysis for complex systems and provides a case study of a development project for a genomics research software system. The system had to accommodate the needs of major genomic projects, including support for researchers, physicians, statisticians, and system administrators. While performing field studies and contextual inquiries, the development team learned that they would have to initiate more rigorous user-centered design practices, including radical methods such as hiring a user, conducting extended contextual inquiries, and performing tests with smaller units of the system. The team learned that it had to design a system that did the following:

- Imposed a single-user interface on the thirty-odd software packages used in the system
- Promoted collaboration among researchers
- Allowed the "state" of a system to be saved at any point and stored
- Provided a linear structure that was simultaneously flexible enough to work with what is often very nonlinear work processes
- Enabled customization so that it would work well with the wide range of both current and future genomic research methodologies

The team's technical communicators likewise faced the challenge of designing information products to accommodate the multivariate, nonlinear nature of genomic research.

Introduction

Of the four aspects of usability posited by Gould and Lewis (1985), usability test methods have been successfully developed to test whether systems are "easy to learn," and "easy and pleasant to use." However, except for the smallest and simplest of devices, usability testing has not proven to be successful for affirming the fourth of those aspects, the usefulness of a product. As the product gets larger and more complex, usability-testing methods prove increasingly inadequate for testing usefulness. This has recently been pointed out (Albers, 2003; Dicks, 2002; Mirel, 2002, 2003; Redish, 2007), but most of us have experienced it on our own when performing tests of larger systems. Our traditional, one- or two-hour usability tests seem to indicate that everything about a system is working fine; yet end users are not able to use the system to complete tasks successfully. For example, Mirel (2002) describes the hospital pharmaceutical program, the smaller parts of which all work well and would no doubt pass muster on conventional usability tests that focus on their effectiveness, learnability, and ease of use, but which, when used as part of the larger hospital system, do not provide the overall usefulness needed by hospital personnel.

In designing the traditional usability test, we look at tasks that end users will perform, although it is surprising how much difficulty even seasoned usability personnel have in distinguishing between system tasks and user tasks. Assume, however, that we properly choose representative user tasks and design a test to measure the effectiveness, ease of learning, and user attitudes about those tasks. We do the best we can to select tasks that, in the typical one- to two-hour test session, will allow us to learn as much about various aspects of the system as we can. The test results can indicate that users could easily complete the tasks, easily learn the tasks, and have highly positive attitudes about the system. However, when they try to use the whole system to accomplish their larger, real-world tasks, the system simply will not support them. That is because very few usability tests measure the usefulness of the overall system. We are good at tests that measure small, incremental, linear sets of steps, but we have not been good at developing tests to measure the usefulness of large, complex systems that are often used collaboratively by users from multiple disciplines and with multiple levels of experience (Mirel, 2002, 2003; Rubin & Chisnell, 2008). Similarly, we know how to provide help and documentation systems that cover well-defined and relatively contained series of steps for completing increasingly larger tasks, but we do not have as much success or much in our literature about how to develop support for large, complex systems that defy well-defined processes.

I experienced this firsthand recently when working as a visiting professor at a large, statistical software organization. My assignment was to write a user guide for a genome solution, which was a collection of many smaller software packages combined with an overarching user interface to allow scientists, researchers, medical doctors, lab technicians, statisticians, and system administrators to conduct collaborative research in genetics. The product was better designed than most such systems, and it included many of the necessities for successfully completing such work that Mirel (2003) points to, including means for nonlinear, multipath studies; for collaboration among disparate professional groups; for saving, labeling, and storing partial work efforts; for supporting collaborative sharing and communication; for accommodating the often nonlinear, ad hoc nature of such research; and so on. Nonetheless, while each of the smaller parts seemed to work quite well, it proved extremely difficult to determine how useful the system was in an overall sense, and, in initial field testing, users needed much hand holding.

Designing a test to determine the usefulness of such a system presents major challenges, including the following:

- A system as complex as the genome solution, a collection of some three or four dozen software programs, cannot be adequately tested with the typical one- to two-hour usability test.
- Even half-day or full-day tests would not cover all of the possible features that researchers might employ over the course of months of actual research.
- Getting adequate sample sizes is always a problem for usability testing. When multiple groups use a product such as the genome solution, audience sample size becomes even more difficult, as one would have to recruit groups of researchers, physicians, statisticians, lab technicians, and systems experts. With such highly educated and specialized users, finding volunteers is extremely difficult and affording to compensate them for testing becomes prohibitively expensive.
- A further difficulty with such systems is getting enough time from the professionals involved; few researchers and physicians are going to agree to the hours or days of testing required to examine the usefulness of such a product.

As Albers (2003) points out, some design problems are complex simply due to the sheer size of either the developers' load or the users' load. The genomics research software is large and complex for both groups. It required an extended, complicated development effort for the programmers, and it must accommodate the extensive number and variety of tasks that its users must perform. Doing anything like traditional usability testing on such a system is simply not possible.

This chapter will present a case study of the genome project and the usability and documentation challenges its developers faced during design and development. Those challenges far exceeded what occurs on more mundane design and development projects. They affected all aspects of the project, including personnel decisions, methodologies for defining and developing the product, field testing, marketing, installation, and testing.

Usability Issues in Designing Usability into the Genome Project

Genomics research constitutes a burgeoning field that almost defies explanation in its complexity. A genomics research institute may include all of the following: PhD researchers in biology, chemistry, botany, zoology, biochemistry, and any other life science; physician researchers in all areas of medical research, from those doing general, pure research to those looking for specific disease causes and cures; statisticians who analyze and interpret the often vast data arrays involved with genomics; a wide array of lab personnel from scientific and technological experts to data entry clerks; administrators, including representatives of funding organizations; and various other support staff. For any single genomics research project, there may be one person working alone or dozens working on the same problem simultaneously.

The Necessity for a Single, Unified User Interface

The genome product on which I worked included, depending on the configuration purchased by a particular customer, some twenty to thirty individual software products working together to allow customers to perform the complicated steps necessary for genome research. Each of those software products is sold by the company in other venues as standalone products or as parts of larger systems with multiple interacting software products also sold by the company. Each of the twenty to thirty products has its own user interface and instructions for use. It would be utterly impossible for an individual customer to learn them all. Therefore, in the initial design for the genome solution, it was realized that a single, unified interface would have to be developed to serve as a front end for the collection of software products.

A further requirement for a unified front end is that this product can be combined with other analytical software packages to allow variations on the types of research possible. So, ultimately, the product might be used in multiple research settings where very different types of research are going on. For example, a commercial lab could be using the product to perform gene expression for the purpose of aiding its research in potential cures for cancer, while a university lab might be doing "pure" research into the genetic composition of a previously unstudied species. With the possibility for linking to other software packages, the need for a single front end became even more compelling.

Another compelling reason for a unified structure was the very nature of the complex work of genetic research and its often nonlinear, ad hoc processes. Paradoxically, the product had to accommodate unstructured, nonlinear work processes and yet had to provide some kind of structure and process if it was going to be usable at all. Analyzing user models for systems in biological research and finding a way to unify them into an approach that would accommodate the various models were crucial.

Thus, the design of a unified front end would be critical in making an otherwise disparate collection of software products usable (and useful) for multiple audiences for multiple purposes. Even though the product team included outstanding, veteran software designers, programmers, and testers, it was realized that the team would have to vary from prior design practices. The first area where changes needed to be made involved personnel.

The Need for Domain Expertise and Field Testing and Observation

Despite their collective experience and expertise in software system design and development, the team did not have the discipline-specific domain knowledge necessary to specify the front-end requirements for the genome solution software. They knew quite well how to combine the various smaller systems to build a powerful statistical analysis engine, but they did not know how to present that power to end users so that they could succeed in using it to do the massively complex statistical and analytical tasks they faced. Further, while they knew how to make the statistics work, they did not have the expertise to know what outputs would be useful for genomic researchers.

Statistical software must perform two primary functions: It must first "crunch" the data supplied to it and do so accurately and within a reasonable time period. This alone is a great enough challenge with genetics research. But the even greater challenge is in how the processed data will be displayed so that they tell researchers what they need to know. There are many ways to view genetic coding data, most often requiring both tabular and graphical representations. To be of value, the output must be easily manipulated to show varying visualizations. In some cases, researchers may need to look at hundreds of different views of the data analyzed by the statistical program.

The software developers for the genomic project realized that they did not have the expertise to design the front end of the proposed product, and, further, that they also did not have sufficient domain knowledge to design the output end either. Nor would conventional methods for performing user and task analysis suffice. Despite numerous field interactions with researchers, the team realized that it simply had to have the expertise "in house," on the team at the company's home location. The field observations and contextual inquiries they performed were necessary but not sufficient for giving them the usability information they needed to successfully design the system.

While technical communicators were not part of the earliest field studies performed by the development team, they did join the team well before the first version (version 1.0) release of the product. Because the overall system had so many components and because the nature of the work users would do with it was so complex and varied, it was decided early to implement a topic-based architecture for the information products supporting the system, which would include an online help system, a user's guide, and some specialized guides aimed at system administrators. This approach was complemented by the XML authoring system that was employed, which complemented the incremental, topic-based information development, fitting nicely with the incremental, iterative nature of the overall system's development. As parts of the overall system successfully passed initial verification and usability testing, technical communicators could develop topics covering them without the concern of excessive rewriting and editing.

Acquiring the Necessary Domain Expertise

The solution for getting the necessary domain expertise was unusual, radical, and expensive. They hired an experienced genomics researcher. They justified this with two main arguments. First, they would not be able to design a reasonably useful system unless they had domain expertise in house. While field researchers were happy to help someone trying to develop software that would allow them to perform their analyses more quickly, accurately, and efficiently, those experts simply could not take the time from their schedules necessary for the software developers to glean sufficient information. Further, the software developers often did not even know what questions to ask the researchers when they did get access to them. So a domain expert was needed for the team to succeed in design and development (Redish, 2007). The second justification for the hire involved marketing and sales. The marketing and sales teams had the same difficulty as the developers; their lack of domain knowledge made it difficult for them to target appropriate audiences and to develop appropriate sales materials and approaches. Having the domain expert work directly with them would solve those problems.

The Necessity for Field Studies and Observation

It should be pointed out that even with the expert in house, the team had to continue to employ conventional field studies and observations to gather sufficient user and task analysis information. This was largely because usability considerations had been secondary in early design to simply getting all of the complex parts of the overall system to function together and to work in the first place. Once there was a working prototype, it could be field tested in a variety of settings to determine how to improve the front-end user interface access to the overall power of the system and to further discover the various affordances that users wanted the system to include.

Developmental Methodology

In addition to its nontraditional staffing practices, the genomic project team also had to follow some unconventional design and development practices. These included all phases of the project, from early data collection and the composing of requirements, through the design, testing, and implementation stages. The team found that they had to gather usability data on an ongoing basis, with frequent iterations of design and testing of large and small design elements. While many project teams give lip service to user-centered design, this project, with a user permanently on the team and groups of users in the field during development, pushed user-centered design to what its proponents claim it should be: an integral part of the initial design of a system. The team, without explicitly discussing it, realized that doing usability testing after the fact and then trying to fix such a large, complex combination of programs simply would not work. They had to get the design, at least in its overall structure and basic mental models, correct the first time. That is not to say that they didn't foresee redesigns and iterations as inevitable, but they realized that the traditional design-it-ourselves-and-then-sell-it-to-the-beta-testers would not work. To ensure that the system was both usable and useful, design and usability studies had to be conducted simultaneously, so that small parts of the system could be iteratively designed and tested until users were successful with them, and then larger elements could be designed and tested, until finally the overall system worked.

Requirements for Design

Despite having a domain expert on the team, it was still necessary to talk to additional genomic researchers. There are so many variations of genomic research arrangements that one person could not possibly be familiar with every model. It was necessary to observe the various types of research and to develop user models for how the researchers structured and carried out their research. This process involved a combination of design and usability study, often involving prototypes of small parts of the system being tested with users in various settings. Hence, the design became user-centered and iterative, as some system parts were tested and retested multiple times until users expressed satisfaction with their usefulness.

Usability Analysis through Field Studies and Contextual Inquiries

While the inner domain knowledge that the team had with a user in place helped with the overall design of the product, it was also essential to incorporate usability studies into the design stages. Any hope of eventually arriving at a useful product required that design and usability assessment be conducted together. The system was simply too large and complicated to use the conventional design-the-whole-thing-and-then-test-with-users approach, which often involves not even testing but selling the product and then watching users essentially do the beta testing that was never done. Because of the complexity of genomic research, the risks were too great to expend the time and resources on an overall system that proved to have usability and usefulness shortcomings. It was during these field observations and contextual inquiries that the development team learned about the need for the overall system to include affordances contributing to successful complex work, such as the following:

- Customizable front-end input "engines"
- Customizable outputs supporting dozens of methods for visualizing data

- Product structure that provided a "standard" research methodology but that allowed for some steps to be skipped or done out of order, for some steps or combinations of steps to be iterated multiple times, and, in general, for making the structure of the system useful for genomic researchers
- Collaboration tools allowing multiple researchers to communicate findings effectively and to share results, reports, data sets, data visualizations, and so on
- Reservoirs for saving various permutations of particular studies for later analysis and cross analysis with other permutations

Primary Design and Usability Challenges

Based on the discoveries from their field studies, the development team faced two primary challenges in attempting to create a unified, coherent system that supported the researchers they had studied. First, the system had to support the collaborative nature of large genome research projects, sometimes involving dozens of personnel who worked on projects at different times and, in some cases, in different locations. Second, the system had to support the often unstructured, nonlinear nature of genomic research while providing some structure of its own.

The Necessity for Supporting Collaborative, Complex Work

First was the collaborative nature of genomic research. The developers realized that the software had to provide support for three main groups: information technologists who would install and administer the multiple programs used in the genomics product, scientists who would design and conduct experiments, and statisticians who would assist scientists in designing experiments and in interpreting the results. These are three groups from different discourse communities for whom cross communication will not be easy in the best of conditions, much less in the intense pressure of a complex research environment. The developers also realized that much of the communication was asynchronous. That is, researchers were sometimes located in different labs, sometimes in different buildings, and sometimes in different geographical locations. Also, they often worked on different schedules, which meant that they needed a way to store information and communicate its importance to other research team members. This would require that the program provide a common storage "repository" where team members could store saved data tables, visualizations, reports, notes, observations, and so on. Further, they needed to be able to understand the nature of folders and files developed by other researchers at other times, so it was important to provide detailed metadata labeling for each folder or file created. Additionally, the researchers needed a way to annotate electronic documents (the equivalent of yellow stickers) to remind themselves or other researchers about important facts or data.

The Necessity for Supporting the Paradox of Structure for Complex System Design

The second biggest problem was attempting to design a product that had to have some kind of usable structure to work on a process that often seemed so complex as to defy any single structure. With dozens of possible ways to approach and conduct genomic research, how could designers create a product and an interface that would accommodate all of them, and, further, allow for other, unforeseen new approaches? This is perhaps the biggest paradox of all facing developers of systems to accomplish complex tasks. They must examine an enterprise that appears to be unstructured, ad hoc, nonlinear, ill-defined, and, often, poorly understood even by experts (Albers, 2003), and somehow design a system to control that enterprise. Of necessity, the system must have some structure, both physically and temporally. In other words, it must have a physical interface with which researchers interact, and it must have some linear process that the researchers complete over time. With complex systems, it is likely that no single physical structure in an interface and certainly no single, linear process can meet the needs of all the users and their tasks. So, while the system must have a basic structure, it must also allow for the flows, eddies, and backwaters that inevitably occur in the flow of a complex process such as genomic research. How did the genomic development team solve this complex developer's paradox? They did two basic things.

Use of a Basic, Unifying Structure

First, in all of their field studies and observations of users, they found a basic structure that genomic research projects follow. This structure involved eleven basic steps. While proprietary interests preclude listing all of the steps here, suffice it to say that the process was basically the empirical, scientific method from hypothesis formulation through data collection, processing, analysis, and interpretation, with several additional steps of data manipulation at the beginning and analysis using visualization at the end. Not all genomic studies would necessarily go through all eleven steps, but they would always go through at least some of them. Also, the process had to allow for iteration, so that after completing some of the steps the researchers might loop output data back into earlier data manipulation and data-processing steps to look at it in different ways. Some of the steps could become small studies in and of themselves. For example, visualizing processed data using numerous different visualization methods could become a series of forming a hypothesis, processing data to provide output visualizations, studying the visualizations, forming new hypotheses based on the results, and so on. Even though the eleven steps did not reflect the structure of all genomic research projects, they provided a system that would accommodate almost any research scenario. The process thus accommodated the numerous, complex, and nonlinear methodologies used by genomic researchers, and provided usefulness both for their overall projects and for smaller segments of the projects. Because design and usability studies, mostly through field observation, were done simultaneously, the overall system became increasingly usable and useful, containing an overall structure that allowed the researchers to follow what to them was a logical work flow and providing a front end that made it easy to make their way through the process.

There was, however, a basic problem with imposing a linear structure on what is often a nonlinear, ad hoc, constantly changing process with loop-backs, divergent paths, dead ends, and no immediately clear, linear path. So, the steps had to be designed to allow for skipping ahead, for looping back to an earlier step, and, particularly in the middle of the process, for moving from any step to any other. The designers built in a structure that has the flexibility to allow for the process to be performed in a nonlinear manner, thus solving the complex system paradox with what was simultaneously a linear, structured system that allowed nonlinear, unstructured processes. The designers further accommodated nonlinear processes by allowing anything anywhere in the process to be saved and stored. Hence, a researcher could store partial outputs and return to them later, could save several permutations of the same data for later analysis, or could save an outlier to study later.

The use of the unifying process also informed the design of the technical communicator's information products for the system. The user's guide would cover the eleven steps in the overall system's interface and would explain how to perform each of the steps, while making it clear that some of them could be omitted and/or performed iteratively. Further, the user's guide would provide an extended example of a complete genomic study to show how a researcher would work through all of the system's stages.

Because the tutorial and exemplary information was to be provided in the user's manual, it was decided that the online help system should function primarily as reference information, providing explanations of the function of each of the system's components. Because many of the help topics were developed prior to the development of the user's guide, many of them were reused in the guide, when appropriate. The online help system was the logical place to provide information about every component and affordance of the system, including each part of the user interface. While this approach may have been more system oriented than task oriented, the overall information library had to supply such information somewhere, and it was determined that the help system was the most likely place where users would look for such information. The task orientation was provided by the user's guide, which explained the overall eleven-step process and provided an extended example of a genomic study.

Use of Customization

The second thing the designers did to solve the complex paradox was to make the program customizable. They knew they had to do so based on user input and on the very nature of genomic research, which must allow for numerous inputs into a system and numerous methods for studying outputs from the system. To reduce the eleven steps down to their most basic form, genomic researchers collect data, put it into a computer to be processed, view the results through various visualization methods, and interpret those results. The two key points for researchers are inputting the data and viewing the results. So, for those two aspects of the process, the developers made it possible to create infinite "engines" that could input data in various ways and that could create visualizations in any way desired by the researchers. Hence, even though the initial design of the software might not accommodate every possible genomic research scenario, the customizations would allow for such accommodation and for future, unanticipated research methodologies.

Design and Usability Methodologies

To achieve the designs that their in-house domain expert and their field observations indicated were necessary, the development team performed three basic variations on the overarching principle of user-centered design: field observation, direct user participation, and extended contextual analysis.

Field Observation

It was necessary to observe researchers going through the processes they employ to collect data, run it through computer analyses, and then analyze the results by viewing them through numerous graphical and tabular presentations. The requirements had to reflect both the processing power required by the massive demands imposed by genomic research and the unified interface that users would interact with to perform all of their tasks. To understand what kind of interface was necessary, team members had to observe researchers over time. Simple one- or two-hour interviews were not sufficient to reveal the variety of tasks the researchers performed using a variety of tools, often a baffling combination of paper and pencil, computer programs, tables, charts, graphical displays, notes scribbled on scrap paper, lab reports, formal written reports of results, folders stuffed with data and results for colleagues to look over, and so on. In some settings, many of the researchers were working on multiple projects simultaneously, such that they had dozens of sets of information, which they needed to be able to combine in a variety of ways, as the same set of data or notes or reports might be used across more than one project. It was impossible to learn what would fulfill these users' needs from observing them briefly.

User Involvement

Because of the necessity for increased user input, it was necessary to arrange for heavier user involvement than on most software projects. That involvement came from having a user (the genomic researcher) join the project team and from interacting continually with users in the field. This was done both by way of e-mail and telephone conversations and with field visits. This project not only involved user-centered design, but also employed user participation in that design. Without that participation, it is highly doubtful that the resulting product would have turned out to be as usable and useful as it is.

Extended Contextual Analysis

Field visits with genomic researchers essentially became extended contextual analyses, wherein team personnel studied not only the processes that researchers followed but also the entire context within which they worked. Such analyses went on across multiple visits, which extended the contextual analysis across weeks rather than the more normal few hours. For example, one of the strengths of the genomic product is that it allows for any kind of file to be saved and linked to one or more folders in a central repository. With so many researchers working on so many files and projects simultaneously, successful collaboration required that they have a method to store everything from the smallest notes to the largest genomic databases. The power of the repository that was designed allows multiple views of the files, such that a researcher can quickly find all of the files created by one assistant or all of the files associated with a particular genomic analysis. It was from observing researchers at work that the team realized they would need to enhance the collaborative features of the program. The experience was similar to that reported by Merholz (2005) in relation to development of a time management tool:

> A fairly standard process would have involved prototyping of this product, and then bringing people in to "test" the prototype. What we did, however, was field research. We went into 12 homes, and saw how people currently managed their stuff. And, believe me, it's messy and complex. One participant used: a church address book, a week-at-a-glance, a Palm-style PDA, a simple address-storing-PDA, and an Access database to manage this task. Had we brought her in to test our prototype, we could have found out all kinds of stuff about how she used this prototype in isolation and away from her tools. But we would have learned nothing about how this tool could possibly have integrated itself into this complex web.

Similarly, the team learned from working with researchers directly that the original design for a key part of the software would not be sufficiently useful for actual research. The original design had called for a handful of input engines designed to handle specific file types for inputting genomic data for analysis. They soon discovered from interacting with and observing researchers in the field that they would need to develop additional engines and that they would need to make all of the engines flexible and customizable. In fact, they discovered that throughout the design of the entire system, they would have to build in much more flexibility, so that they added preference settings for almost every aspect of the interface. The nature and variety of genomic research methodologies are so great as to mitigate against fixed, static designs. Because programmers favor clean, linear designs with straightforward progressions through well-defined procedures, they had to learn from users that the genomics product needed to have flexibility everywhere, because the process is rarely linear and straightforward.

Rather than creating finished designs and trying those with users, the design team learned to test pieces of the software and to go through a series of iterations that were labeled as less than version 1.0, indicating that they were pre-release versions of the code. Hence, usability studies were performed on a continual basis during many of the stages of development. Waiting to do such usability analysis until a finished product was produced would not have led to the necessary flexibility and collaborative features that were required to make the product truly useful for researchers. And trying to test the finished product would be, for the size and complexity reasons cited earlier, extremely difficult and time-consuming if not impossible.

Conclusion

The finished genomic research product is an impressive system for performing complex work. It includes nearly all of the capabilities that Mirel (2002, 2003) stresses as being necessary to perform such work. Despite the cost, it has sold well and been used successfully by many U.S. and international customers.

Had the genomics research development team followed more conventional design methods and released a product that had not been tested with and, in part, developed by its users, the product would have been woefully unusable for its intended audience. No amount of usability testing after the fact would have been able to salvage the product. In a system so large, once certain design decisions have been made, it is often not possible to change them later.

The technical communicators accommodated the complex nature of the product by studying the various rhetorical requirements of the audience when first approaching the system and later when fully engaged in using the system to perform complete genomic studies. The communicators developed appropriate information to guide users in learning the system, where it was concluded that treatment of the unifying eleven-step process and an extended example would guide users in overall system use and understanding. They further provided online help topics on all system components that users could consult as references throughout their ongoing use of the product. The communicators definitely found that topic-based architecture fit better with the nature of the system and of its design than would more traditional document-based architectures.

The product certainly answers one question about usability and complex systems: What we have all been saying about user-centered design for the last twenty years is true. We can design a usable, useful, complex system if we follow the user-centered principles in all stages of design, development, and deployment. Doing so adds to the initial cost and time of development, but leads to a system that meets user needs.

Reader Take-Aways

- Involve users in design throughout the process, and employ multiple methods for doing so, including methods such as hiring a user for the team, extended contextual inquiries, field studies, and user testing.
- Design so that the current state of a system can be saved at any time for later study or restarting.
- Promote collaboration by allowing storage of any type of file in a central location that all users can access.
- Look for the fundamental structure that users follow, even if they rarely perform all of its steps.
- Design so that steps in a procedure do not have to be carried out in a linear fashion but can be completed "out of order."
- Provide information products that give users an overall view of the system and that explain each part of its user interface.
- Make the complex systems as customizable as possible to allow for varying user configurations and requirements.

References

Albers, M. (2003). *Communication of complex information.* Mahwah, NJ: Lawrence Erlbaum.

Dicks, R. S. (2002). Mis-usability: On the uses and misuses of usability testing. In *Proceedings of the SIGDOC 2002 ACM Conference on Computer Documentation,* Toronto (pp. 26–30). New York: Association for Computing Machinery.

Gould, J. D., & Lewis, C. (1985). Designing for usability: Key principles and what designers think. *Communications of the ACM, 28*(3), 300–311.

Merholz, P. (2005). Is lab usability dead? *Peterme.com*. Retrieved from http://www.peterme.com/archives/000628.html

Mirel, B. (2002). Advancing a vision of usability. In B. Mirel & R. Spilka (Eds.), *Reshaping technical communication* (pp. 165–187). Mahwah, NJ: Lawrence Erlbaum.

Mirel, B. (2003). *Interaction design for complex problem solving: Developing useful and usable software*. San Francisco: Elsevier/Morgan Kaufmann.

Redish, J. (2007). Expanding usability testing to evaluate complex systems. *Journal of Usability Studies*, 2(3), 102–111.

Rubin, J., & Chisnell, D. (2008). *Handbook of usability testing: How to plan, design, and conduct effective tests* (2nd ed.). Indianapolis, IN: Wiley.

11

Incorporating Usability into the API Design Process

Robert Watson

Microsoft Corporation

CONTENTS

Introduction 224
API Usability.. 226
    What Is an API?... 226
        API Terms.. 227
        API Examples 228
        Using an API in an Application 228
    Why Is API Usability Important? 230
    What Makes an API Usable? 233
    Measuring API Usability 234
    What Makes an API Difficult to Use? 235
API Design, Development, and Usability 240
    Technical Writers Can Help with Usability.. 241
        Usability Fundamentals.. 241
        Heuristic Evaluations 242
        API Usability Peer Reviews.. 242
    Technical Writers Should Help with Usability 244
Conclusion 245
Reader Take-Aways 246
References.. 246
Additional References 248

Abstract

Application programming interfaces (APIs) define the interface between software modules and provide building blocks that software developers use to create applications. Traditionally, APIs have been considered to be only a software interface; however, if you consider an API as the interface between a software developer and a computer application, you can get some new insight into the complex process of developing software. Applying usability

223

principles to APIs can help API designers make their APIs easier to use, and easier to use correctly. This chapter describes the aspects of usability that are unique to APIs, and how they can be applied in the context of a software development project. The chapter reviews API usability from the point of view of a technical writer who documents APIs, and illustrates how technical writers can work with API design teams to improve API design and documentation by bringing usability into the API design process.

Introduction

Software developers must successfully manage complex, multidimensional interactions between the application's requirements, the application's architecture and design, the programming language or languages, the various APIs used to implement the design, and the people who have an interest in these components. The usability of the APIs used in an application is one of the many factors that influence the effectiveness and the efficiency of these interactions. While an API might not appear to be a user interface at first glance, it is the human–computer interface for a software developer who is writing an application (Arnold, 2005). As a human–computer interface, an API's usability can be measured and improved in the same way as the usability of the more familiar graphical human–computer interfaces.

At the same time, studying an API's usability is complicated by the environment in which the API is used. While using an individual component of an API might be a straightforward task, software developers writing real-world applications face a more complex task—one that must consider the interactions between the API components they use and the interactions between the software they are writing and the end users of that software. The software developer must constantly balance these interactions to ultimately develop a reliable, usable, and ultimately successful software product. Consequently, a study of API usability must take this environment into consideration.

The impact of API usability on software development is growing in importance to software developers and their end user customers as APIs are used more widely in the development of new applications. In the past, there were not as many APIs as there are today, and they were usually used by only a comparatively small group of specialists (software developers). For example, in 1991, when I wrote Microsoft® Windows® programs as a software developer, the Windows API had less than a thousand functions. However, because APIs make it possible for software developers to include the features of other software in their applications, their popularity and ubiquity have increased and show no sign of abating in the foreseeable future. Today, the Windows API includes many individual features, some of which are larger than the entire API of 1991. The Microsoft .NET Framework is another example—it grew from providing 124 namespaces (feature groups) in 2002 to 309 in 2007. These namespaces offered software developers 35,470 different members in 2002 and grew to 109,657 by 2007 (Abrams, 2008). And that is just one software product of the many that software vendors are bringing to market.

At the same time, the number of software developers continues to grow. The U.S. Bureau of Labor Statistics (BLS) reports that there were about 1.3 million computer software engineers and programmers in the United States in 2008, and that number is expected to increase by 32 percent by 2018 (BLS, 2010). The number of computer software engineers and programmers in China and India is also quite large and growing.

As a programming writer (that is, a technical writer who writes text and programming examples that explain and demonstrate how to use APIs), I'm one of the first people outside of the test and development teams to experience a new API as a user. In my work, I'm sensitive to the usability of an API because it directly influences how easy or difficult it is to explain to the user. In the best case, a highly usable API can almost document itself, which should be an API design goal because the API definition is often the first, and sometimes the only, documentation a software developer reads (Cwalina & Abrams, 2009). If the API is well designed, the software developer can understand and use the API easily, and I can provide concise documentation and clear examples. If an API has usability issues, however, it is more difficult for the software developer to understand (Bloch, 2005; Henning, 2007) and requires more explanation to describe how it works—that is, what it really does and why it doesn't do what the user might expect. Usability issues can also make the examples more complicated and difficult to understand.

To a software developer, the consequences of using a difficult API can be very serious (Henning, 2007). In the best case, she might need to spend additional time to try to make sense of a cryptic API. While trying to understand each individual challenge might take only a short period of time, the effect of these minor interruptions is cumulative. Brooks (1995) reminds us that a project becomes a year (or more) late one day at a time—or, in this case, one additional puzzle at a time. Unfortunately, if the software developer has equally cryptic API documentation to work with, he will be forced to experiment with the API to understand how a method or a function actually works, which only adds to the delays. Software developers like to solve puzzles, but the puzzles they should be solving are those related to completing their primary goal of writing their applications. Trying to reverse engineer an API or a programming library to fully understand the API is a time-consuming distraction from the software developer's primary goal. In the worst case, from the customer's perspective, a software developer might unintentionally introduce errors (bugs) into an application when using a difficult API, not realizing that the API doesn't behave as expected until the customer reports a problem (Henning). In the worst case, from the software developer's perspective, the bugs that result from an API with usability issues might drive customers to competing products.

Ideally, usability issues are identified early in the design of an API, when changing the design is comparatively inexpensive, instead of having the testers, technical writers, or, worse, the customers find them later when change is very costly or impractical (McConnell, 1993). To detect and correct usability issues early in the design, usability must be included in the API design process from the earliest specification of requirements and must continue through design, implementation, testing, documentation, and release. This chapter describes what an API is, how to identify usability issues in an API, and how to make usability an organic and integral part of API design. Whether you're a software developer, project manager, usability researcher, test engineer, or technical writer, this chapter describes ways to measure API usability and offers some ideas for applying "downstream" resources earlier in the design process to improve the API's usability.

API Usability

API usability involves the study of a complex system of interactions and uses some domain-specific vocabulary; however, it still builds on the traditional principles of usability. The API usability jargon might appear foreign, but the principles should be familiar to anyone who has studied usability. In spite of the complex interactions, the usability of an API can be measured and reported, as can the usability of any other human–computer interface.

What Is an API?

An API is an interface that makes it possible for an application to use the features provided by other software, such as the features provided by a programming library (Daughtry, Stylos, Farooq, & Myers, 2009; Robillard, 2009; Tulach, 2008). The terms *application* and *programming library* will be used in this chapter to distinguish between the user of an API and the creator of an API. An application is a computer program that is written to perform a task, and that application accesses the features provided by a programming library through the library's API. A programming library, often provided by a software vendor, is software that usually doesn't run by itself but provides functions for applications to use. However, software developers can, and frequently do, write their own programming libraries. Typically, these private, internal programming libraries are developed so that frequently used functions can be shared by several applications.

Tulach (2008) describes an API in terms of a contract: "The API is everything that another team or application can depend on: method and field signatures, files and their content, environment variables, protocols, behavior." Tulach's definition illustrates that what an API includes will vary depending on the programming language, environment, and features, and is not restricted to just the program code that an application calls.

API Terms

Like the scope of an API, the terms used to discuss APIs can vary with the software vendor, programming language, and development environment in which they are used. Depending on the context, the definitions of these terms can also vary. To simplify the discussion in this chapter, I'll use the following descriptions.

- *API elements.* The individual components of an API that make it possible for an application to access the specific functions of a programming library. Examples of API elements include the functions, methods, data structures, file formats, message formats, and symbolic constants that are defined by the API.

- *Function* (also *method*). A subprogram that can be called by another application to perform a specific task and then return control to the calling application.

- *Class.* A programming object that consists of properties that describe the object's characteristics, and methods to manipulate the object's properties and state.

- *Interface.* The signatures (definitions) of the methods, properties, and other elements of a class or structure. An interface does not include the implementations of the defined elements.

- *Structure* (also *struct* or *data structure*). A data type that consists of one or more data elements that are grouped together to create a single programming entity.

- *Constant* (also *symbolic constant*). A constant, numeric value that has a specific meaning to the application or programming library and is given a symbolic name for the programmer to use in an application.

- *Parameter* (also *argument*). A value that an application passes to a function or method to alter the behavior of the function or method, or a value that a function or method returns to the application.

- *Header file.* A file that contains the definitions of an API in a format that can be used by the programming language. Header files can usually be read by software developers and thereby serve as a form of API documentation.

API Examples

Many software vendors publish programming libraries and APIs. Oracle® (previously Sun Microsystems), for example, publishes APIs for Java® software developers, Apple® publishes APIs for Mac® and iPhone® software developers, and Microsoft publishes APIs for Windows and Windows Mobile® software developers. Table 11.1 contains a small sample of the many APIs that are available to software developers. The APIs listed in Table 11.1 were selected to show that an API can provide access to a specific feature or to a diverse collection of features. For example, the Live Search API and the Audio Extraction API provide interfaces to specific features, while the Java FX API and the Windows API are examples of complex APIs.

Software vendors sometimes group related programming libraries into a software development kit (SDK) that contains the programming libraries and other tools that a programmer will need to create an application. Table 11.2 contains a list of some of the SDKs that are available to software developers.

Using an API in an Application

The C Runtime Libraries are examples of programming libraries provided by a software vendor and used by software developers when they write an

TABLE 11.1

Examples of APIs Provided by Software Vendors

API Name	Description
Audio Extraction API in QuickTime 7	Enables software developers to extract and mix the audio from the sound tracks of a QuickTime® movie to create raw pulse code modulation (PCM) data
Live Search API	Enables software developers to programmatically submit queries to and retrieve results from the Windows Live™ Search Engine
JavaFX 1.0	Enables software developers to create rich Internet applications (RIAs) that look and behave consistently across devices
Windows API	Enables software developers of Windows applications to access operating system features such as the following: • The graphical user interface • evices and system resources • Memory and the file system • Audio and video services • Networking and security features
C Runtime Libraries	Enables C and C++ programmers to access operating system features

TABLE 11.2

Software Development Kits (SDKs) Offered by Software Vendors

SDK Name	Description
Java 2 Platform, Standard Edition, v. 1.4.2 (J2SE)	Contains the tools and libraries necessary to develop applications in Java, such as the following: • Java Compiler • Java Virtual Machine • Java Class Libraries • Java Applet Viewer • Java Debugger and other tools • Documentation
Apple iPhone 3.0 SDK	Contains the APIs necessary to write applications that run on Apple's iPhone
Microsoft Windows SDK for Windows 7 and .NET Framework 3.5 Service Pack 1	Contains the tools, code samples, documentation, compilers, headers, and libraries to create applications that run on the Windows 7 operating system

application in C or C++. These libraries contain programming functions and data structure definitions that are frequently used by C programmers (software developers who program in the C language). These libraries include functions to open and close files, read data from and write data to files, accept input from a user, display text to a user, and do many other common tasks.

A Hello World application is a simple application that demonstrates how to write the words "Hello World" on the user's screen. Hello World applications are used by software developers as a test application when trying a new language or feature. Figure 11.1 shows a Hello World application written in C that uses the _cputs() function from the C Runtime Libraries to display the text on the user's screen.

The interface of the _cputs() function is defined in the conio.h header file, shown in the first line of code in Figure 11.1./Figure 11.2 shows the interface definition of the _cputs() function as it is described in the conio.h header file and in the reference documentation of the function (Microsoft, 2010a). From the definition shown in, Figure 11.2, a software developer can determine the type of data to pass to the function (const char *string) and the type of data

```
#include <conio.h>
int main()
{
        _cputs ("Hello World");
        return 0;
}
```

FIGURE 11.1

A simple application written in C that calls a library function.

```
int _cputs(
      const char *string
);
```

FIGURE 11.2
The definition of the _cputs() function.

that are returned by the function (int). However, beyond the data types, the definition in Figure 11.2 does not include any details of what the function does with the data that are passed to it, or any limitations of the function. The software developer must find this information in the reference documentation for this function.

The details of how an API is described and used in an application depend on many factors, such as the programming language being used and the nature of the function provided by the API. Microsoft Visual Basic® applications, for example, use a different programming library than C applications do. Figure 11.3 shows a Hello World application that is written in Visual Basic, and the Console.WriteLine() method, underlined in Figure 11.3, is used to write text to the user's screen instead of the _cputs() function that was used in Figure 11.1.

Programming languages describe the syntax and structure of the commands used to create an application. Therefore, the format, syntax, and structure of an API depend on the programming language that the API supports. Some languages also include functions or programming libraries that enable an application to interact with the operating system or peripheral devices as part of the language definition. While a programming language might include an API or elements of an API, an API does not define any type of program structure. An API defines only how an application that is written in a specific programming language interacts with the features and services provided by its underlying programming library.

Why Is API Usability Important?

The same reasons for studying and improving the usability of a product or website apply to APIs. The International Organization for Standardization (ISO; 1998) defines *usability* as "the extent to which a product can be used by specified users to achieve specified goals with effectiveness, efficiency and

```
Module Module1
   Sub Main()
         Console.WriteLine("Hello World")
   End Sub
End Module
```

FIGURE 11.3
A sample application written in Visual Basic that calls a library function.

satisfaction in a specified context of use" (p. 2). Clearly, designing an API such that software developers can achieve their goal of writing applications with "effectiveness, efficiency and satisfaction" would be better than designing one that is ineffective, inefficient, and unsatisfying. Realistically, of course, it would only be better if the costs of including usability as a feature are sufficiently offset by the benefits provided by improved usability—and they are.

What does ineffective, inefficient, and unsatisfying software cost, and who pays that cost? Cooper (2004) lists some of "the hidden costs of bad software," and includes such things as decreased end user productivity, decreased end user satisfaction, and increased customer support costs. While these costs are real, two of the three are borne by the customer and, in many cases, none of them is seen by the API designers. In a large software company, for example, the customer service engineers can be physically and organizationally distant from the product development engineers. Even in companies that have their customer support engineers close to the product development engineers, there is still some conceptual distance between the two groups of engineers. Customer support engineers focus on the current and previous versions of software—that is, the software that the customers are using. Product development engineers, on the other hand, focus on the next and future versions of the software—the software that has not been written yet. While API developers might have an intellectual sense of the usability issues, the conceptual distance can still make it difficult for them to have a tangible or visceral sense of their product's usability. Bringing product development engineers into a usability test, however, can quickly change this view (misperception) and erase whatever conceptual distance might have existed between the API designer and the customer (Cwalina & Abrams, 2009).

Nielsen (1993) also observes that "the cost savings from increased usability are not always directly visible to the development organization since they may not show up until after the release of the product," making usability engineering difficult to sell on a cost-saving basis alone. Henning (2007) lists many costs of APIs that are difficult to use, and most of them are costs that the customer bears. Stylos et al.'s (2008) case study and Clarke's (2004) examples that are cited in Cwalina and Abrams (2009), however, describe the customer benefits of improving API usability. These cases provide empirical evidence of how improving the usability of an API produced measurable improvements in task completion. While the results described by Stylos et al. and Clarke provide concrete examples, the benefits are still distant from the development team. That the costs of implementing usability engineering are local and visible while the savings are distant and invisible only adds to the challenge of cost justification.

To help make the case for usability engineering to the development team, it is worthwhile to present the team with some immediate and tangible benefits of usability engineering in API design and, at the same time, look for ways to lower the cost, or the perceived cost of usability engineering in API design. Bloch (2005) addresses the development cost–benefit issue when he points out that software development organizations can benefit directly from usable APIs in that "good code is modular," "useful modules tend to get reused," and "thinking in terms of APIs improves code quality." Bloch is suggesting that internal software design (that is, the design of the application that doesn't present a public interface) can benefit when it takes an API-oriented perspective because such an approach encourages a cleaner and more easily maintained implementation, which invites reuse. Such an advantage is something the API developers will benefit from directly—if not in the first release, then certainly in subsequent versions. Henning (2007) observes that while "good APIs are hard [to design and build]," they are a "joy to use" whether you are the API's designer or its target user (p. 25).

Bloch (2005) points out one very important reason that API usability should be critical to designers: "Public APIs are forever—[you only get] one chance to get it right." The cost implication of this observation is that any mistakes made in the design that add costs for the end users will be around and making their work difficult for a long time. Then again, in the worst-case scenario, if the usability issues prevent the API from being widely adopted or cause it to be abandoned completely, they might not be around for very long at all.

Changes to APIs and programming libraries are different from changes to application software. While new elements can be added to the API of a programming library, existing elements cannot change after they've been published without disrupting the applications that use them. Because an API and its programming library are tightly integrated with the applications that use them, a change to an existing API can require software developers to review all of the program codes in their application and update the places that are affected by this change. They must then rebuild the application, retest the updated application, and send the new application to the customer, who must then reinstall it. All of these steps are time-consuming, are expensive, and run the risk of introducing new problems such as new programming errors or annoyed customers.

Having usable APIs to document certainly makes my job as a technical writer easier and more satisfying; however, more to the point, making it easier to write the documentation can actually help improve the API's usability. Robillard (2009) studied the obstacles to learning how to use APIs and identified "insufficient or inadequate examples" as the most frequently reported obstacle to learning an API. If the technical documentation is easier to write, there should be more time to create or improve the documentation and provide better examples of how to use the API. Tied with "insufficient or inadequate documentation" for the most reported obstacle in Robillard's study was "issues with the API's structural design." Robillard's study reported several other documentation-related API learning obstacles, which could have been found in usability reviews during the design phase. Improving an API's usability should directly and indirectly reduce some of the most frequently reported obstacles to learning an API.

The benefits of usability engineering can also be realized when the API is tested and documented before the library is released. Unfortunately, tracking and documenting the cost savings that are realized by the engineering, test, and documentation functions are difficult unless these areas have some type of cost and error tracking already in place before adding usability engineering to the design process. Without a baseline to compare against, any reports of improvement are likely to be mostly anecdotal. In spite of the cost-tracking challenges, Nielsen's (1993) discount usability methods provide some ideas for applying usability economically and incrementally, and achieving improvements in usability without incurring the costs or delays that are often anticipated with such a change to the engineering process (Nielsen, 1993, 2009b).

What Makes an API Usable?

Usability is defined in many places. ISO (1998) provides a general definition of usability, as quoted above. Nielsen (1993) added more detail to the ISO definition in his definition, and Quesenbery (2004) modified Nielsen's list further. These definitions provide a general set of usability components that can be applied to a variety of products but aren't detailed enough to be sufficient for an API usability study. Bloch (2005) listed the following elements of a good API, which bring the factors of API usability into a little sharper focus.

- Easy to learn
- Easy to use, even without documentation
- Easy to read and maintain code that uses it
- Hard to misuse
- Sufficiently powerful to satisfy requirements
- Easy to extend
- Appropriate to audience

Henning (2007) provides some more details of a good API. Henning says that a good API:

- Must provide sufficient functionality
- Should be minimal
- Cannot be designed without understanding its context
- Should be "policy-free if general purpose"
- Should be "policy-rich if special purpose"
- Should be designed from the perspective of the caller
- Should be documented before it is implemented

Table 11.3 summarizes the preceding usability definitions and illustrates where they overlap.

Zibran (2008) reviewed the preceding literature and other sources to create a list of twenty-two design factors that influence API usability. He summarized his findings thus: A useful API is usable if it has five characteristics: (1) easy to learn, (2) easy to remember, (3) easy-to-write client code, (4) easy-to-interpret client code, and (5) difficult to misuse.

Measuring API Usability

If the preceding properties describe a good (i.e., usable) API, then how can they be measured? Green, Blandford, Church, Roast, and Clarke (2006) describe twelve cognitive dimensions that Clarke (2004) used to describe the different components of API usability and the corresponding user preference in the development of the Microsoft .NET Framework 3.0. The usability of an API or API element is then determined by how well the user's preferences match the properties of the API. Table 11.4 lists the twelve cognitive dimensions used by Clarke to evaluate the usability of the Microsoft .NET Framework. The table also lists the questions that each dimension is intended to answer from the point of view of an API and from the point of view of the API's user.

The cognitive dimensions provide a rich, multifaceted view of an API's usability. They also highlight the importance of knowing the user by having parallel dimensions for both the user and the API. Microsoft has used the cognitive dimensions to evaluate the usability of the Microsoft .NET Framework and other APIs. One advantage of using many dimensions is that they provide a more precise description of a usability issue, and design

TABLE 11.3

Usability Definitions

ISO	Nielsen	Quesenbery	Bloch	Henning
Effective	N/A	Effective	Sufficiently powerful Easy to extend	Provides sufficient functionality
Efficient	Efficient	Efficient	N/A	Minimal
Satisfying	Satisfying	Engaging	Easy to use Hard to misuse Easy to read Appropriate to audience	Designed in user context
N/A	Learnable	Easy to learn	Easy to learn	Documented before implemented
N/A	Memorable	N/A	N/A	Caller's perspective
N/A	Low error rate	Error tolerant	N/A	N/A

decisions can be evaluated in terms of how a change in one dimension can create a corresponding change in another dimension.

Evaluating the cognitive dimensions of an API can, however, be a subjective and labor-intensive process. Bore and Bore (2005) object to using the cognitive dimensions for evaluating API usability, citing that "there are too many cognitive dimensions, their interpretation is often not obvious, measuring them is too time-consuming and assigning quantitative measures involves subjective judgment" (p. 155). Bore and Bore used the following, simpler set of dimensions to evaluate an API.

Specificity: What is the percentage of API elements that address application functionality?

Simplicity: How easily can the user translate application functionality to uses of API elements?

Clarity: How obviously is the purpose of an API element reflected in the element's name?

Bore and Bore (2005) chose a small set of predominantly quantitative dimensions to study rather than the more subjective, qualitative, and numerous cognitive dimensions. Nielsen (2004) points out, however, that while quantitative observations have their place, they are often problematic. Qualitative observations provide richer data and insights that provide information that can be used to improve the product's usability. Qualitative observations not only tell how good the usability is, but also often tell why it's that way.

I've found the objective application of the cognitive dimensions to be somewhat challenging. As Bore and Bore (2005) describes, mapping the dimension to a particular situation isn't always clear or objective. Successfully evaluating many of the cognitive dimensions requires a high degree of domain knowledge and some experience. The method used by Bore and Bore, in contrast, was much more objective. The quantitative approach provides more repeatable data collection but provides little in the way of useful insights about how to improve the observed usability. When the cognitive dimensions can be evaluated, however, they provide a specific vocabulary and detailed feedback that can help explain the usability effects and consequences of a design decision. While evaluating all of the cognitive dimensions provides a deep insight into an API's usability, when there are time, resource, or expertise limitations, useful usability information can still be obtained even if only a few dimensions are studied.

What Makes an API Difficult to Use?

In simple terms, referring back to the ISO definition of *usability*, any aspect of the API design that impedes users as they try to achieve a specified goal with effectiveness, efficiency, and satisfaction in a specified context detracts from the API's usability. The cognitive dimensions characterize the degree to which an API design helps or impedes the user—the more closely that the cognitive dimensions of an API match those of the user, the more usable the API. Likewise, the worse the match between the cognitive dimensions of an API and those of the target user, the less usable the API.

One specific example of an API element with usability issues is the socket. Select() method of .NET Framework 1.1 that Henning (2007) reviewed. The socket.Select() method determines the status of network communication sockets for an application, and returns a list of the sockets that meet specified criteria. Figure 11.4 shows the interface definition of the socket.Select() method for use in a C# application.

To use the socket.Select() method to check six sockets and determine whether any of them have data to read, an application would create a list of those six sockets and pass that list to the socket.Select() method. When the function returns, the list contains only the sockets that have data to read. The code example in Figure 11.5 demonstrates this in an application derived from the code example used in the socket.Select() reference documentation (Microsoft, 2010b).

Henning (2007) observed that the socket.Select() method overwrites the contents of the lists that are passed in the checkRead, checkWrite, and check-Error arguments. While that's not a serious problem in the code example shown in Figure 11.5, that example doesn't accurately reflect real-world usage. Henning observes that an application is normally going to keep a list of the sockets that it is using or monitoring for other purposes than simply checking their status, and the list will typically contain hundreds of sockets instead of just the six used in this example. If checking the status of the sockets in a list changes the contents of the list, the application must create a copy of the list each time it wants to prevent losing the contents of the original list. For a small list, this is a nuisance but does not negatively influence performance appreciably. For a large list or several large lists, these copy operations will noticeably degrade performance.

Henning's observations illustrate an example of a work-step unit mismatch, where the user did not expect to need extra copies of the socket lists simply to check their status. As it is implemented, however, the socket. Select() method requires that the user create and manage extra socket lists in order to complete the programming task of checking socket status. Henning

```
public static void Select(
    IList checkRead,
    IList checkWrite,
    IList checkError,
    int microSeconds
)
```

FIGURE 11.4
Definition of the socket.Select() method for C#.

```
static void Main(string[] args)
{
        // Get IP address to use for sockets.
        IPAddress ipAddress = getIpAddress("");
        // Create a list of sockets to monitor.
        ArrayList listenList = new ArrayList();
        int numSockets = CreateNewSocketList
                (listenList, ipAddress);
        // Set the maximum time in milliseconds
        // to wait for a response.
        int timeoutMs = 1000;
        // Call the Select method to find out
        // which sockets are ready to be read.
        Socket.Select(listenList,
                null,
                null,
                timeoutMs);
        // Only the sockets that contain a
        // connection request will be in the
        // listenList array when Select returns.
        // Go through list of returned sockets.
        // and look at each one
        for (int i = 0; i < listenList.Count; i++)
        {
        // Do something with this socket.
        }
}
```

FIGURE 11.5
Code example that illustrates how the socket. Select() method is used in a C# application.

observes that this duplication requires additional code and a duplication of operations that "is inconvenient and does not scale well: servers frequently need to monitor hundreds of sockets ... the cost of doing this [repeated duplication] is considerable" (p. 26).

Henning's observations also illustrate various cases of the cognitive friction that he experienced while trying to use the method—examples of situations where he expected one thing but the method did something else and caused him to review and revise his use of the method. These observations also identify problems with role expressiveness and domain correspondence that could have been identified in usability tests or peer reviews conducted during the design phase of the API.

The preceding analysis is an example of a "microscopic" or detailed study of a usability problem in that it is specific to a single API element. Such a detailed analysis, while thorough, is also very labor intensive. To accommodate a large API (or a small usability staff), the API elements should be matched to the scenarios that they support. Prioritize the scenarios and study the elements that support the most important or most critical scenarios first or in more detail than those that support the less critical scenarios.

Some API usability issues, however, appear only in a larger, macroscopic context. API consistency, for example, includes internal and external consistency. Internal consistency checks an API against itself, and external consistency checks an API against other related APIs. The scope of an external consistency study depends on the users and on the usage scenarios. An API should be checked against any existing related APIs in the same development environment or feature set to identify transferability issues where something that the user already knows from other APIs might cause confusion or contradict something in the API being studied. To identify other areas of potential confusion, an API should also be checked against the other APIs that the user might use to complete a task.

In Watson (2009), I studied an API for internal consistency issues by studying the names of the elements in an API as a group. According to the cognitive dimensions listed in Table 11.4, an API with full consistency is one in which the user can apply what he or she learns in one part of an API to another part of the API. Some of the ways that an API can support a consistent interface include the following:

- Similarly named elements should have similar meanings and uses.
- Similar operations and functions should have similar names and uses.

TABLE 11.4

A Summary of the Twelve Cognitive Dimensions Used to Evaluate API Usability

Cognitive Dimension Name and Scale	API Question	User Question*
Abstraction Level		
Primitive Factored Aggregate	At what level does an API component map to a user's task?	At what level of abstraction does the user prefer to work when attempting a task?
Learning Style		
Informal Task Driven Structured	What learning style does the API afford the user?	What learning style does the user prefer when learning how to use an API?
Working Framework		
Local Global System	What level of information does a user need to keep track of when using the API to complete a task?	What level of information does the user feel comfortable keeping track of when using the API?

(continued)

TABLE 11.4 *(continued)*

A Summary of the Twelve Cognitive Dimensions Used to Evaluate API Usability

Cognitive Dimension Name and Scale	API Question	User Question*
Task Focused Functional Atomic API Focused	What type of code must a user write to accomplish a task using the API?	What type of code does the user prefer to write to accomplish a task using the API?
Progressive Evaluation Local Functional Parallel	At what level can a user evaluate (test) partial implementation of a task or scenario?	At what level does the user prefer to evaluate a partial implementation?
Premature Commitment Minor Differences, Easily Reversed Significant Differences, Easily Reversed Significant Differences, Not Easily Reversed	To accomplish a task, what options does an API present to the user and how visible are the consequences of each option?	What is the lowest level of premature commitment the user can tolerate and still feel comfortable using the API?
Penetrability Minimal Context Driven Expansive	How much does a user need to know about the API's internal workings, and how easy is it for a user to learn that information?	How much does the user want to know about the API's internal workings in order to accomplish a task?
API Elaboration Use It As-Is (Do Nothing) Fine-Tune Replace	What must the user do to extend (elaborate) the API to accomplish a task?	What level of elaboration does the user want to do to an API to accomplish a task?
API Viscosity Low (Easy) Medium (Moderate) High (Difficult)	How easy is it for a user to make changes to code that uses the API?	What level of effort is the user willing to tolerate when changing code that uses the API?
Consistency Full Core Arbitrary	How much of what the user learns about one part of an API can be used in another, similar part of the API?	What level of consistency does the user need in order to feel confident about using the API?
Role Expressiveness Transparent (Easy) Plausible Opaque (Difficult)	How easily can a user tell what an API does by reading the code?	How easy to read does the user need an API to be?

(continued)

TABLE 11.4 *(continued)*

A Summary of the Twelve Cognitive Dimensions Used to Evaluate API Usability

Cognitive Dimension Name and Scale	API Question	User Question*
Domain Correspondence		
Direct	How closely do the classes and	How closely do the classes
Plausible	methods exposed by the API	and methods need to match
Arbitrary	match the user's conceptual	for the user to use the API
	objects?	comfortably?

The analysis method described in Watson (2009) makes it easy to check for adherence to naming and style guides while also checking the internal consistency of an API. While consistency also includes the programming idioms and patterns used by the API, API element naming is still a critical factor in API usability. Cwalina and Abrams (2009) devote an entire chapter to detailing the naming guidelines of the .NET Framework. These guidelines can also be adapted to other environments and languages when existing guidelines are not available.

The API-wide scope of the study in Watson (2009) made it possible to find examples where different methods in the API used a parameter for the same purpose. In one example, ten different methods used a parameter to pass the same information. Nine of the methods used the same name for this parameter, but a different name was used in the tenth method. The difference that was identified by this analysis turned out to be unintentional and was easily corrected early in the development process. Had this inconsistency not been caught, a software developer using the API might wonder how or why that parameter differed from the others, because, in a generally consistent interface, a software developer is likely to assume that any inconsistencies or differences have meaning. Consequently, a software developer will spend time looking for significance and meaning in inconsistencies, even when there are none.

API Design, Development, and Usability

Having dedicated usability engineers working on an API design team will keep usability represented in the requirements, design, and development of an API. Having dedicated usability engineers working with a design team, however, is not the only way to identify usability issues. Nielsen (1993, 2009a) describes *discount usability* as a low-cost alternative for studying usability. Nielsen (2009b) observes that while usability tests with a good usability methodology tend to find more issues than tests with a poor methodology, even usability tests with a poor methodology still identify more usability issues than no testing whatsoever.

Nielsen (2009a) also points out that anyone can do a usability study. He says that while usability experts add value, anyone can improve usability to some degree, and you don't need an expert for every usability research task. He continues, "Everybody on the team needs to take responsibility for improving the user experience" (2009a). Obviously, the API designers have considerable responsibility for the usability of their design—Henning (2007) and Bloch (2005) suggest that they will also benefit from a usable design. The downstream tasks, such as testing and technical writing, will also benefit directly from a usable design even before the target users do. Testing and technical writing have a customer perspective and a vested interest in usability, which makes them good places to start when looking for usability resources beyond the usability experts.

Technical Writers Can Help with Usability

Nielsen (1993) describes four elements of "Discount Usability Engineering":

- User and task observation
- Scenarios
- Simplified thinking aloud
- Heuristic evaluation

While not all technical writers have formal usability training, technical writers will recognize most of the tasks in Nielsen's description of discount usability, and their work benefits directly from usability improvements. In the course of documenting APIs, technical writers are accustomed to working with scenarios, getting subjects, such as subject matter experts, to think aloud, and working with a style guide, which is a lot like a heuristic evaluation. Nielsen (2009) describes how "anybody can do usability," so surely technical writers can.

Ideally, the usability problems in a design will be found early enough in the development process to be corrected before they are implemented. The cost to fix usability issues increases as the design is implemented because, in spite of the flexibility implied by the name, software becomes increasingly inflexible once it's been written. After a development team starts sharing an API among the other developers or development teams, it becomes very difficult (i.e., costly) to change the interfaces. This is where technical writers and discount usability techniques can help a project that does not have the benefit of usability experts.

Usability Fundamentals

Some of the prerequisite information that is required for evaluating usability include scenarios, tasks, and performance goals. Eventually, when they write the API's documentation, technical writers will write about the scenarios that an API supports and the tasks that the user is expected to perform in those scenarios. The scenarios are often used as part of the conceptual and overview help content for an API, and the design tasks from the specification form the basis of many task-oriented, "how-to" topics. Having the technical writers help write those tasks earlier in the specification phase of a design gives them a chance to review these items from a usability perspective, and it relieves the software developers of these tasks.

Heuristic Evaluations

Technical writers can perform a heuristic evaluation of an API specification by comparing the design against an established framework or coding standard. The coding standard to be applied might be internal if the development team has studied usability issues and created a coding standard or coding style guide to use, or an external coding standard might be used if the API must comply with an established coding standard or framework. Deriving or adopting a heuristic or coding standard is a separate and prerequisite task that must be performed before a heuristic evaluation can be accomplished. The process for developing a heuristic is outside of the scope of this chapter.

Evaluating an API heuristically is similar to a heuristic evaluation of a graphical user interface in many ways. For example, in both cases, there must be an established standard or heuristic to evaluate against and the evaluator must have some expertise in the medium being evaluated. To evaluate a graphical user interface, that expertise includes some training in visual design. To evaluate an API, the requisite expertise includes an understanding of programming and programming languages. Technical writers who document APIs should have sufficient expertise to evaluate a specification using a heuristic. Having a technical writer evaluate an API uses no development resources and gives the technical writer a chance to become familiar with the API—something she will need to do before she starts writing anyway.

A programming context isn't always necessary to evaluate all aspects of an API's usability. The approach used in Watson (2009), for example, evaluated the lexical consistency of an API by looking only at the elements of an API and considering them as text with very little programming context. The only programming context was applied when selecting and organizing the terms to evaluate. After the terms were selected, they were compared simply as words. The inconsistencies were easily identified and corrected or explained.

API Usability Peer Reviews

Usability testing can be conducted early in the design phase, even when the API exists only in the specification, by using API usability peer reviews. Farooq and Zirkler (2010) describe API usability peer reviews as "an adapted version of cognitive walkthroughs," and they are also a practical example of discount usability evaluation for API designs. In an API peer review, the API's program manager (the person who is responsible for the specification of the API's design and who coordinates the development resources) conducts a short, 1.5- to 2-hour review of an API feature or a scenario supported by the API. Two to four reviewers who are representative of the API's target software developers provide feedback to the program manager. After the review, the program manager summarizes the feedback as actions for the design team to take.

Nielsen (2009a) says that "it takes only three days to complete a small usability project," while Farooq and Zirkler's (2010) API usability peer reviews require only 3.5 hours spread over a few days. The low cost and flexibility of API usability peer reviews make them cost-effective for use throughout the design and development of an API. Farooq and Zirkler illustrate how API usability peer reviews compare quite favorably to API usability tests in terms of finding usability problems, even though an API usability peer review requires only between 6 and 20 percent of the resources of an API usability test. API usability peer reviews can be conducted very early in the design process by using a paper prototype of the API that is derived from the API specification. API usability peer reviews can be conducted throughout the development process, and they can shift from paper prototypes to actual program code when the programming library contains working code, even if it's only a prototype or beta-test version of the programming library.

In an API peer review, the program manager does most of the active facilitation of the review process, while the usability engineer's role is more advisory in nature. Before the review, the usability engineer instructs the program manager on the peer review process and on how to conduct the review meeting. During the review meeting, the usability engineer observes the interaction between the program manager and the reviewers as the program manager facilitates the review. After the meeting, the usability engineer helps the program manager articulate the observations and record them as work items. For an API peer review, someone who is familiar with the process, such as a technical writer, can fill the role of the usability engineer even if he has only limited usability experience.

The API usability peer review format encourages both macroscopic and microscopic examination of an API. The reviewers naturally apply their prior experience with other APIs and will point out cases where the API under review differs from their experience. At the same time, using the sample API to solve a programming task encourages the reviewers to take a microscopic look at the API being reviewed. To complete the task, the reviewers must successfully fit all the API elements together correctly as they would if they were actually using the API in a program. This exercise shows whether the target users (the reviewers) will fit the API elements together as the designers expected. If not, the think-aloud nature of the review will give the designers the necessary insight into the user's perspective to enable them to identify the gaps between the assumed behavior and actual behavior.

Technical Writers Should Help with Usability

API design is typically and reasonably an engineering responsibility and includes evaluating and understanding the requirements, selecting the programming idioms and design patterns, designing the internal architecture and implementation, and designing the external interface, which includes naming the programming elements. The design of an API and the documentation of that API, however, are typically separate processes that produce separate and different products. Software developers produce the header files that define the API and the programming library's executable files, while technical writers write the documentation that is released as some form of help library, Web content, or hardcopy book. While this might accommodate the development organization's structure, to the user, the API (that is, the header and interface definition files of the API) constitutes the first source of the API's documentation (Cwalina & Abrams, 2009). The API's help documentation might actually be a distant second, or third, source of documentation after Web search results and community content (information obtained from forums and other unofficial, Web-based sources), depending on how the help content is accessed (Cwalina & Abrams; Nykaza et al., 2002). The help documentation is also viewed somewhat skeptically by software developers either because the documentation is insufficient (Robillard, 2009) or because the software developers have had negative experiences with documentation in the past (Nykaza et al.). If there is any contradiction between the header file and the help documentation, the header file will be the more credible source of documentation because the application gets the API definition information from the header file and not from the documentation.

Robillard (2009) identifies obstacles that software developers encounter when learning an API. The obstacle categories reported most frequently in Robillard include "Obstacles caused by inadequate or absent resources for learning the API (for example, documentation)—Resources" and "Obstacles related to the structure or design of the API—Structure" (p. 30). Usability testing and specification reviews would identify "issues with the API's structural design." Involving technical writers in the design process helps improve the documentation aspect of an API, and documentation deficiencies can be identified and corrected earlier, removing the key obstacles to learning an API as identified by Robillard.

Because API reference documentation is necessary (Cwalina & Abrams, 2009) and conceptual topics and examples are important to learning an API (Robillard, 2009), it might be tempting to focus on better documentation alone instead of better usability. Providing better documentation is certainly not a bad thing, but any usability issue with an API is best addressed at the source—that is, the API design, not just the documentation. Henning (2007) lists some of the consequences of poor APIs, and he observes that "programmers take longer to write code against poor APIs than against good ones" (p. 29). He also observes that poor APIs require additional program code that "makes programs larger and less efficient." Improving documentation won't improve these aspects of a poor API; they must be fixed in the API design.

Conclusion

API usability has gotten very little attention since McLellan, Roesler, Tempest, and Spinuzzi (1998) published what could be the first API usability study in 1998 (see also Daughtry et al., 2009). Henning (2007) attributes this lack of attention to the programming environment of the 1970s and 1980s, when APIs were limited to specialized functions or operating system services. Back then, the focus of computer science curriculam was data structures and algorithms because, as Henning observes, "if I wanted to create software, I had to write pretty much everything from scratch" (p. 35). Henning believes that with the advent of open source software, this is no longer the case and that software engineering today consists more of integrating existing functionality than writing software from scratch. But it's not just open source software that puts increased attention on API usability; Daughtry et al. (2009) attribute it "to the increase in the number and size of APIs." Daughtry et al. continue, "It is now rare for anyone to write code that does not call APIs, and some tasks, such as user interfaces and graphics or the use of network resources and web-based services, cannot be performed without using APIs" (p. 2).

The importance of API usability is increasing, and while it might be getting more attention than it has in the past, that doesn't necessarily mean that it's getting a lot of attention or even enough attention. Henning (2007) observes that the shift from creating software from scratch to integrating existing software "has gone largely unnoticed [by many universities]." Daughtry et al. (2009) also mention a "lack of role models in … API design." Consequently, adding usability as an API design feature is likely to involve some degree of usability education along with some degree of persuasion.

Demonstrating the benefits of usability for the customer and the development team can be done at any point in the product's development cycle. Demonstrating these benefits earlier in the process is better than later; however, demonstrating them later is still better than not at all—if not for the current design, then for the next product or version. A key element in demonstrating these benefits is collecting the necessary data. The data can come from the categorization of software defect reports (bugs), customer service calls, and product feedback.

While studying API usability involves elements of computer science and software engineering, it is still based on the same principles of usability that are used in any other usability study. In spite of the fact that API usability tasks involve writing program code instead of navigating a Web page, API usability still requires knowing the user, developing scenarios and tasks, defining usability goals, observing users, and all the other usability research methods such as those Barnum (2002) describes.

Having usability engineers working with an API design team is the gold standard; however, usability support can also come from other sources. Technical writers, for example, are also in a position to assist with usability. On projects that don't have usability engineers, technical writers can provide the usability support using Nielsen's discount usability techniques. Applying technical writing resources to usability tasks early in the design can improve not only the design's usability from the start but also the API's documentation later in the project, both of which help improve the end user's experience with the resulting product.

Reader Take-Aways

- The cost and importance of API usability issues are increasing. What might have been an acceptable cost with a comparatively few interfaces that were used by a comparatively small population is becoming increasingly unacceptable as the numbers of programming interfaces and users of those interfaces grow.

- Improving the usability of an API provides design and development benefits in the short term in addition to the longer-term benefits afforded to the users of the API.

- Including usability as an API feature does not need to dramatically increase the design and development costs. Usability can be phased into a development effort gradually as the resources permit, and usability research does not require dedicated usability engineers.

- API design and usability use a specialized vocabulary that is specific to software design. Understanding this domain and vocabulary helps articulate and communicate usability issues to software developers who are designing an API.

- Domain knowledge of API design is helpful, but it is not necessary for every aspect of API usability studies. The domain knowledge of others, such as the software developers who are designing an API, can be leveraged and applied to API usability studies, for example, through API usability peer reviews.

References

Abrams, B. (2008). *Number of types in the .NET framework.* Retrieved from http://blogs. msdn.com/brada/archive/2008/03/17/number-of-types-in-the-net-framework.aspx

Arnold, K. (2005). Programmers are people, too. *Queue, 3*(5), 54–59.

Barnum, C. M. (2002). *Usability testing and research.* New York: Longman.

Bloch, J. (2005). *How to design a good API and why it matters.* Retrieved from http://lcsd05.cs.tamu.edu/slides/keynote.pdf

Bore, C., & Bore, S. (2005). Profiling software API usability for consumer electronics. In *International Conference on Consumer Electronics* (Eds.), *2005 Digest of Technical Papers: International Conference* (pp. 155–156). Gibsonia, PA: International Conference on Consumer Electronics.

Brooks, F. P. (1995). *The mythical man-month: Essays on software engineering, anniversary edition* (2nd ed.). Boston: Addison-Wesley Professional.

Clarke, S. (2004, May). Measuring API usability [electronic version]. *Dr. Dobb's Journal Windows/.NET Supplement,* S6–S9. Retrieved from http://www.ddj.com/windows/184405654

Cooper, A. (2004). *The inmates are running the asylum: Why high-tech products drive us crazy and how to restore the sanity.* Indianapolis, IN: SAMS.

Cwalina, K., & Abrams, B. (2009). *Framework design guidelines: Conventions, idioms, and patterns for reusable .NET libraries* (2nd ed.). Indianapolis, IN: Addison Wesley.

Daughtry, J. M., Stylos, J., Farooq, U., & Myers, B. A. (2009). API usability: Report on special interest group at CHI. *ACM SIGSOFT Software Engineering Notes, 34*(4), 27–29.

Farooq, U., & Zirkler, D. (2010). API peer reviews: A method for evaluating usability of application programming interfaces. Paper presented at CHI 2010, April 10–15, Atlanta, GA.

Green, T. R. G., Blandford, A. E., Church, L., Roast, C. R., & Clarke, S. (2006). Cognitive dimensions: Achievements, new directions, and open questions. *Journal of Visual Languages & Computing, 17*(4), 328–365.

Henning, M. (2007). API design matters. *Queue, 5*(4), 24–36.

International Organization for Standardization (ISO). (1998). *Ergonomic requirements for office work with visual display terminals (VDTs)—part 11: Guidance on usability* (ISO/9241-11/1998). Geneva: Author.

McConnell, S. (1993). *Code complete.* Redmond, WA: Microsoft Press.

McLellan, S. G., Roesler, A. W., Tempest, J. T., & Spinuzzi, C. I. (1998). Building more usable APIs. *Software, IEEE, 15*(3), 78–86.

Microsoft Corporation. (2010a). _cputs, _cputws (CRT). *MSDN Library.* Retrieved from http://msdn.microsoft.com/en-us/library/3w2s47z0(VS.71).aspx

Microsoft Corporation. (2010b). Socket.Select method. *MSDN Library.* Retrieved from http://msdn.microsoft.com/en-us/library/system.net.sockets.socket.select(VS.71).aspx

Nielsen, J. (1993). *Usability engineering.* Boston: Academic Press.

Nielsen, J. (2004, March 1). Risks of quantitative studies. *Jakob Nielsen's Alertbox.* Retrieved from http://www.useit.com/alertbox/20040301.html

Nielsen, J. (2009a, December 21). Anybody can do usability. *Jakob Nielsen's Alertbox.* Retrieved from http://www.useit.com/alertbox/anybody-usability.html

Nielsen, J. (2009b, September 14). Discount usability: 20 years. *Jakob Nielsen's Alertbox.* Retrieved from http://www.useit.com/alertbox/discount-usability.html

Nykaza, J., Messinger, R., Boehme, F., Norman, C. L., Mace, M., & Gordon, M. (2002). What programmers really want: Results of a needs assessment for SDK documentation. In *Proceedings of the 20th Annual International Conference on Computer*

Documentation (pp. 133–141). Toronto: ACM. Retrieved from http://portal.acm.org.offcampus.lib.washington.edu/citation.cfm?id=584955.584976&coll=ACM&dl=ACM&CFID=27598684&CFTOKEN=34478726

Quesenbery, W. (2004, February). Balancing the 5Es: Usability. *Cutter IT Journal*, 4–11.

Robillard, M. P. (2009, November/December). What makes APIs hard to learn? Answers from developers. *Software, IEEE*, 27–34.

Stylos, J., Graf, B., Busse, D. K., Ziegler, C., Ehret, R., & Karstens, J. (2008, September). A case study of API Redesign for improved usability. Paper presented at the 2008 IEEE Symposium on Visual Languages and Human-Centric Computing (VL/HCC).

Tulach, J. (2008). How to design a (module) API. *NetBeans*. Retrieved from http://openide.netbeans.org/tutorial/apidesign.html

United States Bureau of Labor Statistics (BLS). 2010. Computer software engineers and computer programmers. In *Occupational outlook handbook* (2010–2011 ed). Retrieved from http://www.bls.gov/oco/ocos303.htm

Watson, R. B. (2009). Improving software API usability through text analysis: A case study. Paper presented at the Professional Communication Conference. Retrieved from http://ieeexplore.ieee.org/search/srchabstract.jsp?arnumber=5208679&isnumber=5208666&punumber=5191363&k2dockey=5208679@ieeecnfs&query=%28%28ipcc+2009%29%3Cin%3Emetadata+%29&pos=19&access=no

Zibran, M F. (2008). What makes APIs difficult to use? *IJCSNS International Journal of Computer Science and Network Security, 8*(4), 255–261.

Additional References

Apple Computer. (2010). *Develop for iPhone 3.0*. Retrieved from http://developer.apple.com/iphone/program/sdk/

Apple Computer. (2010). *Using the audio extraction API in QuickTime 7*. Retrieved from http://developer.apple.com/quicktime/audioextraction.html

Bias, R. G., & Mayhew, D. J. (2005). *Cost-justifying usability: An update for an Internet age*. Boston: Morgan Kaufman.

Clarke, S. (2003, November 14). Using the cognitive dimensions framework to design Usable APIs. *Stevencl's WebLog*. Retrieved from http://blogs.msdn.com/stevencl/archive/2003/11/14/57065.aspx

Clarke, S. (2003, November 24). Using the cognitive dimensions, continued: Learning style. *Stevencl's WebLog*. Retrieved from http://blogs.msdn.com/stevencl/archive/2003/11/24/57079.aspx

Clarke, S. (2003, December 3). Using the cognitive dimensions: Working framework. *Stevencl's WebLog*. Retrieved from http://blogs.msdn.com/stevencl/archive/2003/12/03/57112.aspx

Clarke, S. (2003, December 22). Using the cognitive dimensions: Progressive evaluation. *Stevencl's WebLog*. Retrieved from http://blogs.msdn.com/stevencl/archive/2003/12/22/45143.aspx

Clarke, S. (2003, December 22). Using the cognitive dimensions: Work step unit. *Stevencl's WebLog*. Retrieved from http://blogs.msdn.com/stevencl/archive/2003/12/22/45142.aspx

Clarke, S. (2004, January 22). Using the cognitive dimensions: Premature commitment. *Stevencl's WebLog*. Retrieved from http://blogs.msdn.com/stevencl/archive/2004/01/22/61859.aspx

Clarke, S. (2004, February 3). Using the cognitive dimensions: Penetrability. *Stevencl's WebLog*. Retrieved from http://blogs.msdn.com/stevencl/archive/2004/02/03/66713.aspx

Clarke, S. (2004, February 10). Using the cognitive dimensions: API elaboration. *Stevencl's WebLog*. Retrieved from http://blogs.msdn.com/stevencl/archive/2004/02/10/70897.aspx

Clarke, S. (2004, March 10). Using the cognitive dimensions: API viscosity. *Stevencl's WebLog*. Retrieved from http://blogs.msdn.com/stevencl/archive/2004/03/10/87652.aspx

Clarke, S. (2004, March 17). Using the cognitive dimensions: Consistency. *Stevencl's WebLog*. Retrieved from http://blogs.msdn.com/stevencl/archive/2004/03/17/91626.aspx

Clarke, S. (2004, April 23). Using the cognitive dimensions: Role expressiveness. *Stevencl's WebLog*. Retrieved from http://blogs.msdn.com/stevencl/archive/2004/04/23/119147.aspx

Clarke, S. (2004, May 17). Using the cognitive dimensions: Domain correspondence. *Stevencl's WebLog*. Retrieved from http://blogs.msdn.com/stevencl/archive/2004/05/17/133439.aspx

Clarke, S. (2005, March 29). HOWTO: Design and run an API usability study. *Stevencl's WebLog*. Retrieved from http://blogs.msdn.com/stevencl/archive/2005/03/29/403436.aspx

Conger, J. L. (1992). *The Waite Group's Windows API bible: The definitive programmer's reference*. Corte Madera, CA: Waite Group.

Daughtry, J. M., Farooq, U., Stylos, J., & Myers, B. A. (2009, April 9). API usability: CHI'2009 special interest group meeting. Meeting held at CHI 2009, April 4–9, 2009, Boston, MA.

Department of Trade in Services of MOFCOM. (2008). *China's software industry: Current status and development strategies—trade in services*. Retrieved from http://tradeinservices.mofcom.gov.cn/en/i/2008-09-03/54732.shtml

Kreitzberg, C. B., & Little, A. (2010, May). Useful, usable and desirable: Usability as a core development competence. *MSDN Magazine*. Retrieved from http://msdn.microsoft.com/en-us/magazine/dd727512.aspx

Microsoft Corporation. (2010). C run-time libraries. *MSDN Library*. Retrieved from http://msdn.microsoft.com/en-us/library/abx4dbyh.aspx

Microsoft Corporation. (2010). Class (C++). *MSDN Library*. Retrieved from http://msdn.microsoft.com/en-us/library/w5c4hyx3.aspx

Microsoft Corporation. (2010). The #define directive. *MSDN Library*. Retrieved from http://msdn.microsoft.com/en-us/library/teas0593(VS.80).aspx

Microsoft Corporation. (2010). Function procedures. *MSDN Library*. Retrieved from http://msdn.microsoft.com/en-us/library/6xxtk8kx(VS.80).aspx

Microsoft Corporation. (2010). Interface (C# reference). *MSDN Library*. Retrieved from http://msdn.microsoft.com/en-us/library/87d83y5b(VS.80).aspx

Microsoft Corporation. (2010). Live search API, version 1.1b. *MSDN Library*. Retrieved from http://msdn.microsoft.com/en-us/library/bb251794.aspx

Microsoft Corporation. (2010). Methods (C# programming guide). *MSDN Library*. Retrieved from http://msdn.microsoft.com/en-us/library/ms173114(VS.80).aspx

Microsoft Corporation. (2010). *SDK*. Retrieved from http://msdn.microsoft.com/en-us/windows/bb980924.aspx

Microsoft Corporation. (2010). struct (C++). *MSDN Library*. Retrieved from http://msdn.microsoft.com/en-us/library/64973255(VS.80).aspx

Microsoft Corporation. (2010). Windows API. *MSDN Library*. Retrieved from http://msdn.microsoft.com/en-us/library/cc433218(VS.85).aspx

Moskalyuk, A. (2005, April 19). Number of software developers in China and India to grow at 25.6% and 24.5% a year. *ZDNet Research Blogs, IT Facts*. Retrieved from http://blogs.zdnet.com/ITFacts/?p=7667

Oracle. (2010). *Download Java 2 platform, standard edition, v1.4.2 (J2SE)*. Retrieved from http://java.sun.com/j2se/1.4.2/download.html

Oracle. (2010). *Frequently asked questions about Java*. Retrieved from http://java.sun.com/products/jdk/faq.html

Oracle. (2010). *JavaFX 1.0 API overview*. Retrieved from http://java.sun.com/javafx/1/docs/api/

Section IV

Practical Approaches: Methods for Evaluating Complexity

Theoretical concepts for conceptualizing usability and complex systems are not enough. Neither are demonstrative case studies, however helpful they may be. We need to see practitioners in action, formulating hypotheses that they address with innovative but replicable methodologies. The authors in this section of the book present practical approaches to evaluating complexity.

First, Carol Barnum and Laura Palmer highlight the shortcomings of traditional post-test questionnaires. Employing as an alternative product reaction cards, along with other forms of feedback, Barnum and Palmer demonstrate through sound findings that the cards are highly effective at expanding understanding of user experiences, especially that found in complex systems environments. David Golightly, Mirabelle D'Cruz, Harshada Patel, Michael Pettitt, Sarah Sharples, Alex W. Stedmon, and John Wilson explore what they perceive as two critical dimensions of complex usability: technical complexity and social and organizational complexity. Through their case studies, they argue that a comprehensive grasp of both dimensions of complexity requires a novel mix of both quantitative and qualitative methods. Next, Donna Kain, Menno de Jong, and Catherine Smith use "plus-minus" document-based usability evaluation to show how audiences of risk and emergency communications often interpret and use information differently than expected. Finally, Jason Cootey describes the creation of an interactive test scenario, Otis in Stageira, meant to offer more effective testing for systems, such as computer games, that are more complex in nature. In particular, Otis in Stageira was developed to measure whether players can learn rhetorical strategies of persuasion during gameplay.

12

Tapping into Desirability in User Experience

Carol M. Barnum and Laura A. Palmer

Southern Polytechnic State University

CONTENTS

Introduction 254
You Can Lead a Horse to Water, but You Can't Make It Drink 254
    Satisfaction Ratings in Self-Rated Questionnaires 255
    Post-Test Questionnaires in Comparative Studies 256
How (and Why) Microsoft Created (and Used)
    Product Reaction Cards 257
    Creation and Use in 2002 257
    Application and Use in 2004 259
How Others Have Used the Product Reaction Cards 261
Other Methods Used to Study Affect 263
Our Use of Product Reaction Cards 263
    Learning to Use the Cards 264
    Designing Methods to Show the Results 265
Results from Our Studies, from Simple to Complex Systems 265
    Moving toward Complexity 265
    Hotel Group Study on Loyalty Program Enrollment 267
    Website for Teacher Professional Development 268
    Network Monitoring and Management 269
    A Major Hotel Group Reservation Study 270
    Call Center Application 274
Conclusion 275
Reader Take-Aways 277
Acknowledgment 278
References...278

Abstract

In this chapter, we discuss our use of Microsoft's product reaction cards to tap into the elusive quality of desirability in user experience. Desirability, as an affective response to a product, has been difficult to measure through instruments such as traditional post-test questionnaires. Since 2006, we have used the product reaction cards to expand our understanding of user experience. Coupled with other feedback mechanisms, we have found the results obtained from the cards help users express their positive and negative feelings about a product. Our work demonstrates that when the findings from participants' choices are collated, a shared experience emerges. In presenting our findings from a variety of studies, we hope to give readers the tools and the incentive to use them in complex studies and thereby expand our understanding of their usefulness.

Introduction

Evaluating complex systems generally requires complex testing protocols (Redish, 2007). The need to use complex testing protocols results from the nature of complex systems, which increase the cognitive load for users, who are expected to perform tasks that are different "in kind, not just degree" (Mirel, 2003). Not only does the complexity of the testing protocol increase, but also user experience practitioners often find themselves in domains beyond their expertise, requiring reliance on a domain-specific subject matter expert to set up realistic scenarios for testing and to assist in analyzing the findings and making recommendations. As well, some of the common qualities associated with usability, such as ease of use, are not necessarily important factors in evaluating complex systems. Other factors, such as ease of learning, ability to interact with the system over time, and reduction of human error (Redish), take precedence. Designers of these complex systems are left with "major headaches" in determining how to design a system that meets users' needs (Albers, 2004, p. 128). However, if complex systems designs fail to account for users' *feelings* about the system—a nebulous group of factors relating to "desirability"—users may be reluctant to adopt a system, or, in situations where they have no choice, forced to use a system they don't like.

In this chapter, we focus on the need to probe and the means to understand the nature of desirability of complex systems. We report on our use of Microsoft's product reaction cards, created to elicit the elusive desirability factor from users, and how the results from our studies address this critical factor in users' satisfaction with new products.

You Can Lead a Horse to Water, but You Can't Make It Drink

The International Organization for Standardization's (ISO, 1998) definition of *usability* has three major elements for gauging the usability of products for specified users with specified goals in specified contexts of use: effectiveness, efficiency, and satisfaction. *Effectiveness* refers to the "accuracy and completeness with which users achieve specified goals"; *efficiency* refers to the "resources expended in relation to the accuracy and completeness with which users achieve goals"; and *satisfaction* refers to the "freedom from discomfort, and positive attitudes toward the user of the product" (ISO, p. 2; as cited in Hornbaek, 2006, p. 82). Usability evaluations generally do a good job of uncovering effectiveness and efficiency issues within complex systems, but may not do so well in addressing satisfaction.

While effectiveness and efficiency are certainly important to overall system usability, satisfaction may be the element that makes or breaks successful adoption. If users do not experience satisfaction when using a system, their attitude about using the system may affect the efficiency and effectiveness of their interaction with the system. If the cognitive load put on the user is too great, the user may end tasks prematurely, resulting in suboptimal effectiveness and even mistakes. Low satisfaction levels may result in higher than expected burnout or employee turnover, requiring more training in system use. Even worse, when system design is perceived as "too complex," users may balk at using it, causing companies to abandon expensive purchases in search of something deemed more user-friendly. Numerous documented cases of systems being scrapped because of resistance or rejection point up the problem and recall the old adage, "You can lead a horse to water, but you can't make it drink."

For this reason, it's critically important to gauge the user's reaction to a complex system, not only in terms of whether it has the functionality the user needs but also in terms of the user's sense of satisfaction with the system. Designers of complex systems must design for all three factors. The question is how to uncover the third factor—satisfaction.

Satisfaction Ratings in Self-Rated Questionnaires

When concerns about satisfaction are addressed, they are typically solicited through a post-test questionnaire, which asks questions or poses statements that users respond to, using a rating scale of five points, seven points, or something similar. However, usability practitioners are well aware of the inflationary nature of users' responses to questions about their satisfaction with a system. This tendency to give pleasing responses in self-rated questionnaires (called the *acquiescence bias*) often produces incongruities between the ratings that users give and the users' experiences witnessed by observers. This bias extends even to the relationship between people and computers. Reeves and Nass (1996) report in their groundbreaking book, *The Media Equation: How People Treat Computers, Television, and New Media like Real People and Places,* on a number of such studies investigating the way in which people react to technology. As one example, they report that people evaluated a computer's performance more favorably when they were asked to complete a questionnaire on the same computer they used for the task. In contrast, they gave lower ratings when asked to complete the questionnaire on another computer. In other words, people's desire to be pleasant extends not just to other people, but also to the computer on which the interaction takes place.

All of this leads us to the question "How do we elicit reliable feedback from users about their experience?" As reported by Hornbaek (2006), who reviewed 180 published usability studies, "The challenges are to distinguish and empirically compare subjective and objective measures of usability; to focus on developing and employing measures of learnability and retention ... [and] to extend measures of satisfaction beyond post-use questionnaires" (p. 79).

Post-Test Questionnaires in Comparative Studies

We are not suggesting that post-test questionnaires have no value. Despite the inherent flaws noted above, post-test questionnaires can be useful in providing *relative* rates of user satisfaction, particularly when the scores are used in comparative evaluations. One such comparative evaluation (Tullis & Stetson, 2004) was designed to measure the effectiveness of a variety of feedback instruments in identifying user preferences for a financial website. Participants in the study were given one of the feedback instruments and asked to use it to compare their experience with two competitor websites. With a total sample of 123 participants, each feedback instrument was used by nineteen to twenty-eight participants. The five instruments used were as follows:

- SUS (System Usability Scale) developed at Digital Equipment Corporation
- QUIS (Questionnaire for User Interface Satisfaction) developed at the University of Maryland
- CSUQ (Computer System Usability Questionnaire) developed at IBM
- Fidelity's questionnaire, the study authors' instrument developed for internal testing purposes
- Words (adapted from Microsoft's product reaction cards)

The researchers found that SUS and CSUQ provided the most consistent responses at twelve to fourteen respondents. They further noted that SUS, which at ten statements is the shortest, framed all of the statements on the user's reaction to the website *as a whole*. This instrument proved the most reliable in indicating user preference for one website over another.

Regarding the applicability of the results in a broader context, Tullis and Stetson (2004) provided several caveats:

- They evaluated only two websites.
- They chose what they considered two typical tasks and tested only these two tasks.

- They focused their study on the *comparison* of two websites, not on a study of a single website.

Because one of the feedback instruments used was Microsoft's product reaction cards, their comments on the use of the cards are especially pertinent to the focus of this chapter:

> When evaluating only one design, possibly the most important information is related to the diagnostic value of the data you get from the questionnaire. In other words, how well does it help guide improvements to the design? That has not been analyzed in this study. Interestingly, on the surface at least, it appears that the Microsoft Words might provide the most diagnostic information, due to the potentially large number of descriptors involved. (Tullis & Stetson, 2004, p. 7)

We concur. We have found that Microsoft's product reaction cards unlock information regarding the user's sense of the quality of satisfaction in a more revealing way than any other tool or technique we have tried.

We have used the cards in studies ranging from the simple to the complex. In our experience, the reason for the success of product reaction cards is simple: This tool provides a way for users to *tell their story* of their experience, choosing the words that have meaning to them as triggers to express their feelings about their experience. While it is always important to understand the user's experience, it becomes even more important in evaluating complex systems because the very complexity of the system makes it harder to know what users like and dislike, and, most importantly, why. The cards open the door to this understanding in ways that other methods often do not.

This chapter describes the development and use of the product reaction cards, the adaptations of use by others, and our use of them. We then present several case studies to show the consistency of findings we have obtained in our studies of products across a range of industries and applications. We illustrate various ways to present the findings in reports or presentations. What this approach provides is a method to capture and document user satisfaction with a system.

How (and Why) Microsoft Created (and Used) Product Reaction Cards

Creation and Use in 2002

Product reaction cards were developed by Microsoft (Benedek & Miner, 2002) as part of a "desirability toolkit" created to get at the quality of desirability, a key component in user satisfaction. The motivation for creating the desirability toolkit was driven by the limitations inherent in standard feedback mechanisms, such as Likert-type questionnaires, as well as semi-structured interviews. The problem with questionnaires, as stated earlier, is that study participants tend to give similar, generally positive responses to all questions. An additional problem, as Benedek and Miner point out, is that the user experience practitioner defines the range of questions, restricting the participants' responses to the specific questions or statements asked. The problem with interviews, they point out, is that they are time-consuming to give and even more time-consuming to analyze. Microsoft concluded that these barriers made it unlikely that usability practitioners would assess desirability when conducting product evaluations. Thus, they concluded that the need existed, but a practical mechanism did not.

To create a mechanism for getting at the missing desirability factor, a Microsoft product team kicked off the project with a brainstorming session involving eight usability engineers. The challenge was to come up with ideas beyond the traditional methods for collecting feedback on users' experiences. The brainstorming session began with this challenge: "You are a usability engineer and you want to find out what makes something desirable. What can you do?" (Benedek & Miner, 2002). The criteria for offering up a new process were that (1) it had to be quick to administer, and (2) the results had to be easy to analyze. The outcome was the "desirability toolkit," which comprised two parts: (1) a faces questionnaire, in which participants would be asked to choose a photograph of a face whose expression matched their experience; and (2) product reaction cards, in which participants would be asked to choose descriptive words or phrases from a large set of cards.

The faces questionnaire was intended to give participants ambiguous stimuli in the form of photographs of faces. The goal was to get feedback from participants on their emotional responses (such as happiness or frustration) to using the product.

Figure 12.1 shows a sample face and a sample statement, in which participants were asked to rate how closely the facial expression matched their

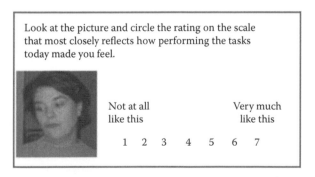

FIGURE 12.1
Example item (one of six faces) from the faces questionnaire.

experience. In use, some participants responded well to this approach; others found it difficult to interpret a meaning and assign a rating to the face.

The other tool in the desirability toolkit was the product reaction cards. The words on the cards represented a broad spectrum of dimensions of desirability. The words were obtained from market research, prior user research, and team brainstorming. Based on their observations of the higher-than-average positive response from participants in completing post-test questionnaires, the team set the ratio of positive to negative words at 60 percent positive and 40 percent negative or neutral words.

The product reaction cards were used in the following way:

- At the end of a testing session, participants were asked to select cards that reflected how they felt about their experience with the product. Their choices were recorded.

- Next, participants were asked to narrow their selection to their top five choices.

- Then they were asked to explain their choices. In this step, participants told the story of their experience, picking their storyline and its outcome from the prompt provided by the word or phrase on each card.

This method was used by four user experience engineers in four different lab studies. Everyone who was involved liked it: the engineers and observers for the rich qualitative feedback they received from the users, and the participants for seeing how their feedback was valued.

Refinements of the number of cards and the words on the cards resulted in a toolkit of 118 words, while also maintaining the ratio of 60 percent positive and 40 percent negative or neutral words. Table 12.1 shows the final set of 118 cards, which was used in three studies by the Microsoft Windows group. Coupled with the four initial pilot studies, these study results proved beneficial to the teams, regardless of whether they were conducting formative or summative studies of a single product, comparative evaluations of different versions of the same product, or comparative evaluations of different products.

Application and Use in 2004

For the launch of MSN Explorer 9, the product reaction cards were used in a different way (Williams et al., 2004). In this study, the team's objective was to "cut through this complexity" of diverse users and diverse user preferences for a visual design of an integrated browser by providing a feedback mechanism for users to articulate their preferences among four design choices. Requiring the same criteria as the earlier Microsoft studies, the method had to be fast, it had to complement existing methodologies, and it had to be effective in understanding user preferences about visual design.

TABLE 12.1

Complete Set of 118 Product Reaction Cards

Accessible	Creative	Fast	Meaningful	Slow
Advanced	Customizable	Flexible	Motivating	Sophisticated
Annoying	Cutting-edge	Fragile	Not secure	Stable
Appealing	Dated	Fresh	Not valuable	Sterile
Approachable	Desirable	Friendly	Novel	Stimulating
Attractive	Difficult	Frustrating	Old	Straightforward
Boring	Disconnected	Fun	Optimistic	Stressful
Business-like	Disruptive	Gets in the way	Ordinary	Time-consuming
Busy	Distracting	Hard to use	Organized	Time-saving
Calm	Dull	Helpful	Overbearing	Too technical
Clean	Easy to use	High quality	Overwhelming	Trustworthy
Clear	Effective	Impersonal	Patronizing	Unapproachable
Collaborative	Efficient	Impressive	Personal	Unattractive
Comfortable	Effortless	Incomprehensible	Poor quality	Uncontrollable
Compatible	Empowering	Inconsistent	Powerful	Unconventional
Compelling	Energetic	Ineffective	Predictable	Understandable
Complex	Engaging	Innovative	Professional	Undesirable
Comprehensive	Entertaining	Inspiring	Relevant	Unpredictable
Confident	Enthusiastic	Integrated	Reliable	Unrefined
Confusing	Essential	Intimidating	Responsive	Usable
Connected	Exceptional	Intuitive	Rigid	Useful
Consistent	Exciting	Inviting	Satisfying	Valuable
Controllable	Expected	Irrelevant	Secure	
Convenient	Familiar	Low maintenance	Simplistic	

Source: Developed by and © 2002 Microsoft Corporation. All rights reserved. Permission granted for personal, academic, and commercial purposes.

Based on MSN branding efforts and market research, the key words driving the Explorer study were "fast, fun, innovative, desirable." Because the time frame allotted for usability studies was only a few weeks and the team wanted larger responses than could be obtained in single-user-testing studies, they created a number of testing and feedback techniques to elicit user preferences in larger numbers. Among these methods was a group-sorting process, using the product reaction cards, followed by a focus group discussion.

To accomplish their goal, they ran ten 2-hour studies with approximately eight people per session and two facilitators, divided into three phases:

- *Phase 1: Individual measures without discussion.* Participants were instructed to review the four design options, presented in various formats, spending one minute in front of each design in each format. They were asked to complete a form indicating their choice of the best design and also to rank the other three choices.

- *Phase 2: Individual mark-up with group discussion.* Based on each participant's first choice, participants were grouped with others who had made the same choice and taken to a separate room to engage in several activities, including marking up pages for each design, using a red pen for negative comments and a black pen for positive comments. Following this task, they were asked to write down three words they thought best described each of the four designs, then complete a Likert-type questionnaire.

- *Phase 3: Group consensus using product reaction cards.* Each group was asked to reach consensus on the top three cards that reflected their preferred design, using 110 cards from the deck of product reaction cards. The group then completed a form that asked for a brief explanation of the reason for each word chosen. The sessions were videotaped. The sessions ended by combining groups who favored a particular design for a focus group discussion of the one they liked best, with time for discussion, as well, about the ones they liked least.

The resulting design direction was based on the data collected from this study. Project success was measured by customer feedback from the beta release and subsequent launch to the general public. A large number of spontaneous, positive comments from the twenty-five thousand beta users were supported by survey response data, which showed that users consistently rated the *appearance* of MSN 9 above even those features rated most *useful* in the product. Among all the methods used, the team concluded that future testing would focus more on the product reaction cards and less on traditional Likert-type statements and the other methods used in this study.

How Others Have Used the Product Reaction Cards

When the results of these Microsoft studies using the product reaction cards were presented at conferences (the Usability Professionals' Association [UPA] conference in 2002 and CHI 2004), others very likely picked up the technique. However, the number of published studies on how they have been used is quite small and mostly anecdotal. One illustration is provided by David Travis (2008), who reports on his use of approximately one hundred words from the deck of product reaction cards (or substitutes to suit the study). His studies indicate that participants tend to speak more negatively (and honestly) when using the cards, as compared to their responses when completing questionnaires.

His methodology for using the words is different from their original use by Microsoft. The participants in his studies do not sort through cards;

instead, they use a paper checklist of the words. Participants are asked to read through the list and select as many words as they like to reflect their experience. As is the case with Microsoft, Travis then asks the participants to narrow their choices to a top five. For each participant, the order of the words on the list is scrambled (using an Excel spreadsheet to generate the randomized list). For some studies, words reflecting the key brand values are used in place of some of the other, less relevant words. The guided interview that follows is based on the participant's explanation of the meaning of the selected words.

In addition to recording the words chosen, Travis conducts a verbal protocol analysis based on the participant's narrative, using a sheet with two columns: one for positive reactions and one for negative reactions. The positive and negative scores are added up to compute the percentage of positive comments.

Another reported use of the product reaction cards comes from Christian Rohrer (2009) on his studies of desirability at eBay and Yahoo!. In both cases, his team created cards that matched brand values. Users were instructed to select cards to express their responses to various designs. One approach was a qualitative usability study, in which individual participants in a lab setting were shown different visual designs for Yahoo! Personals, and asked to select the cards that best matched their responses to each design. Participants were then interviewed to understand their reasons for selecting the cards they did.

This method was adapted for use in a quantitative study, in which various images were embedded into a survey and participants were asked to respond with their choice of a word or phrase from the list that matched their feelings about each design. This process generated a lot of data for analysis and compilation into quantitative results.

In the desirability studies conducted at eBay, the team used paired opposites in a survey to assess responses. Examples of paired opposites include the following:

- Uninteresting–interesting
- Makes me feel unsafe–makes me feel safe
- Forgettable–captivating
- Cluttered–clean

As with the MSN 9 study, the goal of these studies was to identify the most favorable design, based on users' initial responses. As Rohrer (2008) attests,

> This is most useful when you are trying to make a good first impression with your target audience and invite them to interact more deeply with your site or product so they can discover whether it will meet an underlying need.

Other Methods Used to Study Affect

Other studies have attempted to understand people's emotional responses to user experience, or *affect*, devising various experimental methods (de Lera & Garreta-Domingo, 2007; Hazlett & Benedek, 2007; Kaltenbacher, 2009; Kim, Lee, & Choi, 2003; Mahlke, Minge & Thuring, 2006; Nurkka, 2008). In a study about attractiveness, Tractinsky, Katz, and Ikar (2000) used a pre-test and post-task questionnaire to determine a correlation between users' perceptions of the aesthetic quality of a computer system and its usability. They conclude that such a correlation exists: The perceived beauty of design and perceived usability held up in participants' ratings after usability testing, and users' perceptions had no relation to the actual usability of the system. They further conclude that "user satisfaction is important even for involuntary system use" and that "the positive affect created by aesthetically appealing interfaces may be instrumental in improving users' performance as well" (Tractinsky et al., 2000, p. 141). The authors state the importance of further study of the aesthetic aspects of design for their influence on perceptions of usability.

Another study (Thuring & Mahlke, 2007) used a more detailed questionnaire to explore the relationship between people's preferences for some systems over others. They, too, conclude that factors such as aesthetic qualities and emotional experiences play an important role in feelings about system usage and per-ceived usability. Similar findings were reported in a study of visual appeal and usability conducted by the Software Usability Research Laboratory (SURL), in which the authors (Phillips & Chaparro, 2009) found that the impact of positive first impressions based on visual appeal has a greater influence on users' per-ceptions of usability than the actual usability of the sites tested. These studies indicate that behavioral affect, or emotional responses to system design, influ-ences perceptions of usability, which, in turn, influence performance.

Predictions for continuing interest in learning about desirability are indicated in the September 2009 issue of *Usability News* with its attention-grabbing headline: "HCI 09: Interaction Gets All Emotional." Describing emotion as a property to take into account during both design and usability, author Joanna Bawa (2009) predicts from recent studies that emotional design may be the next big field. Another indicator of increased interest is a report by Forrester Research called *Emotional Experience Design* (Rogowski, Temkin, & Zinser, 2009). Current research trends show that what may best gauge users' true assessments of a system is the quality of their experiences.

Our Use of Product Reaction Cards

Our interest in understanding user experience and our sense that we were not getting the feedback we wanted led us, naturally, to the product reaction cards. We began using the cards in 2006, and we have found them highly valuable in penetrating the aspect of user experience that defies the regular measurements of mouse clicks, optimal paths, or help desk calls. We have found that they uncover how users *feel* about their experience interacting with a system. And we have found that using the cards adds an element of methodological plurality to our studies. Adding these results to our other methods of data gathering allows us to triangulate our findings and produce more meaningful and substantive results for our clients.

Learning to Use the Cards

Implementing the cards in our studies proved, at first, to be challenging. There was no detailed explanation about how the cards were used. Only the paper presented by Benedek and Miner in 2002 and the follow-up paper presented by Williams et al. in 2004 gave any framework for how the cards could be incorporated into a study. Our search for studies that claimed to be measuring emotional responses provided few useful results, beyond those we have reported earlier in this chapter. One study by Kim et al. (2003) used descriptive terminology (adjectives) to measure preferences. However, this study was difficult to administer and time-consuming to interpret; it was well out of our reach as usability practitioners who seek more immediate and relevant feedback. For our purposes, the cards were an easy-to-use tool that would provide rich and accessible findings for our studies.

What follows is an analysis of our work with the cards. We have used the cards to gauge our participants' emotional reaction to the product or process they have just evaluated. We have found that within each study and across a wide spectrum of users and tasks, the expressions of participants' experiences are surprisingly similar, as indicated by the cards. Participants choose the same words with such regularity that it is clear to us, as usability professionals, a common and shared experience exists that can be understood qualitatively and measured quantitatively.

In our early work, we began with only a basic understanding of how the cards were used and what they might provide. We were, for example, unaware that there were 118 cards in the full set. We mistakenly presumed the full set was the 55 words contained in Benedek and Miner's (2002) original published proceedings. However, after several studies, we located the full set of cards and started to use all 118 cards. Regardless of whether the set was 55 or 118 cards, we found the cards provided participants with a tool through which they could express what they felt in a way we had not heard before, as they finished an individual scenario or concluded a full usability session. As early as our first study, we saw that overall feelings, thematic word clusters, and repeated word choices made up a new qualitative data set conveying negative, neutral, or positive feelings about our users' experience. Additionally, many of the selected words cluster around themes or concepts, such as *speed* or *ease of use* or *complexity* or *stressfulness*. Thus, the cards have provided us with a tool to probe with our users to understand what they like and what they dislike about their experience with a system. From this insight, we are better able to offer recommendations for specific changes to improve the user experience.

Designing Methods to Show the Results

These studies have also given us the opportunity to explore how to best present the results of card choices. Although the cards are a qualitative methodology, presenting the results in visual displays offers a way to show the clustered results quantitatively. Qualitative results require researchers to be significantly more dynamic with their analytical approach and move past traditional modeling such as Venn diagrams. As the field expands beyond word counts and other numerical measures, researchers have to be prepared to understand semantic nuances and create visualizations of language-based outcomes. No single design fits every situation; instead, in our work, we've been flexible and experimental in the ways we visually present our findings.

In this chapter, we present numerous representations of results, so as to show various ways to present the results. Our goal is to encourage other usability practitioners to consider presenting outcomes with visuals that tell the participants' stories in the most clear and compelling ways possible. It is important to note, however, that a single method should be selected for a particular study so that the presentation of findings is consistent throughout the report or presentation of results.

Results from Our Studies, from Simple to Complex Systems

The following presents highlights of our work with the cards in usability studies of simple to complex systems. In first showing their promise in simple studies, we then show how the same patterns emerge in complex studies.

Moving toward Complexity

Our early studies with the cards in simple usability tests let us know right away that there was value in the method and that we wanted to add it to our data-gathering processes. In a study for a major cable TV provider, we saw what the cards could deliver in the way of emotional responses. The product was simple—the documentation for the self-install kit for a digital cable home setup—but the study, in this case, was complex, as we were testing four different designs of the instruction set and four different setup configurations with different types of televisions, cables, and accessories. We include the study results here because they show the results that can be obtained in learning user preferences in comparative evaluations, and, of particular interest to readers of this book, because the focus of the study was on the documentation.

The company wanted to learn which of four designs for the self-installation of a digital cable box resulted in the highest success rate and greatest sense of satisfaction. In the study, twenty participants (plus a pilot user) were randomly assigned to one of four instruction designs; for each, they were required to unpack the materials from the carton and follow the printed instructions. Although the combination of equipment and television connections differed in each scenario, the overarching similarity for all participants was the requirement to follow the instructions to connect cables and a digital box to the television set. Once the task was completed, participants were asked to select three or four or five cards from the set of product reaction cards.

Across the board, the instructions were viewed positively as *straightforward*, *usable*, and *easy to use* by the participants, as shown in Figure 12.2. However, for the basic setup of the cable box to a TV, a negative card—*confusing*—was selected four times. As well, the word *inconsistent* was selected three times. Viewing the cards in these clusters helped us understand user preferences for an effective instruction set. None of the four options was a clear winner; however, the technical communicator heading the team was well informed about what needed to be addressed to make the experience a positive and successful one for users.

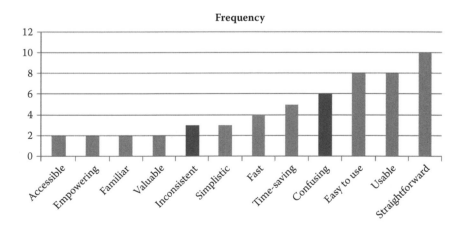

FIGURE 12.2
Card choices across all documentation designs tested.

Hotel Group Study on Loyalty Program Enrollment

A large international hotel group was redesigning its credit card enrollment process for customers signing up for enhanced benefits from a subscription-based loyalty rewards program. It wanted to compare the current version of its credit card enrollment process for preferred customers to join a loyalty rewards program with two different prototype sites for the same process. Among the six participants in the study, seventeen words were selected to describe the current site, with ten of these being positive words (58 percent). Prototype 1 also elicited seventeen word choices from participants, fifteen of which were positive (88 percent). Prototype 2 elicited only thirteen word choices, nine of which were positive (69 percent).

It was clear from the card choices that participants felt only marginally positive about their experience with the current site. With the exception of one participant, everyone had at least one negative word in his or her collection of card choices. The words *easy to use* and *usable* were selected twice to describe the current site; however, *confusing* and *time-consuming* were also selected twice. For Prototype 1, only two participants included negatives in their choices; both chose the word *confusing*. Like Prototype 1, Prototype 2 had only two participants select negative words. The term *confusing* was selected twice, with *inconsistent* and *cluttered* selected once each. On the basis of these cards, no clear winner had yet to emerge.

However, a difference was noted when it came to ease of use. Although all three sites were described by the same four terms—*easy to use, organized, familiar,* and *usable*—Prototype 1 outranked the others, as shown in Figure 12.3.

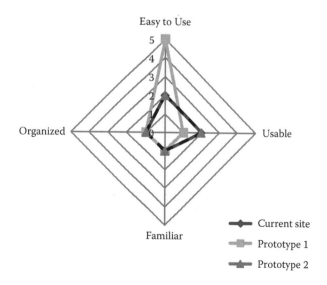

FIGURE 12.3

Comparison of card choices related to ease-of-use category for current site, P1 and P2.

When we looked at Prototype 2 in more detail to see what card choices differentiated it from Prototype 1, the *speed* category was the one significant area of difference. Prototype 2 was described as *time-saving* (2) and *fast* (1); only one reference to speed—*time-saving*—was found in the choices for Prototype 1.

As expected, the compilation of participant preferences indicated that the current site was the least preferred of the three versions. We expected that the remaining five participants would select Prototype 1, as the cards indicated it was more positively rated and the ease-of-use language was strong. However, our results were split.

Two participants selected Prototype 1 as their preferred site, while three selected Prototype 2. This slight preference toward the second version was interesting. Prototype 2 was less easy to use than Prototype 1, but it was seen as time-saving. This difference in the category of speed might have been the variable that tipped the decision slightly in favor of Prototype 2. Without the cards, however, we might never have seen this breakdown in participants' thinking.

Website for Teacher Professional Development

Our first use of the cards for a complex website was a study of a prototype website for teachers in professional development programs. The website was being developed as an online tool to support professional development activities for teachers in designated leadership programs. It employed many elements for communication, including common social networking features, to connect teachers with career coaches, principals, supervisors, and others. This website was a complex system in that it required the users to make many decisions in different parts of the site and establish ways to interact with a number of constituencies who would also be interacting with the users on the site. Following their interaction with the site prototype, five participants—all teachers—were asked to select the cards that best described their experience. While their experiences on the site weren't without difficulties, the cards indicated their positive feelings about the site's potential.

In total nineteen words were selected by the participants to describe their experience, as shown in Table 12.2. Of those nineteen, fifteen words were positive, which, without further analysis, could have led us to conclude that participants had a mostly positive experience with the site. Positive words chosen more than once included *professional* (twice), *customizable* (twice), and *impressive* (three times). However, four of the five users experienced some element of negativity with the website. The site was *frustrating* (chosen twice), *confusing*, and *hard to use*. Participants liked the potential of the website, but because the networking features of the site didn't function exactly like the social networking sites they used regularly, they found many of the functions unclear and in conflict with their established mental model.

TABLE 12.2

Card Choices for Teachers' Professional Development Website

Pilot	User 1	User 2	User 3	User 4
Frustrating	Professional	Impressive	Frustrating	Organized
Professional	Hard to use	Meaningful	Impressive	Impressive
Collaborative	Innovative	Comprehensive	Powerful	Customizable
	Useful	Approachable	Customizable	
		Confusing		

Note: Four of the five participants chose a negative card from the deck, indicating their level of frustration with the site.

Network Monitoring and Management

As an example of a highly complex system, we share our use of the cards in a usability study for a software company that produces computer network monitoring and management applications. In this study, six participants— all IT professionals—were asked to perform a series of complex but typical tasks with the new Web-based application. The study had several goals: The client wanted to gain feedback about features such as the navigation and user interface of the new system, and the client wanted to know if the participants could easily learn the new system. Our scenarios were constructed accordingly, and, at the end of the testing session, we used the cards to gain a sense of the users' feelings about this new application.

Among the cards chosen, the breakdown of positive versus negative language was thirty cards and thirteen cards, respectively, which was very close to Benedek and Miner's (2002) finding of 60 percent positive and 40 percent negative choices as the basis for their card creation. The negative card selections were our immediate focus, as we wanted to know the aspects of the product that evoked a negative experience. Figure 12.4 is a tag cloud of the negative language.

Four of the six participants chose negative words, with *time-consuming, frustrating, and inconsistent* chosen more than once by participants. Although we expected more positive language choices than negative choices, we didn't expect that the four participants who chose negative cards would also have

FIGURE 12.4

Tag cloud of negative language for network monitoring system.

positive reactions to the application. While the product's strongest negative dimension was *time-consuming*, the product was also *usable* according to four participants. Cards including *flexible* and *accessible* were selected by three of the six participants; the similar or exact same card selections showed us a trend developing with participants and their impressions.

A Major Hotel Group Reservation Study

In a multiphased study for a hotel group's online reservation system, the hotel group sought to understand the user experience with activities related to guest services. Although a task as seemingly straightforward as booking a hotel room seems simple, the decision-making process is dependent on the user's ability to acquire information from different sources within the website, which makes the task complex. In a highly competitive space, such as hotel reservations, if the competition makes the task a better experience for the user, the hotel may lose some potential sales to a website with a better booking experience for users. For this comparison, our client wanted to understand the user experience of making these decisions, as compared to that of a competitor. Twelve participants used the client's site and the competitor's site, selecting cards after using each one. Then the results were compared.

Our client's hotel reservation site received a strong positive rating from the participants, with twenty-eight out of thirty words reflecting a positive experience. Of the thirty-one words used to describe the competitor's site, only twenty-one of the words reflected a positive experience. Thus, the comparison of the cards showed that the client site was preferred by the participants. More telling, though, were the themes emerging from the individual word choices.

As with our other studies, selected words demonstrated the development of themes in the cards. *Speed, ease of use,* and *value* were again categories that emerged from the participants' card choices. A new category—*appeal*—also emerged from the selections. The word *appealing* was chosen twice; the words *inviting* and *desirable* were also selected. Clearly, the client's site projected an ethos the participants found pleasing.

Many words occurred more than once to describe both sites in this study. Figure 12.5 compares words that occur more than two times for both the major hotel group and its competitor. The repetition of language, as shown visually, depicts that our client's site differentiates itself in many positive ways, including being usable, straightforward, and time-saving. But, our client's site was also rated as less easy to use and not as fast as the competitor's site.

In the second phase of this usability study with the major hotel group, twelve participants compared the current site with a prototype site. Participants were asked to make a reservation and include add-on options, such as tickets to a local sports event, using both versions of the site.

Through the use of the cards, their feelings toward the current site were clear immediately, with eleven (68 percent) positive words selected as compared to five (32 percent) negative words selected. As represented in Figure 12.6,

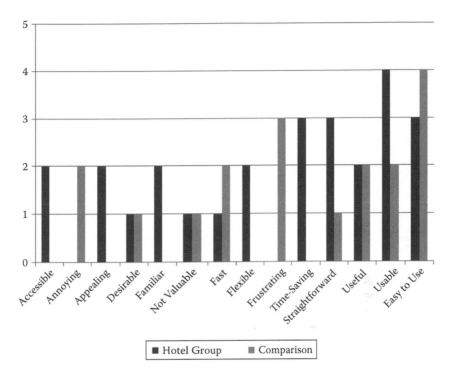

FIGURE 12.5
Frequency of cards chosen for client's hotel reservation site compared with competitor's site.

several words were selected repeatedly by participants to describe their experience with the current site. *Efficient* and *time-saving* were selected three times each. They clustered with the individual selections of *straightforward*, *accessible*, and *inviting*. When visually represented, we see this cluster of positive language and gain a big-picture view of factors contributing most strongly to participants' positive assessments. The words *useful, predictable, usable,* and *valuable*—chosen once each—also demonstrate, via their small clusters, important secondary influences on the positive ratings.

For the prototype, the number of positive cards totaled seven (46 percent), while the negative cards chosen numbered eight (53 percent). Thus, the participants were somewhat negative about their experience with this site. *Easy to use* was the only positive card selected twice for the prototype site. Clearly, there was a noticeable difference in how the participants were responding to the two versions of the site, choosing cards that showed a more positive response with the current site.

When reviewing the negative words chosen for the current site, we found that just 14 percent of the total words chosen reflected an undesirable experience. The word *time-consuming* occurred twice; *annoying, cluttered,* and *inconsistent* were selected once each. Figure 12.7 shows the word clusters for negative language for the current site and prototype site.

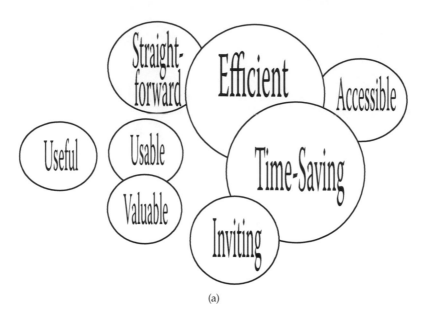

(a)

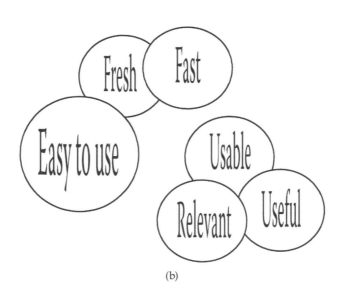

(b)

FIGURE 12.6
Positive language for the current site (a) versus the prototype site (b).

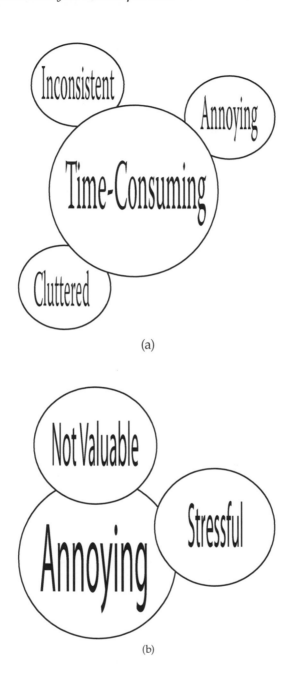

FIGURE 12.7
Negative language for the current site (a) versus the prototype site (b).

The cards' overall negative language rating for the prototype site was significantly different at 57 percent, which became clear to us when we looked at the selected words, their frequencies, and their cluster effects. As *annoying* was selected three times, we knew participants were not happy with the prototype. While *time-consuming* was a negative for the current site, it was a less severe choice than *annoying*. When we saw that *annoying* was paired with multiple selections of *not valuable* and *stressful*, we quickly understood that our participants' affective response to the prototype was not positive.

In the next phase of this series of comparative usability studies, twelve participants compared the newest prototype site with the current site, as well as two competitors' websites. In this study, as we have seen in some of our other studies, there was no clear favorite. The cards told us that participants were, at best, mixed-to-negative about their experience with all four sites. The prototype site and Competitor 2 were overwhelmingly disliked, with 66 and 75 percent negative language used as descriptors.

Figure 12.8 shows the results obtained from the twelve participants where their word selections are repeated one or more times across the evaluated sites. Individual card selections for all of the sites indicate that each site was rated as frustrating by at least one participant. The major hotel group's current site was the most frustrating, with participants making three selections of that word. The prototype site was also *frustrating* (2), *confusing* (3), *busy* (2), and *not valuable* (2).

Call Center Application

A call center was transitioning from a very old, green-screen application to a Web-based, graphical interface. It was critical to gain insight about the employees' reactions to the radically different look and feel as well as their interest in learning to use the new system. Additionally, the call center employees would be working with one part of a complex system that represented the upcoming consolidation of multiple data sources into a centralized information source.

Eleven participants used the new interface in a series of typical telephone-based interactions with customers. Of the total cards chosen by participants, fifty-one of the fifty-six choices were positive (93 percent); only five words (7 percent) reflected a negative assessment of working with the new interface. As the current system involved switching back and forth between databases to answer customer queries, the participants chose cards that expressed their pleasure with the speed and ease of use that this new system provided. That the system felt familiar and that it was easy to understand also resonated positively with the participants, especially as they were using the system without any training. Five of the eleven participants selected *easy to use* from the 118 cards (as shown in Figure 12.9), among other words picked more than once.

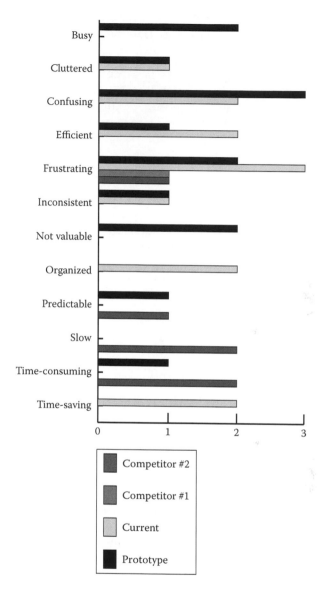

FIGURE 12.8
Frequency selections for prototype, current, and two competitor sites.

Conclusion

As far back as 1989—which, in the history of usability research, is the early days—Jakob Nielsen wrote an article called "What Do Users Really Want?" In it, he reported on the results of a study in which the respondents

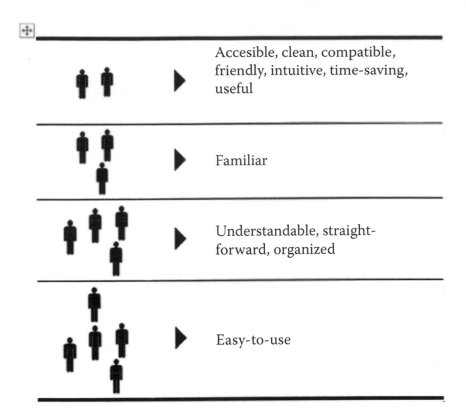

FIGURE 12.9
Number of participants selecting the same cards to describe their experience with the new call center interface.

(characterized as an experienced group of users) were asked to evaluate the importance of a number of interface characteristics when thinking about programs they use. The goal was to understand participants' priorities regarding product usability. The top-rated characteristic was *pleasant to work with*, which was considered more important in its effect on overall "user friendliness" than ease of learning. Nielsen says that he doesn't know precisely what *pleasant to work with* means, but it was clearly important to the respondents in rating a product user-friendly. Now, more than twenty years later, we find ourselves still grappling with that hard nut to crack: What does *pleasant to work with* mean to users, and how do we find out?

We have found that we are taking a big step forward in getting at this nebulous, critical feeling of pleasure, or desirability, through the use of the product reaction cards created by Microsoft. Whether the study is small and straightforward or large and complex, the cards have given our users the means to share their stories of their experiences with us. When we analyze the cards and cluster the similar words, we see that users reveal both positive and negative aspects of their experiences. Although the positive almost always outweigh the negative, as the 60:40 percent ratio adopted by Benedek and Miner (2002) in their creation of the product reaction cards reflects, we can gain particular insights into the negative words chosen, despite the fact that they tend to be fewer, because the users explain what they mean by their choice of the cards. And they freely choose the cards without any direction from us or a standard questionnaire.

Of course, even with the richness of the results we have received in using the cards, there are limitations to our work:

- Our studies are primarily small in scale, even when testing complex systems, so that the results are based on as few as six participants. Larger studies are needed to learn whether our findings can be replicated when more participants are involved.

- Our use of the product reaction cards does not negate the influence of the acquiescence bias seen with other satisfaction scales used in usability studies.

- The time to administer the card choice process and allow for users to share the relevance of their card selection may restrict the use of the cards in studies that are tightly controlled by time constraints.

- The cards should not be used as the sole means of getting participant feedback regarding their experiences. They work best when used along with other satisfaction survey instruments or when used as a baseline for comparison in iterative studies.

Even with these limitations, we believe that our experience in using the cards can help others testing complex systems to employ this method to gain insight into user experience. We hope that through our analysis of the results of some of our studies, we will advance the conversation and also promote further use by our fellow usability practitioners.

Reader Take-Aways

Complex systems focus on the common qualities of usability, just as simple systems do. However, complex systems, by their very nature, put greater demands on the cognitive load of users. Testing the usability of complex systems often requires complex testing protocols. In putting together a complex testing protocol, it is important to include the element of engagement, also called *desirability*. The key points presented in this chapter address a method to elicit users' sense of desirability in their interaction with a system. These include the following:

- The need to elicit feedback from users about "satisfaction," one of the cornerstones of the ISO 9241-11 definition of usability
- The limitations of self-reported questionnaire responses, which tend to show higher-than-observed responses to users' interactions with the system
- The development and use of product reaction cards by Microsoft
- The ongoing exploration of desirability by others, but little being reported by others on the use of product reaction cards
- Our use of the product reaction cards in a number of studies, presenting outcomes in various visual displays
- The value of using product reaction cards, in conjunction with other feedback mechanisms, to triangulate study findings and thereby increase our understanding of the users' experience

Acknowledgment

Portions of this chapter appeared in a conference paper entitled "More than a Feeling: Understanding the Desirability Factor in User Experience," published in the *Proceedings for CHI 2010*, April 10-15, 2010, Atlanta, Georgia.

References

Albers, M. J. (2004). *Communication of complex information: User goals and information needs for dynamic web information*. Mahwah, NJ: Lawrence Erlbaum.

Bawa, J. (2009, September 15). HCI 09: Interaction gets all emotional. *UsabilityNews.com*. Retrieved from http://www.usabilitynews.com/news/article5963.asp

Benedek, J., & Miner, T. (2002). *Measuring desirability: New methods for evaluating desirability in a usability lab setting*. Retrieved from http://www.microsoft.com/usability/UEPostings/DesirabilityToolkit.doc

de Lera, E., & Garreta-Domingo, M. (2007, September). Ten emotion heuristics: Guidelines for assessing the user's affect dimension easily and cost-effectively. In D. Ramduny-Ellis & D. Rachovides (Eds.), *Proceedings of the 21st BCS HCI Group Conference HCI 2007*, Lancaster University. Swindon, UK: British Computer Society.

Hazlett, R., & Benedek, J. (2007). Measuring emotional valence to understand the user's experience of software. *International Journal of Human-Computer Studies*, 65(4), 306–314.

Hornbaek, K. (2006). Current practice in measuring usability: Challenges to usability studies and research. *International Journal of Human-Computer Studies, 64*, 79–102.

International Organization for Standardization (ISO). (1998). *Ergonomic requirements for office work with visual display terminals (VDTs)— part 11: Guidance on usability* (ISO/9241-11/1998). Geneva: Author.

Kaltenbacher, B. (2009). From prediction to emergence: Usability criteria in flux. *Journal of Information Architecture, 1*(2), 31–46.

Kim, J., Lee, J., & Choi, D. (2003). Designing emotionally evocative homepages: An empirical study of the quantitative relations between design factors and emotional dimensions. *International Journal of Human-Computer Studies, 59,* 899–940.

Mahlke, S., Minge, M., & Thuring, M. (2006). Measuring multiple components of emotions in interactive contexts. In *CHI 2006 Extended Abstracts on Human Factors in Computing Systems* (pp. 1061–1066). Atlanta, GA: CHI.

Mirel, B. (2003). Dynamic usability: Designing usefulness into systems for complex tasks. In M. J. Albers & B. Mazur (Eds.), *Content and complexity in information design in technical communication* (pp. 232–262). Mahwah, NJ: Lawrence Erlbaum.

Nielsen, J. (1989). What do users really want? *International Journal of Human-Computer Interaction, 1*(1), 137–147.

Nurkka, P. (2008). *User experience evaluation based on values and emotions.* Retrieved from wiki/research.nokia.com/images/5/5e/Nurkka_ValuesEmotions.pdf

Phillips, C., & Chaparro, B. (2009, October). Visual appeal vs. usability: Which one influences user perceptions of a website more? *Usability News, 11*(2). Retrieved from http://www.surl.org/usabilitynews/archives.asp

Redish, J. (2007, May). Expanding usability testing to evaluate complex systems. *Journal of Usability Studies, 2*(3), 102–111.

Reeves, B., & Nass, C. (1996). *The media equation: How people treat computers, television, and new media like real people and places.* New York: Cambridge University Press.

Rohrer, C. (2008, October 28). *Desirability studies.* Retrieved from from http://www.xdstrategy/com/2008/10/28/desirability-studies

Rogowski, R., Temkin, B. D., & Zinser, R. (2009). *Emotional research design: Creating online experiences that deeply engage customers.* Forrester Research. Retrieved from http://www.forrester.com/rb/Research/emotional_experience_design/q/id/47918/t/2

Thuring, M., & Mahlke, S. (2007). Usability, aesthetics and emotions in human-technology interaction. *International Journal of Psychology, 42*(4), 253–264.

Tractinsky, N., Katz, A., & Ikar, D. (2000). What is beautiful is usable. *Interacting with Computers, 13,* 127–145.

Travis, D. (2008, March 3). *Measuring satisfaction: Beyond the usability questionnaire.* Retrieved from http://www.userfocus.co.uk/articles/satisfaction.html

Tullis, T. S., & Stetson, J. N. (2004). A comparison of questionnaires for assessing website usability. Paper presented at the Usability Professionals' Association conference, 2004. Retrieved from http://home.comcast.net/~tomtullis/publications/UPA2004TullisStetson.pdf

Williams, D., Kelly, G., Anderson, L., Zavislak, N., Wixon, D., & de los Reyes, A. (2004). MSN9: New user-centered desirability methods produce compelling visual design. In *Proceedings of CHI Conference* (pp. 959–974), Vienna. Austria: CHI.

13

Novel Interaction Styles, Complex Working Contexts, and the Role of Usability

David Golightly
University of Nottingham

Mirabelle D'Cruz
University of Nottingham

Harshada Patel
University of Nottingham

Michael Pettitt
Orange PCS

Sarah Sharples
University of Nottingham

Alex W. Stedmon
University of Nottingham

John R. Wilson
University of Nottingham and Network Rail

CONTENTS

Introducing Complexity 282
    Defining Complexity 282
        Technical Complexity 283
        Contextual Complexity 283
Case Domain 1: Collaborative Work Environments 286
Case Domain 2: Virtual Environments 289
Case Domain 3: Sociotechnical Systems 294
The Implications of Complexity 297
Conclusion 299
Reader Take-Aways 300
Acknowledgments 300
References 301

Abstract

This chapter presents a series of case domains demonstrating two dimensions of complex usability. The first dimension is the *technical* complexity of systems, which moves away from a traditional "input–output" view of interaction to consider the use of multiple modalities, tasks, and interfaces during a single interaction with a system. Second is the complexity associated with the social and organizational *context* in which the technology is used. These dimensions of complexity, in particular contextual complexity, require a triangulation of sometimes novel methods, both quantitative and qualitative, in order to fully design for user needs. Each case domain (collaborative work environments, virtual environments, and sociotechnical systems) represents a different setting in which complex usability emerges, and where different methods are required.

Introducing Complexity

The pressures of real software development mean that usability input often consists of a usability inspection against agreed heuristics or standards, or a limited usability "test" with a small sample of users. While both of these approaches have their limitations (Barnum et al., 2003; Cockton & Woolrych, 2002; Johnson, Salvo, & Zoetewey, 2007), they may be adequate in situations where the problem space is well known, where the platform for the technology is well understood, and where there is a history of design successes and failures to draw on.

As human factors specialists, however, our current and recent work involves assessing systems that are complex—we are studying the intersection where multiple people simultaneously interact with diverse technologies and systems. Our contribution to this collection discusses how, in the field of human factors, we have had to move beyond traditional testing protocols and standard usability measures to meet the needs of such systems. Complex systems require us, as researchers, to both adopt and develop flexible multimodal assessment methodologies in order to understand the nuances of interaction between multiple persons and complex, layered activities. Our work in three distinct areas—collaborative work environments, virtual environments, and sociotechnical systems—demonstrates how usability methods and practices must expand to provide meaningful and significant assessments of complex systems and users.

Defining Complexity

Our experience is that the construct of usability itself is complex and multifaceted. We can describe the three application areas presented in this chapter as typical examples of complex usability, defining *complexity* over two dimensions: technical complexity and contextual complexity.

Technical Complexity

Technical complexity refers to either the fundamental interaction characteristics (input and output) of the technology being complex, or where the underlying system architecture is complex, linking a variety of different systems, architectures, agents, databases, or devices. In both cases, at some point a decision must be made regarding the system image that is presented to the user—if a system is made transparent, and the underlying complexity of the system represented, this may support some users in forming an appropriate mental model of the system architecture and operations, but may hinder others by providing them with detail that is not actually needed in order to effectively work with the technology and achieve their goals.

This challenge has been known and addressed for many years—an early example being the use of a metaphor of files and folders to represent document location on a windows interface. While issues of technical complexity may appear at first to be a "solved problem" for interface design, the rapid pace of change in information and communication technologies (ICT) means we are now required to design for a whole new range of technologies that are emerging into common use and demand the development of new interaction styles and new guidelines. A recent example is the mobile phone, where the user is required to learn novel, sometimes nonintuitive forms of interaction, or new models of how technology works (e.g., the relationship between on-device data versus data provided for applications over a network), in order to manage and manipulate the handheld device.

Therefore, for this first dimension of complexity, it is critical that human factors methods are developed that enable the designer and evaluator to accurately assess the extent to which the user interface presents the capabilities and limitations of a technology. This is achieved either directly through the interface in terms of icons, labeling, interaction models, metaphors, and so on, or through appropriate support and documentation, to the user in a clear manner. If this presentation is achieved effectively, the user is more likely to harness the potential of a device or system. However, it is crucial that determining what clarity or appropriateness might mean for any given user in any situation is shaped by the second dimension of complexity—the context in which the technology is implemented.

Contextual Complexity

Contextual complexity includes the broader tasks, roles, or jobs the technology is proposed to support, especially when tasks are open-ended or unstructured (Redish, 2007). A task is not viewed merely as a series of inputs and outputs with a technology, but the work, leisure, or lifestyle goals the user wishes to achieve and how a technology reflects and integrates with this work. For example, imagine that our mobile phone user is now on a train, and wishes to communicate with a colleague using his or her mobile phone. The method and efficiency of interaction are first influenced by technical complexity—how well does the user understand how to make a call, send a message, or use e-mail via mobile Internet? However, the contextual complexity issues include the extent to which they are familiar with the person he or she wishes to call—if it is a close friend or colleague, he or she may be willing to risk making a voice call, knowing that the reception is weak and he or she may lose connection during the call. If the person being contacted is a potential client or employer, the user may choose to use e-mail, to avoid the risk of being cut off (and thus appearing unprofessional), and may not want fellow passengers on the train to overhear. A well-designed system will consider this more contextually complex type of scenario when developing a system, to ensure that the appropriate information required to support these contextual goals is incorporated into the system.

This appreciation of the broader role of context has brought with it a theoretical shift in human factors away from a typical scenario where an individual has a purely "cognitive" problem-solving view of the world (influenced by Newell & Simon, 1972, though their original work does not actually disregard the context where problem solving takes place), to one that sees sense making, problem solving, and interaction as distributed between people and artifacts (Hutchins, 1995) and shaped by the "ecology" or environment in which the interaction takes place (Vicente, 1999). The result is that interaction often takes on a highly situated nature, being shaped by an emerging and developing understanding of tasks and constraints, rather than restricted to a canonical "plan" (Suchman, 1987). Because we are now looking at the wider factors of user's goals situated within a task environment, we need to support all the potential uses, anticipated and even unanticipated, that a technology may be applied to, making it very difficult to design around prescriptive input and output sequences. This is as relevant to safety-critical systems, where systems (and their operators) should be able to function during rare and unexpected events (Garbis & Artman, 2004), as it is to mobile devices where it may be more challenging to predict all the contexts (e.g., on a train, in the home, at the office, and in a bar) that may have an influence on how technology is used. Contextual complexity is also heavily influenced by the culture in which the technology is being implemented—this could be a geographical culture, such as how local social norms can influence the naming of places in navigational aids (Diamon, Nishimura, & Kawashima, 2000); or organizational culture, such as how the norms one company may have for communication and collaboration may differ from another.

One implication of the need to reflect both technical and contextual complexity in design is an emphasis on multiple methods. While laboratory and quantitative work still has an important part to play, most often addressing the issues of technical complexity, qualitative methods such as ethnography, interviews and content analysis, and detailed observation are also vital for capturing the factors that contribute to contextual complexity. In practice, we find that methods need to be used in combination so that all aspects of usability can be addressed.

With that in mind, the rest of this chapter presents examples from our work in three specific domains that encompass both technical and, in particular, contextual complexity. We concentrate on considering the applicability of methods in complex situations, how existing measures can be tailored, and where new, bespoke methods are required.

Table 13.1 describes the three case domains. The first of these is collaborative work environments, where the challenge of design and evaluation is being able to effectively create multi-user systems, and capture the contextual characteristics of real and virtual working environments. The

TABLE 13.1

Chapter Summary: A Summary of Technical and Contextual Complexity, plus Methodological Challenges for Each Case Domain

Case Domain	Complex Usability Dimension		Methodological Challenges
	Technical Complexity	Contextual Complexity	
Collaborative Work Environments	Conversational turn taking, document management, and user status	Embedding technology in multiple organizations with differing cultures Changing the way in which people communicate	Deciding which aspects of collaboration to measure Capturing real-time interaction from multiple actors
Virtual Environments	Combination of the user or users in the real and virtual world with visual, auditory, and haptic feedback	Multiple user types, experience of technology, skills of task	Capturing salient issues from the real world and understanding requirements of future systems Understanding and selecting multimodal technologies
Sociotechnical Systems	Presentation of the interface for complex control systems Use and adaptation of the interface in unexpected ways	Communication between actors (within and beyond the control room) Safety-criticality of work Multiple tasks and priorities	Observing and quantifying interaction in the real world Understanding collaborative activity Building meaningful representations of complex domains

second domain is multimodal virtual environments, where the variety of inputs and outputs, the challenge of working in three dimensions, and the lack of design guidelines or standards provide a wealth of challenges for the usability expert. Our third example is sociotechnical systems, where the aspects of multiple constraints and goals within complex control environments, and the challenges of both capturing and representing the factors that shape interaction in safety-critical domains, are considered. At the end of the chapter we review the emergent themes from these three examples of complex usability, and discuss some of the methodological and design implications of addressing complex usability requirements.

Case Domain 1: Collaborative Work Environments

The ability of organizations to foster and enhance cooperation and collaboration through collaborative work environments (CWEs) is of increasing importance as industries move toward more distributed ways of working, involving multidisciplinary teams within multiple organizations. In this case domain, *collaboration* is defined as involving two or more people engaged in interaction with each other, within a single episode or series of episodes, working toward common goals (Henneman, Lee, & Cohen, 1995; Mattessich & Monsey, 1992; Meads, Ashcroft, Barr, Scott, & Wild, 2005; Montiel-Overall, 2005; Schrage, 1990; Wilson, 2006). CWEs incorporate the use of collaborative technologies (e.g., videoconferencing, virtual environments, augmented reality, and shared virtual whiteboards) into the workspace to support individuals and teams who are working together in the same or different physical locations.

Collaboration is a key contributor to a high level of contextual complexity of a technology—a system that supports collaboration may simply support an individual task but is more likely to support goal-directed behavior such as collaborative decision making. In a situation such as this, there may be multiple actors involved, each with either complementary or competing goals that are, in turn, shaped by the work context or aims of their respective collaborating organizations. Users may switch between different tasks and modes of communication and use different devices depending on their current (and possibly unpredictably changing) context. Users may be working in different organizations, in multiple locations, and across different time zones, and they may be engaged in synchronous or asynchronous collaboration. There are also issues such as responsibilities, power, and seniority within the work group. In addition to these contextual complexity factors, there are fundamental technical complexity issues in terms of managing turn taking in communications, presenting the status of users, or handling permissions and locking for shared documents, as well as complexity in terms of presenting the same information appropriately on different display systems (e.g., PDAs, desktop computers, and powerwalls).

CWEs also bring with them their own methodological challenges for design and evaluation. For example, multiple users must be observed at the same time and it may be difficult to gain access to required users simultaneously. It has previously been stated (Neale, Carroll, & Rosson, 2004) that the general lack of evaluation of collaborative systems and lack of evaluation strategies and tools to capture the factors that influence collaboration are hindering the progressive development of technology to facilitate collaboration (we propose that this statement can be extended to many cases of complex usability). Therefore, it is important to develop methods that can be applied across different system types and domains to yield generalizable data.

Using existing heuristics or cognitive walkthrough for initial evaluation of collaborative systems can prove difficult as there are few that address issues related to supporting group work as well as task work (Ereback & Hook, 1994). Systems with a high level of contextual complexity will be likely to have multiple users; therefore, it would be beneficial to examine how such heuristics could be extended to consider multiple roles. Baker, Greenberg, and Gutwin (2001) adapted Nielsen's (1992) heuristic evaluation method for evaluating collaborative systems by using heuristics relevant to teamwork, for example, to identify usability issues that could arise in distributed teams who share a visual interface. This approach could be extended to capture the similar or competing goals of different individuals within a team to consider the contextual complexity issues further.

Typical usability inspection methods for assessing collaborative systems fail to address the cultural or organizational aspects of collaborative group work, focusing instead on the low-level characteristics of collaboration that are deemed to be common to group interactions, such as turn taking and user status (Gutwin & Greenberg, 2000). Although this context-free evaluation allows basic usability problems to be identified, and may address many of the issues associated with technical complexity, it is important to reintroduce the importance of the context in which tasks are carried out during further evaluation, to ensure the issues relating to contextual complexity are captured (Neale et al., 2004; Pinelle & Gutwin, 2001).

An example of complex usability in the context of CWEs can be shown in our current work in a European Commission (EC)–funded Integrated Project, CoSpaces (IST-5-034245), with industrial, research, and business partners from twelve European countries developing innovative collaborative working solutions that are responsive to industrial needs. The project is exploring how advanced technologies (virtual, augmented, tele-immersive, mobile, and context aware) can be deployed to create collaborative engineering workspaces for supporting planning, design, assembly, construction, and maintenance activities in the aerospace, automotive, and construction sectors.

To articulate some of the different usability elements contributing to contextual complexity, we have developed a human-centered descriptive model of collaborative work based on a transdisciplinary review of existing research on collaboration and collaborative and team working, supported by a program of user interviews, workshops, and expert brainstorming sessions and user requirements elicitation work with the CoSpaces user partners. Combined with our extensive experience related to collaborative work with industrial and educational teams (see Wilson, Patel, & Pettitt, 2009), our model categorizes the following underpinning factors involved in collaborative work:

- *Individuals*: skills, well-being, and psychological factors
- *Teams*: roles, composition, common ground, shared awareness and knowledge, relationships, and group processes
- *Interaction processes*: communication, coordination, decision making, and learning
- *Tasks*: type, structure, and demands
- *Support*: knowledge management, error management, teambuilding, training, resources, networks, and tools
- *Context*: culture, environment, business climate, and organizational structure

In addition, we identified a further category of *overarching* factors (i.e., goals, incentives, constraints, experience, trust, management, conflict, performance, and time); these factors are often relevant across, and impact on, many of the other main factors.

This model defines the numerous elements that contribute to the contextual complexity of a technology, and while it can inform the requirements process and evaluation methodology, the model itself presents a challenge. It is important to be pragmatic about the role of a human factors specialist during an interface design process—if we are to ensure that our guidance is incorporated into the early stages of system design, it is probably not appropriate to overwhelm a design team with a large amount of information. Therefore, assuming we overcome the challenges of collecting the information regarding some or all of these aspects, at some point this information regarding requirements must be filtered and presented in a form that will be useful to a technology designer.

Having used the model as a framework for the CoSpaces system design, the evaluation approach we have followed to assess CoSpaces technologies is to compile real-world scenarios, developed during the user requirements elicitation phase, to highlight the areas that should be evaluated. We have then used role-play simulations based on these scenarios to test prototype systems in laboratory settings. In order to capture the full richness of the issues that influence contextual complexity, we have applied a range of methods including questionnaires, verbal protocol, task completion time, and observation to first evaluate the system out of a real-life context, before moving evaluation into the real world as development of the technology progresses—first by testing prototype systems and tasks with real users and settings with some degree of control, before implementation of the final system and evaluation with real users completing real tasks in real settings. This work follows a "living labs" philosophy, where interaction with products, systems, and environments is studied in relatively realistic environments, with the goal of retaining some control over participant interactions but also providing some level of ecological validity. This multiphase approach is essential to allow us to capture the contextual complexity elements of the proposed technology and feed this information into the design process as early as possible.

In summary, the CoSpaces case is typical of working in the domain of collaborative work environments. While there are technical complexity challenges, such as in managing turn taking, the major challenges are in fully understanding the wealth of factors associated with collaboration and deciding which factors should be considered for the effective design of CWEs. These challenges are predominantly cases of contextual complexity, and can only be fully understood by taking an approach to design that relies on full user participation, and evaluation that tries to re-create a real working environment as closely as possible.

Case Domain 2: Virtual Environments

Virtual environments (VEs) are computer-generated simulations of real and imaginary worlds created and experienced through virtual reality (VR). These experiences are generally three-dimensional and experienced through a number of senses—sight, sound, touch, and smell—using a range of input devices (e.g., a mouse, a joystick, a glove, gesture recognition, or speech recognition) and output devices (e.g., projection screens, headset technologies, force feedback, and tactile devices). The objects within these environments are "independent" of the user in that they can display real-world behavior or autonomy in terms of gravity, acceleration, or friction so that their behavior is appropriate when interaction occurs with another object or user. Moreover, the user is said to experience feelings of immersion and/or "presence" in the VE (Wilson & D'Cruz, 2006). Figures 13.1 and 13.2 give examples of different technology configurations that may be used for interaction with a VE.

Various key contributions through the years have identified a range of challenges for human factors in designing VEs and the selection of appropriate technologies (Sharples, D'Cruz, Patel, Cobb, & Wilson, 2006). It is often felt that VEs differ from similar computer-based technologies as they are specifically based on the user or users' experience within a VE. That is, their design and selection of appropriate technology components are defined by the "purpose" of the application (e.g., virtual prototyping, training for maintenance, and so on). This means technical and contextual complexity issues are closely intertwined as it is largely the context (and experience of the development team) that determines the VE requirements.

FIGURE 13.1
Example of a multi-user, projected display.

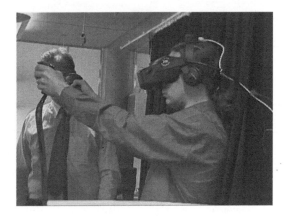

FIGURE 13.2
Example of an early single-user, head-mounted display (HMD).

How to interact within a 3D environment is just one example. There are still no standard interaction metaphors available, although we have carried out research on designing appropriate 3D interaction devices and menus (Patel et al., 2006). The selection of devices and menus is strongly related to context of use, as this determines the variety of actions and behaviors required to be displayed. For example, if the VE is a reconstruction of a building and the only interaction required is for the user to "walk through" and view from a variety of perspectives, then this can be addressed with the use of a joystick or wand or even a simple navigation menu on screen activated by a mouse. The usability issues are minimal as users are more experienced with these devices and type of interaction. However, if the VE were a training application of a maintenance task involving psychomotor skills and required collaboration with another person, then the interaction devices and menus would become more complicated and less defined. Early examples of VE training applications reported that the technology was more a hindrance to training as the complexity of the technology meant that users often spent longer learning how to use the training tool rather than how to do the task (D'Cruz, 1999).

One other specific usability issue associated with the use of 3D technologies is virtual reality–induced symptoms and effects (VRISE)—the combination and varieties of positive and negative effects, notably VR-induced sickness, that may result from a period of use of VR. This is likely to be a consequence of both the technical and contextual factors. Previous work has identified that a combination of factors will contribute to experiences of sickness, including technical factors such as the type of displays, and the sensory conflict experienced due to the differing information presented to the visual and vestibular systems; and contextual factors, such as the variety of user types, and range of environmental conditions present in the workspace in which the technology is implemented (Nichols et al., 2000). While the prevalence and influence of some of these causative factors have been reduced by advances in display and control technology, large data sets (>300 participants) still suggest that approximately 5 percent of the population will experience effects so severe that they are unable to continue with using the technology.

In addition to VRISE, many of the other contextual complexity issues highlighted in the CWE case domain are also relevant here, in particular those associated with the appropriate implementation of VR technology and the way in which its introduction might alter traditional work patterns and communications. Returning to the training example, there is the crucial issue of at what stage of training VEs should be applied. VEs can be used to combine as well as extend traditional training methods by providing an interface to different training media (e.g., text, videos, and digital representation of a number of scenarios). As a computer-based technology, they can be used to support independent learning (as well as collaborative learning), enabling the trainee to access information and practice skills remotely, unless specialist equipment is required such as a large-screen projection or a haptic device. In this case, a specific training room or access to a training center may be required, which will then add to the cost of the training program.

The appropriate methods used to assess both technical and contextual complexity in VEs depend on constraints (e.g., time and resources available) and also on the specific VR or VE system or application. A large number of different evaluation approaches and methods are listed in D'Cruz and Patel (2006). These methods are grouped into seven main categories: general usability issues, user interactions and behaviors, user performance and outcomes, user experience, sickness and other effects, physical and social presence, and psychophysical and psycho-physiological measures.

An example of complex usability, and relevant methods, applied to VEs can be shown in our work on Sound and Tangible Interfaces for Novel Product Design (SATIN). SATIN (IST-5-034525) is an EC-funded project to develop a multimodal interface consisting of an augmented reality environment in which users are able to stereoscopically view 3D objects and both explore and modify the shape of these objects by directly touching them with their hands. This visual–haptic interface is supplemented in an innovative way by the use of sound as a means to convey subtle information (curve shape or curvature) and feedback about the virtual object and the user interaction. More specifically, this application of sonification enables the designer to explore geometric properties related to the surface of the object that are barely detectable by touch or sight. This example illustrates a particular aspect of technical complexity in the presentation of three types of display information (auditory, haptic, and visual) to the user simultaneously, and the need to capture the interaction with all of these senses. The contextual complexity for this case comes from the need to understand users (in this case, product designers), tasks, and the overall product design process, including how it might vary between sectors (e.g., designing a car body as opposed to designing a piece of glassware).

As a starting point for understanding the context of use for designers now and in the future, road-mapping exercises were conducted with SATIN end users and separately with around sixty people at the 2008 CREATE conference. With end users, road mapping focused on "design practice in the future" and "technology to support future design." In focus group–style sessions, end users debated these topics, facilitated by researchers who prompted discussion with key questions such as "How will designers use technology for design in the future?" and "What innovative methods of interaction will be developed?" At CREATE 2008, a range of experts in human–computer interaction for design—including academics, usability consultants, technology developers, and designers—were split into groups of eight and provided with modeling clay to encourage them to physically model the form or context of use in which future design technology may be applied. Participants were provided with some example products to consider like a cup and saucer, door handle, and washing machine. Both road-mapping activities resulted in broad agreement in the types of requirements that future technology should meet, and the types of features that would be desirable. As a result, a number of critical requirements for future systems were identified, such as minimal interfaces, more accurate representations of color and natural materials, "intelligence" in systems (e.g., real-time analysis of component integration, including aesthetic evaluation and retention of character), compatibility and standardization of file extensions, and the ability to easily generate Class A surfaces (currently not possible in polygon-based systems).

During development of the SATIN prototype, four usability evaluations were conducted. These were performed in an offline environment (i.e., the technology was not actively implemented in a company during the evaluation) so the experimenters had some influence over the tasks to be completed. In a similar approach to the "living labs" example in Case Domain 1, elements of the real-world context were replicated in the laboratory—for example, the products represented to the users were similar to those that might be used in their everyday work, and the users themselves were members of the organizations that were identified as future users of the technology (including household product designers and car exterior designers). User trial evaluations combined observations of users completing tasks, and success and failure rates as performance measures, with subjective questionnaires that assessed the participants' perceptions of, and attitudes toward, the technology. It was also important to apply methods less commonly used in usability evaluations such as the assessment of VRISE (Nichols et al., 2000) and also workload. The latter was specifically included following earlier observations that system users may approach a high level of workload when having to combine information from all three sensory displays in order to effectively understand the nature of the curve being presented—an example of the consequences of technical complexity. High workload levels may have been considered particularly unacceptable given the contextual complexity of the designers' tasks.

The third evaluation of the SATIN prototype followed a slightly different approach in that it was primarily based on heuristics, or design principles. Heuristics from Nielsen are viewed as especially powerful since they were derived from a large body of literature and established low-level guidelines (Greenberg, Fitzpatrick, Gutwin, & Kaplan, 2000). Since most established heuristics were borne from work on systems quite distinct from SATIN, primarily desktop-based graphical user interfaces, some reinterpretation was considered necessary. This emphasizes the point made earlier in this chapter that current heuristic approaches do not allow us to capture all of the necessary aspects of complex usability—therefore, they were adapted to address the technical and contextual complexity issues presented in SATIN. Final criteria were categorized under the following headings: intuitive interaction and realism; feedback; user focused; cognitive demand; effective design; error and help; and safety and comfort. A set of prioritized recommendations was produced following the evaluation, with the agreement that high-priority requirements should be resolved by the final evaluation, whereas low-priority requirements were those that would be required of a market-ready system.

In summary, the novel multimodal interaction style implemented by the SATIN project required usability researchers to adapt existing methods so they were appropriate to the system under investigation. We employed a number of established techniques, including interviews, focus groups, user trial evaluations, and heuristics. Each time, we paid particular focus to the novel characteristics of SATIN and extended and adapted the techniques to ensure that all appropriate aspects of technical and contextual usability were captured. Our success in applying these techniques is encouraging for usability practitioners since it demonstrates that, with appropriate adjustments, established practices can be made applicable to scenarios in which we encounter complex usability with very novel interfaces.

Case Domain 3: Sociotechnical Systems

Our final example is in the area of sociotechnical systems. Typically, these are safety-critical or command-and-control-type systems—examples include process and power control, air traffic control, and emergency dispatch (Stanton & Baber, 2006). Many of the underlying principles are also applicable to single-user or single-artifact systems, but this category of sociotechnical system offers unique challenges in terms of the complexity of the underlying domains that the operators are trying to control, and risks associated with error.

One area that we have been closely involved in is rail control, in collaboration with Network Rail, the UK's rail infrastructure provider. This includes not only conventional "signaling" (*train dispatch* in the United States) but also looking forward to future train control systems where an operator may use a high degree of automation and decision-making support to control large territories. There are not only both design and implementation challenges in modifying existing technology, but also more far-reaching challenges in the effective design of wholesale changes to the rail control system, and the roles involved (Wilson, Farrington-Darby, Cox, Bye, & Hockey, 2006).

Rail signaling provides an example of complex usability that contains elements of both technical and contextual complexity. While the actual technology used to control signals in the United Kingdom is not particularly complex in the structure of its individual elements—and, indeed, for purposes of resilience and reliability should almost be as simple and durable as possible—the workstation of the signaler is a combination of a large number of different information sources and controls. Signalers will often use simple artifacts to "distribute" their cognition, for example, in the use of text on magnetic strips placed on the display to act as temporary reminders. This characteristic of distributed cognition is typical of many domains—for example, the importance of the physical "flight strip" as a reference point for air traffic controllers (Berndtsson & Normark, 1999). In this case, technical complexity lies in the combination of interaction opportunities presented to the user and how, in some cases, the user may augment them or use them in unanticipated ways.

Rail control also provides a rich example of contextual complexity—signaling environments are collaborative, with a signaler being required to communicate with both those who are co-located and also those who are situated in remote locations. This interaction is often very subtle, with highly coded, or even non-verbal, cues being used by one signaler to guide the attention of another (Heath, Svensson, Hindmarsh, Luff, & Vom Lehn, 2002). Signalers will also engage in "active overhearing" of what is going on around them, to stay informed of events, which is typical of other domains involving many agents, such as emergency dispatch (Blandford & Wong, 2004). In addition, the safety-critical and time-critical nature of the operating environment presents a level of pressure and motivation to the operator that is likely to influence how he or she chooses to complete tasks. Finally, rail signaling provides a clear example of a situation where the "task" is only one part of the "job" (Golightly et al., 2009). While a main role of the signaler is to interact with the signaling interface to set routes or points to ensure that trains follow the correct path, in fact, this task is only one part of the signaler's job—other elements of the job include communication, planning, and record keeping, for example.

The characteristics of rail control as a sociotechnical system, and the complexity of the usability issues involved, highlight a number of methodological issues. We have experienced many of these within a number of specific human factors technology development and assessment projects. Some of these have looked at the introduction of new supporting systems for signalers and new forms of automation, while others have been evaluating procedural change and how well the existing interfaces support these new working practices.

First, the distributed nature of the system across artifacts means that changing one aspect of the system may have implications for other tasks. Each individual artifact may have its own technical complexity, and this will be amplified by the pressure of the safety-critical context in which it is embedded. For example, the usability issues associated with a new form-filling interface for signalers became insurmountable when placed within the context of a real signal box (Balfe et al., 2009). Piloting and comparison of work activity pre- and post-task have proved a useful way of assessing the impact of new technologies. As part of our work to study the impact of signaling automation, we have developed categorizations of signaler activity that we can use for counting up how signalers exhibit planning, monitoring, and interactive behavior (Sharples, Balfe, Golightly, & Millen, 2009). From this, we can see if signalers change their behavior as a result of automation— early evidence is that the signalers switch from an active monitoring of the system to a much more passive involvement. This is indicative of a high level of trust and, at least in part, the usability of the automated system.

We can also supplement these quantitative observations with subjective measures (e.g., of perceived workload) and also with qualitative comments from the signalers themselves. As an example of the value of this, we see variations in behaviors between signal boxes. Clarification with comments from signalers shows that local context in terms of traffic conditions but potentially also local attitudes (an example of local working culture) influence whether signalers trust the automation, or choose to manually intervene. These data show that the technical and contextual complexity of the system can clearly affect the way in which a signaler completes his or her task, and that it is important to ensure that any methods applied are of appropriate breadth so that the range of these issues are captured.

This brings us to the issue of understanding and describing work that is distributed across actors, whether these are colleagues in the signal box, or drivers, station staff, and the like out in the world. The sheer complexity of the rail context makes it difficult to re-create a "living lab" approach to rail signaling; therefore, the most appropriate approach is often to apply qualitative approaches in situ. Examples in the literature include Roth, Multer, and Rasler (2003), who carried out an extensive ethnographic and interview-based study of rail control in the United States to understand the implications of open-channel communications. This approach revealed key characteristics of the signaler task, particularly the role of active overhearing to help support proactive train planning. Similarly, Heath et al. (2002) used an ethnographic and content analysis–type approach to tease out the subtle nature of cues used within a metro control signal box to guide behavior and information exchange. Through these approaches we are able to tackle contextual complexity arising from multiple agents working together, for instance in recent work to understand how the sociotechnical system as a whole (including the role of multiple interfaces) supports situation awareness (Golightly et al., 2009).

The most pressing methodological concern is that we are considering a safety-critical system, and technology must be supportive to the operator, even (perhaps especially) in nonroutine situations. The difficulty here is designing and evaluating systems to perform in contexts that may not be anticipated, not only in terms of events in the world but also in terms of creative and ad hoc ways that operators choose to interact with systems. To address this, we have recently looked to cognitive work analysis (Vicente, 1999) as a holistic approach to understanding the constraints that shape sociotechnical systems. In particular, we have been using work domain analysis, which results in an abstraction hierarchy of the work domain. Space prevents a detailed discussion of this approach here (see Lintern, 2009, for an excellent introduction), but an example is shown in Figure 13.3. This illustrates that the entire work domain, from high-level purpose down to physical artifacts used to execute the work, can be placed together in a single representation. As a result, it is possible to consider the wider implications of removing, changing, or adding a piece of the domain—not just physical artifacts but also potentially changing work functions or changing work priorities. These are the kind of changes we may need to understand when dealing with wholesale changes to the rail control role, and they are just as applicable to evaluation and training needs as they are to design.

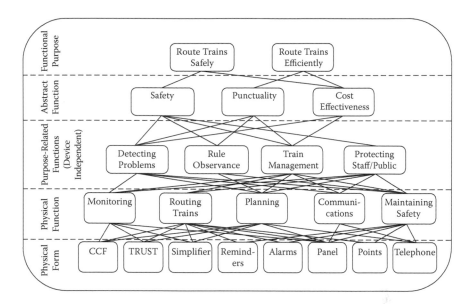

FIGURE 13.3

An abstraction hierarchy example for rail signaling. (From Millen, L., Sharples, S., & Golightly, D. (2009). *Proceedings for the Conference in Rail Human Factors, 2009,* Lille, France. Nottingham, UK: Human Factors Research Group.)

The Implications of Complexity

The three cases presented in this chapter demonstrate two points:

1. They illustrate the importance of both technical and contextual complexity in three different technology contexts—collaborative work environments, virtual environments, and sociotechnical systems.

2. They provide example methods that can be applied to examine the human factor issues associated with complex usability. Because of the inherent complexity, it is often necessary to use multiple methods in combination.

As we have demonstrated, complexity takes on both a technical and contextual dimension. For example, even though the user interface challenges of the SATIN project were highly technical, these challenges had to be addressed from the perspective of the role of the technology within a design task (i.e., the context). That said, we would be as unwise to focus on context at the expense of low-level usability as we would be to emphasize technical usability at the expense of context. In the case of VR systems, it is only through efforts to improve basic usability (e.g., dealing with the problems of wearing headsets, or system response) that the technology is now viable to the point where we have to consider context more closely. Repeatedly in our work, we find that even the best-conceived system that fully appreciates users and their contextual needs can fail if the fundamentals of the user interface are not well implemented.

A number of methods have been discussed, and across all three case domains we place importance on using multiple methods to address usability. We would emphasize that there is no one single method that can be used to ensure usability, and this is particularly apparent in cases of complex usability. Because complexity may exist from a number of different sources, it is necessary to sometimes use different approaches to unravel all the various issues that might lie within. Also, each method has its own strengths and weaknesses. For example, qualitative interviews are open to subjectivity on the part of the usability specialist, whereas more quantitative work is only as valuable as the reliability and validity of the data being collected. Triangulating across a number of methods can improve accuracy and enables us to build a fuller understanding of the most pertinent issues. Also, earlier exploratory approaches can be useful to target issues that need more detailed examination later.

Implicit within this is the need to use different methods at different times. This includes a need to move usability activities from being a "testing" activity (the end-of-line problem) through to being a "requirements-gathering" activity, much earlier in the process. If this is to be achieved, the complex usability methods need to generate outputs that dovetail into the development activities. Practical experience of working in large-scale development projects highlights that usability specialists rarely work directly with developers but, instead, have a more fruitful relationship with marketing functions that are tasked with defining product needs, and business analysts tasked with defining the functional makeup of a product. Also, as we note in the case of the model of CWEs used in CoSpaces, there is an onus on us to transform and present an often vast body of data into something that is directly relevant for other stakeholders. We need appropriate representations to do this, and forms such as use cases, scenarios, personas, and style guides are valuable channels for communicating user needs to other stakeholders in the development process. Professional communicators can make a specific contribution to this activity, as they can often express these user requirements both linguistically and visually in a clear and meaningful manner.

We have touched on the contribution that culture can play in understanding users and user requirements and emphasized the importance of users when considering contextual complexity. Culture is usually seen in terms of geography—how people may have different attitudes or expectations based on their local culture. A powerful example of this is the use of colors, where different colors may suggest different meanings (e.g., red means danger for Europeans but signifies luck in China; Madden, Hewett, & Roth, 2000). Iconography must also be international (Wang, 2008), and one motivation for using icons is that it reduces the need to localize the interface text for each territory where a product may be sold or implemented. The examples we have discussed in this chapter emphasize the importance of local working culture and practice in addition to national culture. For example, a company that regularly uses videoconferencing may have a different attitude (not necessarily positive) to the installation of a CWE than a company with little experience of such technologies.

Our experience of working on international projects (either EU research projects, or for multinational corporations) has shown us there is also a cultural angle to how a usability specialist carries out work, and how he or she applies methodologies (Golightly et al., 2008). For example, certain companies may be more than happy to let real end users—the operators of technology—talk to usability specialists about their needs and experiences. Other organizations may only do so through a representative (often a member of management) who acts (often unreliably) as a proxy for the real users. It may also be difficult to get a clear and consistent understanding of usability terminology when crossing language barriers. Can technical partners who do not have English as a first language understand what our methods are, particularly when, for example, self-report questionnaires are written in English? Can these partners understand what our outputs mean?

Conclusion

Fundamentally, human factors and usability professionals and professional communicators share a common aim in that we are both committed to developing user experiences that are safe, effective, and engaging. In this sense, we often work in the same "space" between users and developers, employing many comparable and overlapping techniques; and, ultimately, we are all engaged in "user-centered design." As part of this user-centered design, we seek out and utilize a detailed understanding of who our users are—their needs, abilities, and limitations. This issue is more acute in the case of complex usability. We have demonstrated here that complexity has a number of aspects that need to be managed—communicating the technical complexity of the technology, and understanding the complexity of the context in which users engage with the technology. We also find that the multiple methods, often qualitative, are required to capture knowledge of users. As technological advances continue and the scope of application areas potentially grows deeper and wider, we believe complexity for the usability specialist is likely to increase, not decrease. Our hope is that by applying some of the practices described in this chapter, this complexity will be better understood and not passed on to the user.

Reader Take-Aways

- Complex usability has two dimensions—a technical and a contextual dimension. The two are interdependent, and both are of high importance to usability specialists and the successful implementation of systems that are relevant to complex usability.

- It is unlikely that a single method will be sufficient to capture all aspects of technical and contextual usability. If methods are used in combination to triangulate across the different aspects of complexity, then we are more likely to identify any critical issues in enough time to address them.

- We need to ensure that new methods are generalizable across applications or domains—current methods tend to be specific and interface focused—and new methods must acknowledge both the technical and contextual complexity of systems. Conversely, one size does not fit all—in many cases, it may be necessary to adapt methods to a specific context in response to issues identified. Therefore, instead perhaps of a single tool set to be applicable to all technology types, it would be more useful to have a toolbox from which we can select tools appropriate to the complex usability issues that we encounter.

- Real-world data are very valuable, but can sometimes be difficult or impossible to obtain. In addition, the value of laboratory or controlled studies means that quantitative and statistical comparisons are often more practicable. The data source needs to be considered carefully on the basis of the eventual requirements of the data analysis, as well as take into account practical, safety, and organizational constraints.

- The involvement and communication with all stakeholders, especially the marketing and business analyst roles, are essential to ensure that complex usability issues are appropriately addressed. Methods to address complex usability challenges will often lead to complex data sets that require management and careful communication. This needs to be considered if these detailed methods are to be effectively integrated into the design process.

Acknowledgments

We would like to thank those former and current members of the Human Factors Research Group who have contributed to the work presented here. In particular, we would like to thank Jacqueline Hollowood for her work in the SATIN project.

References

Baker, K., Greenberg, S., & Gutwin, C. (2001, May). Heuristic evaluation of groupware based on the mechanics of collaboration. Paper presented at the 8th IFIP Working Conference on Engineering for Human-Computer Interaction (EHCI'01), May 11–13, Toronto.

Balfe, N., Lowe, E., Abboud, R., Dadashi, Y., Bye, R., & Murphy, P. (2009). Signaller forms: The ultimate irony of automation. In *Proceedings for the Conference in Rail Human Factors, 2009*, Lille, France [CD-ROM]. Nottingham, UK: Human Factors Research Group.

Barnum, C., Bevan, N., Cockton., G., Nielsen, J., Spool, J., & Wixon, D. (2003). The "Magic Number 5": Is it enough for Web testing? In *Proceedings of CHI '03* (pp. 698–699). New York: Association for Computing Machinery.

Berndtsson, J., & Normark, M. (1999). The coordinative functions of flight strips: Air traffic control work revisited. In *Group '99: Proceedings of ACM Siggroup Conference on Supporting Group Work* (pp. 101–110). New York: Association for Computing Machinery.

Blandford, A., & Wong, B. L. W. (2004). Situation awareness in emergency medical dispatch. *International Journal of Human-Computer Studies, 61*(4), 421–452.

Cockton, G., & Woolrych, A. (2002). Sale must end: Should discount methods be cleared off HCI's shelves? *Interactions, 9*(5), 13–18.

D'Cruz, M. D. (1999). Structured evaluation of training in virtual environments. PhD thesis, University of Nottingham, UK.

D'Cruz, M. D., & Patel, H. (2006, January). Evaluation methodologies review. *INTUITION (IST-NMP-1-507248-2) Milestone M1.10_1*.

Diamon, T., Nishimura, M., & Kawashima, H. (2000). Study of drivers' behavioral characteristics for designing interfaces of in-vehicle navigation systems based on national and regional factors. *JSAE Review, 21*(3), 379–384.

Ereback, A. L., & Hook, K. (1994). Using cognitive walkthrough for evaluating a CSCW application. In *Proceedings of ACM CHI 1994* (pp. 91–92). New York: Association for Computing Machinery.

Garbis, C., & Artman, H. (2004). Team situation awareness as communicative process. In S. Banbury & S. Tremblay (Eds.), *A cognitive approach to situation awareness: Theory and application* (pp. 275–296). Aldershot, UK: Ashgate.

Golightly, D., Stedmon, A., Pettitt, M., Cobb, S., Sharples, S., Patel, H., et al. (2008). English, Slovenian, Swedish: Capturing user requirements on large projects. In P. Bust (Ed.), *Contemporary ergonomics 2008*. Aldershot, UK: Ashgate.

Golightly, D., Wilson, J. R., Lowe, E., & Sharples, S. C. (2009). The role of situation awareness for understanding signalling and control in rail operations. *Theoretical Issues in Ergonomics Science, 11*(1–2), 84–98.

Greenberg, S., Fitzpatrick, G., Gutwin, C., & Kaplan, S. (2000). Adapting the locales framework for heuristic evaluation of groupware. *Australian Journal of Information Systems, 7*(2), 102–108.

Gutwin, C., & Greenberg, S. (2000, June). The mechanics of collaboration: Developing low cost usability evaluation methods for shared workspaces. In *Proceedings of IEEE 9th International Workshop on Enabling Technologies: Infrastructure for Collaborative Enterprises (WETICE '00)*, June 14–16, Gaithersburg, MD (pp. 98–103). Piscataway, NJ: IEEE.

Heath, C., Svensson, M. S., Hindmarsh, J., Luff, P., & Vom Lehn, D. (2002). Configuring awareness. *Computer Supported Cooperative Work, 11*, 317–347.

Henneman, E. A., Lee, J. L., & Cohen, J. I. (1995). Collaboration: A concept analysis. *Journal of Advanced Nursing, 21*, 103–109.

Hutchins, E. (1995). *Cognition in the wild*. Cambridge, MA: MIT Press.

Johnson, R. R., Salvo, M. J., & Zoetewey, M. W. (2007). User-centred technology in participatory culture: Two decades "beyond a narrow conception of usability testing." *IEEE Transactions on Professional Communication, 50*(4), 320–332.

Lintern, G. (2009). *The foundations and pragmatics of cognitive work analysis: A systematic approach to design of large-scale information systems*. Retrieved from http://www. cognitivesystemsdesign.net/home.html

Madden, T. J., Hewett, K., & Roth, S. (2000). Managing images in different cultures: A cross-national study of color meanings and preferences. *Journal of International Marketing, 8*(4), 90–107.

Mattessich, P. W., & Monsey, B. R. (1992). *Collaboration: What makes it work? A review of research literature on factors influencing successful collaboration*. St Paul, MN: Amherst H. Wilder Foundation.

Meads, G., Ashcroft, J., Barr, H., Scott, R., & Wild, A. (2005). *The case for interprofessional collaboration: In health and social care*. Oxford: Blackwell.

Millen, L., Sharples, S., & Golightly, D. (2009). The application of cognitive work analysis in rail signalling. In *Proceedings for the Conference in Rail Human Factors, 2009*, Lille, France [CD-ROM]. Nottingham, UK: Human Factors Research Group.

Montiel-Overall, P. (2005). Toward a theory of collaboration for teachers and librarians. *School Library Media Research, 8*. Retrieved from http://www.ala.org/ala/aasl/aaslpubsandjournals/slmrb/slmrcontents/volume82005/theory.htm

Neale, D. C., Carroll, J. M., & Rosson, M. B. (2004). Evaluating computer-supported cooperative work: Models and frameworks. In *Proceedings of CSCW '04*, November 6–10, Chicago (pp. 112–121). New York: Association for Computing Machinery.

Newell, A., & Simon, H. A. (1972). *Human problem solving*. Englewood Cliffs, NJ: Prentice Hall.

Nichols, S., Ramsey, A. D., Cobb, S., Neale, H., D'Cruz, M., & Wilson, J. R. (2000). *Incidence of virtual reality induced symptoms and effects (VRISE)*. HSE publication 274/2000.

Nielsen, J. (1992). Finding usability problems through heuristic evaluation. In *Proceedings CHI 1992* (pp. 372–380). New York: Association for Computing Machinery.

Patel, H., Stefani, O., Sharples, S., Hoffmann, H., Karaseitanidis, I., & Amditis, A. (2006). Human centred design of 3D interaction devices to control virtual environments. *International Journal of Human Computer Studies, 64*, 207–220.

Pirelle, D. & Gutwin, C. (2001). Group task analysis for groupware usability evaluations. In Proceeding of IEEE 10th International Workshop on Enabling Technologies: Infrastructure for Collaborative Enterprises, June 20-22, Cambridge, MA; IEEE Computer Society, 102-107.

Redish, J. (2007). Expanding usability testing to evaluate complex systems. *Journal of Usability Studies, 2*(3), 102–111.

Roth, E. M., Multer, J., & Rasler, T. (2003). Shared situation awareness as a contributor to high reliability performance in railroad operations. *Organization Studies, 27*(7), 967–987.

Schrage, M. (1990). *Shared minds: The new technologies of collaboration.* New York: Random House.

Sharples, S., Balfe, N., Golightly, D., & Millen, L. (2009). Understanding the impact of rail automation. In *Proceedings of HCI International*, San Diego, CA. New York: Springer.

Sharples, S. C., D'Cruz, M., Patel, H., Cobb, S., & Wilson, J. (2006). Human factors of virtual reality: Where are we now? In *Proceedings of IEA*, July 10–14, Maastricht, the Netherlands, Amsterdam: International Association for the Evaluation of Educational Achievement.

Stanton, N. A., & Baber, C. (2006). Editorial: The ergonomics of command and control. *Ergonomics, 49*(12–13), 1131–1138.

Suchman, L. (1987). *Plans and situated action: The problem of human-machine communication.* Cambridge: Cambridge University Press.

Vicente, K. (1999). *Cognitive work analysis: Toward safe, productive, and healthy computer-based work.* Mahwah, NJ: Lawrence Erlbaum.

Wang, H.-F. (2008). The cultural issues in icon design. In P. Bust (Ed.), *Contemporary ergonomics 2008.* Aldershot, UK: Taylor & Francis.

Wilson, J. R. (2006). Collaboration in mobile virtual work: A human factors view. In J. H. E. Andriessen & M. Vartiainen (Eds.), *Mobile virtual work: A new paradigm?* (pp. 129–151). Berlin: Springer.

Wilson, J. R., & D'Cruz, M. (2006). Virtual and interactive environments for work of the future. *International Journal of Human Computer Studies, 64*, 158–169.

Wilson, J. R., Farrington-Darby, T., Cox, G., Bye, R., & Hockey, G. R. J. (2006). The railway as a socio-technical system: Human factors at the heart of successful rail engineering. *Journal of Rail and Rapid Transport, 221*(1), 101–115.

Wilson, J. R., Patel, H., & Pettitt, M. (2009, April). Human factors and development of next generation collaborative engineering. In *Proceedings of the Ergonomics Society annual conference*, London, April 22–23. London: Taylor & Francis.

14

Information Usability Testing as Audience and Context Analysis for Risk Communication

Donna J. Kain
East Carolina University

Menno de Jong
University of Twente

Catherine F. Smith
East Carolina University

CONTENTS

Introduction 306
Communicating about Risks and Emergencies 307
    Crisis and Emergency Risk Communication (CERC) 308
    Audiences for Risk and Emergency Information 310
Usability Evaluation of Documents 311
    Feedback-Driven Audience Analysis 312
Testing with the *Hurricane Survival Guide* 313
    Pilot Study Participants 314
    Document Usability Evaluation Protocol 315
    Document Analysis 316
        Comments about Use 318
        Comments about Design and Writing 320
        Comments about Information 323
Conclusion: Document Usability Evaluation as Audience Analysis 328
Reader Take-Aways 329
References 330

Abstract

Guidance on professional "best practices" for risk and emergency communication includes advice to "know the audience." But many factors influence audiences' reception of information about risks in complex contexts. These include experience, level of concern, and social norms as well as the information they receive. Typical survey methods used to gauge the public's understanding of risks and emergency communications may overlook crucial insights about how audiences interpret information and make judgments. We discuss using "plus-minus" document-based usability evaluation to examine attitudes and needs of residents in a storm-prone area with the goal of developing an empathetic approach to providing information.

Introduction

In this chapter we do not discuss a large, complex, computer-based information system but rather a fairly ordinary print document that explains what people need to know and do when their region is threatened by a hurricane. The complexity we address is that the brochure functions within a dynamic system of information and interactions in which actors share the overarching goal of safety but respond from different interpretations of risks and different vantage points in the context. For the emergency management professionals who created the document and who are responsible for the safety of residents in their local jurisdictions, the primary purpose of emergency preparedness information is to encourage people to make efficacious choices, often quickly, in potentially dangerous situations. While this may seem like a straightforward information need, the reality is that official messages intended for people in harm's way may not reach them, may not be clear or convincing, and may conflict with other sources and residents' existing knowledge, attitudes, and beliefs. Consequently, developing effective usable information products for conveying risk and emergency messages requires that writers and designers understand not only the information needs of their audiences but also the problems and concerns that influence audiences' reception and use of information, whether print or electronic.

We suggest that useful audience analysis can be accomplished through document usability evaluation. Generally, this form of testing is conducted as part of a design and development process to improve specific messages before they are provided to audiences. However, testing can also serve as a research tool to find out about the audience, provide opportunities for audience input, and help information designers to cultivate empathy by engaging with representative audience members. To illustrate this approach, we report on one part of a pilot study of reception and interpretation of risk and emergency information about hurricane preparedness. As part of the study, we conducted document-based interviews using the "plus-minus" evaluation method (de Jong & Rijnks, 2006; de Jong & Schellens, 1997, 2001). While the interviews provided useful information about the particular document we tested, our primary purpose in conducting the tests was to gain perspective about audience and context that could be applied more generally to improve risk and emergency information for this population.

Our pilot study was conducted in 2007 in one coastal county (referred to hereafter as the County) of North Carolina. The North Carolina coast is a high-risk area for damage from hurricanes, tropical storms, and the winter storms known there as *nor'easters*. The County has coastal areas on both the Atlantic Ocean and a large sound and includes barrier islands, protected seashore areas, many square miles of estuaries, and popular resort locations. The population more than doubles during the peak tourist season, which coincides with the Atlantic hurricane season that runs from June 1 to November 30 annually. Seven other counties located along North Carolina's Atlantic coast, several much more heavily populated, are similarly at risk from the landfall of serious storms. Communication about severe weather presents an ongoing, cyclical challenge for communicators in this area because the local population as well as visitors must be prepared each year for the potential of a major storm.

To provide background for our discussion of the pilot study, we first review crisis and emergency risk communication and the needs of audiences using this type of information. We then review the literature pertaining to usability evaluation of documents that informs our research, describe the pilot study we conducted, and discuss our findings. We conclude by suggesting ways that usability evaluation of documents can serve as audience analysis that provides rich information about audiences' information needs and situational concerns.

Communicating about Risks and Emergencies

Recent costly natural disasters in the United States, including hurricanes Isabelle (2003), Katrina (2005), Rita (2005), and Ike (2008); the 2008 Midwest floods; and the 2009 California wildfires call attention to the necessities and challenges of providing information to the public about risks from natural hazards and messages about emergency preparedness and emergency response. For example, people who may be evacuated from their homes need to know how to prepare and stay informed, when to leave, where to go, and what to expect when they return. If the information people receive is not adequate because it fails to address their needs, concerns, and perceptions of the risks, residents may not leave the area, thus exacerbating a dangerous situation. Inadequate, late, callous, and overly condescending communication may also cause audiences to distrust sources of the information and subsequent messages from them.

Crisis and Emergency Risk Communication (CERC)

The types of communication that natural hazards such as severe weather events invoke are identified by the Centers for Disease Control (CDC), Federal Emergency Management Administration (FEMA), and other government agencies as *crisis and emergency risk communication* (CERC). Definitions of CERC conflate several key terms—*crisis, emergency, risk,* and *disaster*—that each warrants some brief definition (see also CDC, 2002; Lundgren & McMakin, 2004; Seeger, 2006).

Whether the issues are health, environmental, or natural hazards such as hurricanes, communicating with the public about risks, crises, emergencies, or disasters is a complex process. People first need information about their risks—the type, severity, probability, and potential consequences of hazards that may affect them. Risk is the likelihood or probability that a specific event, decision, or behavior may result in a negative outcome or have a negative impact on individuals or populations. Experts understand risk as a measurement of potential loss from hazards and adverse occurrences or from decisions that are made relative to them (Luhmann, 1993; Renn, 2008; Sjöberg, 2000; Taylor-Goolby & Zinn, 2006). Risk communication, which developed rather "arhetorically" in the fields of "communication, cognitive psychology, and risk assessment" (Grabill & Simmons, 1998, p. 416), is aimed at developing and providing information about the risks of hazards to the public and other stakeholders engaged in evaluating and making decisions about potential harm. But risk information generated by experts that relies on scientific, technical, and probabilistic information is not well understood by the public (Gigerenzer et al., 2005; Handmer & Proudley, 2007; Keller, Siegrist, & Gutscher, 2006). Nor are these the only types of information that should be considered in determining risks and communicating about them. Consequently, goals for risk communication researchers and practitioners have shifted over the last twenty years to incorporate, or at least recognize, the affective ways that people think about, respond to, and make decisions about risks.

While risk communication focuses on conveying the probabilities and forecasting consequences—often with the goal of changing behavior—crisis, emergency, and disaster communication focuses on actual events in motion. Though sometimes used together or interchangeably, particularly in preparedness literature aimed at the public, the terms *crisis, emergency,* and *disaster* signify differences in scale, scope, and response type. *Crisis* has generally been associated with reputation management by businesses, organizations, and other groups that have some responsibility for a negative event—either they cause or find themselves dealing with a situation that endangers or causes harm to others (e.g., the Exxon *Valdez* oil spill or the Union Carbide chemical plant accident in Bhopal, India). Emergencies are foreseeable, dangerous events that have the potential to cause serious, localized damage but for which local emergency management agencies are equipped to cope (e.g., fires, localized flooding, and moderate hurricanes).

Disaster generally refers to emergency events that overwhelm the resources of local emergency management, thus requiring the involvement of state, regional, and sometimes national agencies and for which the severity of some consequences are unforeseen (e.g., the terrorist attacks of September 11, 2001, or Hurricane Katrina).

Crisis and emergency risk communication focuses on providing information about events to stakeholders, the media, and the public as quickly and accurately as possible; managing resources and logistics on the ground safely; and instructing the public about life- and property-saving actions they should take. In these situations, events are often serious and the situation uncertain and rapidly changing. People in harm's way can be afraid, confused, and sometimes cut off from information. For communicators, Peter Sandman (2006) argues,

> When people are appropriately concerned about a serious hazard, the task is to help them bear it and to guide them through it. This is the true paradigm of crisis communication. In a crisis, people are genuinely endangered and rightly upset. (p. 257)

Consequently, in areas that are likely to face extreme conditions, effective and empathetic risk and preparedness information is crucial.

Emergencies can become disasters and crises when infrastructures are overwhelmed or when the magnitude of an emergency is unanticipated. The boundaries between risk, emergency, crisis, and disaster are often blurred. Hurricane Katrina serves as a case in point. In addition to the weather emergency becoming a disaster when levees broke on Lake Ponchartrain, Katrina also constituted a crisis for a number of agencies involved in trying to deal with the disaster and its aftermath. Many of the problems that led to the disaster were, in fact, foreseeable. However, the severity of events coupled with inadequate management led to consequences that had not been imagined. The actions, or perceived lack of action, by officials and the poor quality of communication with the people impacted were notable for the apparent lack of empathy (see, e.g., Cole & Fellows, 2008).

As with risk communication, one aspect of CERC that can cause problems for the public is overreliance on expert perspectives in developing and providing information. For example, the CDC (2002) includes the following characteristics in its definition of CERC:

> [C]ommunication [that] encompasses the urgency of disaster communication with the need to communicate risks and benefits to stakeholders and the public.
> [E]ffort by experts to provide information to allow an individual, stakeholder, or an entire community to make the best possible decisions about their well-being within nearly impossible time constraints and help people ultimately to accept the imperfect nature of choices during the crisis.

[E]xpert opinion provided in the hope that it benefits its receivers and advances a behavior or an action that allows for rapid and efficient recovery from the event. (p. 6)

Approaches to communication that center on expert analysis and decisions assume that if people are provided with accurate information, they will understand the problems and follow instructions. They will interpret the situation as it is described to them and make the "right" choices—this is the rational decision model. If the public fails to understand or draw the same conclusions as the experts, then it is assumed they need additional information or education. Communication models that privilege the knowledge of experts over that of the public tend to discount the knowledge that the public actually has. A shortcoming to CERC, as Sandman (2006) recognizes, is that while it "focuses on the need for sources to share what they know, not mentioned is the possibility that sources need to learn what the public knows" (p. 260).

Morgan, Fischhoff, Bostrom, and Atman (2002), in advancing the idea of a mental models approach to risk communication, acknowledge that expert models of risk provide the foundation for risk communication. But expert models often fail to account for factors that the public may include in their decision-making processes. Thus the mental models of audiences, primarily nonexperts in risk and emergency, must also be integrated into effective risk and emergency communication efforts.

Audiences for Risk and Emergency Information

As Albers points out, "[D]ifferent groups of people come to understand a situation differently using different mental routes and assigning different priorities to information elements" (see this volume, Chapter 6). When the audience is "the public," the diversity of audience members and their mental models pose a particular level of complexity. It might be tempting for those preparing risk and emergency information to assume that "the public" is just too large to analyze, or that the audiences' mental models are similar to the communicators', especially when they all live in the same area and experience the same situations. But people make judgments about risk and make decisions about actions in response based on a number of factors including the level of media focus on an issue, trust in information and in the sources providing information, risk–benefit trade-offs, proximity (the closeness of personal experience to a risk), level of uncertainty, and place-based considerations (Baron, Hershey, & Kunreuther, 2000; Grabill and Simmons, 1998; Masuda & Garvin, 2006; Paton, 2007; Ropeik & Slovic, 2003; Rosati & Saba, 2004; Sapp, 2003).

In addition, information about risks and adverse events circulates among audiences from different groups, including people who are directly and indirectly affected by a risk, the media, business communities, public officials, volunteer groups, emergency management personnel, and residents (Haddow & Haddow, 2008; Ward, Smith, Kain, Crawford, & Howard, 2007). Though these groups share the objective of understanding and mitigating risks and managing in emergencies, they have different information needs that depend on their involvement and roles, knowledge and expectations, language, literacy skills, access to information, and other factors (Grabill & Simmons, 1998; Lazo, Kinnell, & Fisher, 2000; Riechard & Peterson, 1998). Moreover, as Masuda and Garvin (2006) point out, people understand risks and emergencies in localized ways, "as embedded in social context" and "situated within the social experiences and interactions of individuals, groups, and institutions" (p. 439).

This is the case for the location and population about which we conducted our pilot study. Coastal areas that are subject to tropical weather systems continually face the threat of storms and related conditions such as flooding and erosion. Thus, the need for risk communication is ongoing and the need for emergency—and possibly disaster—information is periodic, yet inevitable. Emergency management professionals in areas under periodic threat from natural hazards maintain a constant state of readiness, and the public must have ready access to usable and useful information.

Usability Evaluation of Documents

A number of resources about developing and disseminating risk and emergency information recognize the systemic complexities of risk and emergency communication contexts and provide guidance on communicating with various audiences (e.g., CDC, 2002; Haddow & Haddow, 2008; Lundgren & McMakin, 2004; Renn, 2008; Seeger, 2006). As part of planning risk and emergency communication, most of these sources advise that communicators undertake audience analysis by collecting demographic information about the audiences using surveys, conducting focus groups, pretesting messages, and testing messages with users to better understand audiences' information needs (Lundgren & McMakin; Renn; Brown, Linden-berger, & Bryant, 2008). In planning and testing messages, Lundgren and McMakin, for example, advise that the "best way is to go out and talk to your audiences, actually meet them face to face" (p. 129). They also note, however, that communicators often fail to take this step with the public because of the dispersion and numbers of people involved, the costs and time necessary, the potential hostility of the audience depending on the issue, and communicators' own reluctance to meet with audiences (p. 133). While we agree that audience analysis and message testing are essential for effective risk and emergency messages, we suggest that survey methods, focus groups, and document comparisons typically employed may overlook crucial insights about audiences' attitudes and concerns that might help communicators develop the empathy necessary to design effective information for target audiences. *Empathy*, in this sense, means a practice that reflects concern for audiences' experience and needs from their perspective.

Feedback-Driven Audience Analysis

The long-standing tradition in all communication subdisciplines suggests that conducting formative evaluation research optimizes the quality, user-friendliness, and/or effectiveness of functional documents. Several studies have demonstrated that collecting and using reader feedback is an important step toward developing effective written materials (de Jong, 1998; Schriver, 1997; Swaney, Janik, Bond, & Hayes, 1999). Yet it appears difficult for communication professionals or technical writers to anticipate the needs and preferences of document users. Studies by de Jong and Lentz (1996), Lentz and de Jong (1997), and Schriver (1997, pp. 452–455), for instance, showed that experts cannot accurately predict the results of a reader-focused evaluation. The underlying problem is a lack of *empathy*—the willingness and ability of writers and reviewers to fully understand the motives, needs, feelings, prior knowledge, skills, and context of the target audience (de Jong & Lentz, 2007). Conducting a formative evaluation study is a way of repairing design problems in specific documents caused by such a lack of empathy. For people communicating with the public in crisis, risk, and emergency situations, knowledge of the audience and empathy are essential (Seeger, 2006).

Although the usefulness of reader-focused evaluation for the effectiveness of specific documents is virtually undisputed, it is important to also consider its limitations. A significant limitation was mentioned by Weiss (1988), who argued that the benefits of usability testing are overestimated. Especially when evaluation data are gathered toward the end of a design process, there is a substantial risk that the most severe and fundamental user problems are already "wired in" and it is too late to correct them. In more general terms, formative evaluation may help to optimize documents within the boundaries of the overall strategy chosen, but may not be strong enough to put the overall strategy itself to the test. Writers may thus end up fine-tuning a suboptimal solution. This drawback seems plausible, but there are no studies available to confirm or refute this potential pitfall.

The benefits of reader-focused evaluation, however, do not necessarily have to be limited to the optimization of specific documents. Studies by Schriver (1992) and Couzijn (1999) showed that technical writers may be expected to learn a lot about effective written materials when they are exposed to detailed reader feedback on a regular basis. When writers are confronted with specific problems that readers experience while using a document, they will be better equipped to empathize with the audience, anticipate their needs and preferences, and predict the problems they may experience. This learning process may be general, but it is more likely that it will be situational—for a specific type of document (e.g., user manuals or forms), a specific type of subject matter (e.g., government subsidies or alcohol), and/or a specific type of audience (e.g., adolescents or seniors). The specific reactions of readers to functional documents may help to develop tacit or explicit knowledge about the audience in writers.

Consequently, the document is used as a means to explore the knowledge, attitudes, skills, and/or behaviors of individual people in the target audience. The evaluation is not intended to optimize a specific document, but to learn more about the audience. From the perspective of document design processes, the reader-focused evaluation is not used as a method of formative evaluation but as a method of predesign research, aimed at supporting design specifications for policy and communication in the future. Schriver (1997) distinguished between three types of audience analysis: (1) *classification-driven audience analysis*, based on general characteristics of the audience; (2) *intuition-driven audience analysis*, in which an imagined reader is construed; and (3) *feedback-driven audience analysis*, based on detailed observations of the ways that people from the target audience actually use documents. An important drawback of classification-based audience analysis is that the step from results to an actual document is often too large. A drawback of intuition-driven audience analysis is that there is no way of knowing that the imagined readers truly represent the target audience.

Feedback-driven audience analysis appears to be the most promising of the three options. The research described in this chapter can be seen as a specific type of feedback-driven audience analysis. In addition to examining how well the residents we talked to understood the information in the guide, our use of the plus-minus interview technique allowed us to learn about how and why residents make decisions about risks from severe weather events, how they perceive the risks posed by storms, and how they evaluate the trustworthiness of messages and messengers.

Testing with the *Hurricane Survival Guide*

In the pilot study that we report on here, we focused on an example of risk communication to understand the residents' responses to the information and their perceptions of the risks from storms. General information about emergency preparedness is available from many sources, including the U.S. Department of Homeland Security (through FEMA), the CDC, the state of North Carolina (Ready N.C.), nonprofit organizations such as the Red Cross, the media, and local emergency management. But information that people need most to make decisions about risks and to cope with emergencies is location specific. Even more specialized information is necessary for residents with medical needs and for the health agencies that provide assistance to them.

Risk and emergency information must also be available for people who speak languages other than English (primarily Spanish in the area we studied).

A variety of print and online materials about weather risks and storm preparation are available to residents in the County including websites, brochures, checklists for preparedness, and storm-tracking maps. Local emergency management and public information officers also communicate preparedness year-round and especially during hurricane season through public events such as Hurricane Awareness Week. One downside of perennial risk messages for people in at-risk areas is that messages can become cultural background noise. Residents may experience issue and/or threat fatigue, as did Floridians who were given evacuation orders four times in 2004. The danger is that people become complacent about risks and inured to messages about them.

From the sources available, we chose to have residents help us evaluate a print document prepared by the County entitled *Hurricane Survival Guide*. The guide is an 8 ½ × 11, four-page document, the goal of which is to help residents and visitors to the area understand and prepare for hurricanes and tropical storms. The *Guide* provides information about what to do, where to go, and how to stay informed in the event of a hurricane, whether the readers evacuate or stay in the area. It includes Saffir-Simpson descriptions of hurricane categories (1–5); names of past storms; advice for boat and pet owners; contact information for local emergency management; information about where to get official information through local media; a hurricane preparedness checklist; an explanation of the reentry protocol, which involves residents obtaining a permit in the form of a car windshield sticker; a map of evacuation routes; and a list of conditions to expect following a "direct hit."

Pilot Study Participants

We recruited 20 participants in several locations: a popular local coffee shop, a branch of the County public library, and a YMCA fitness center. The number of participants is consistent with research about both mental model approaches and document usability evaluation that 20–30 interviews will uncover the majority of attitudes toward contextual risks as well as the majority of problems with a document (de Jong and Schellens, 2000; Morgan et al., 2002). Participants received a $20 compensation for their cooperation in the research sessions, which lasted, on average, one hour.

We were particularly interested in talking to people who had considerable knowledge of the area and at least some experience with storms. Residents are the people most likely to have to choose staying or evacuating when a hurricane approaches the area and to deal with the aftermath upon returning—vacationers are directed to leave and have fewer reasons than residents to risk staying. In addition, we wanted to determine whether and how residents' situational understanding differed from that of the officials who created the brochure. Visitors would not have enough background to comment on area conditions or residents' behaviors, attitudes, and problems. Consequently, all of the participants selected for the pilot study were either year-round residents (eighteen) or people who spent enough of the year in the County to be familiar with the risks from severe weather (two). All of the participants had at least one experience with a serious storm. More than half reported experiencing six or more storms. Eight of the participants had never evacuated for a storm, five had evacuated once, and the remaining seven had evacuated two or more times.

Otherwise, the sample of participants was diverse and included eleven women and nine men ranging in age from eighteen to eighty with the median age of forty-two. Education levels ranged from no high school diploma to master's degrees, with fourteen participants having completed undergraduate or graduate degrees. All but three participants were employed, with annual incomes ranging from $14,500 to $200,000. Ten participants were married; six were caring for children or parents at home.

Document Usability Evaluation Protocol

During the document testing, participants read and evaluated the content of the *Guide* using "plus-minus" markup (de Jong & Rijnks, 2006; de Jong & Schellens, 2000). Participants were asked to read the document completely. They were further instructed that as they read, they were to mark any content to which they reacted positively for any reason with a plus symbol (+) and to mark any content to which they had a negative response for any reason with a minus symbol (–). Participants were told that they could mark any unit of content from individual words, sentences, and paragraphs to images, sections, or whole pages. Immediately following the reading and marking, researchers interviewed the participants. During the interviews, and using the markings on the guide as prompts, interviewers asked participants to explain the plus or minus marks they noted on the guide.

In Morgan et al.'s (2002) description of the mental models approach to risk communication, they advocate using open-ended interviews to develop the public's mental models of risk. We supplemented the plus-minus method with additional open-ended questions that served a similar function. After reviewing the document with the participants, we asked each participant about

- whether they thought any information was missing in the *Guide*;
- their experiential knowledge about hurricanes;
- ways they have incorporated dealing with hurricanes into their lives;
- motives for people not to evacuate when an evacuation order is issued;
- characteristics of people who evacuate for hurricanes;

- characteristics of people who stay for hurricanes;
- the role television and radio stations play during hurricanes; and
- the role authorities play during hurricanes.

The discussions with each participant took place immediately after he or she had completed reading and marking the guide.

On prepared interview forms, researchers recorded the comments that participants made about the plusses, minuses, and answers to questions during the evaluations. We also recorded and transcribed the interviews to capture additional insights that people provided about their perceptions and attitudes. The document evaluation provided starting points for discussion, though one limitation we should note is that the interviews were shaped topically and sequentially by the structure of the document and the order of the additional questions.

Document Analysis

We collected information about residents' assessments of the *Guide* and, more generally, about their attitudes toward information sources, the area, serious storms, and preparedness. The participants had positive and negative things to say about the *Guide*. Participants' comments were, as expected, useful in isolating problems with the document. Interviewers recorded 180 discrete comments about problems on interview forms, and we assigned those to various topics in three categories. About 11 percent of problem comments (19), which we labeled *Use*, focused on distribution (where people could acquire a copy, or who should receive it) and comments about the usefulness or appropriateness of information for the participants as audience. Another 22 percent of the problem comments (41) focused on topics that we labeled *Design and Writing*, which includes comments about overall design and design elements, organization, and wording. The remaining 67 percent of the comments (120) focused on *Information*. We organized these comments into four topics including comments about the need for more information or less information about a topic, identification of contradictions in information, and elaborations of information for the benefit of the interviewer. Comments are summarized and briefly defined in Table 14.1, "Summary of Comments."

Most of the problem comments can be understood in terms of the salience of the *Guide* and the information in it. Delort (2009) defines *salience* as qualities of an object's features that draw attention leading to choices of items for cognitive processing. Texts "may be visually and linguistically salient" (p. 4). Albers (2007) holds that information salience contributes to how well readers conceptually grasp an overall message, how well they understand relationships among pieces of information, and how effectively they can draw conclusions from information. Factors in texts that interfere with salience include design that fails to assist readers in identifying important information, and incorrect,

TABLE 14.1

Summary of Comments

Category	Description	Topics (Most Cited)	Number of Comments
Comments about Use			
Distribution	Suggestions for where or how the *Guide* could be provided		5
Audience	Comments about ways that different audiences are addressed in *Guide* or how audiences would respond to it		14
		Subtotal	**19**
Comments about Design and Writing			
Design	Comments about overall appearance of the document, graphics, fonts, color, and images		25
Organization	Comments about placement of information; suggestions for moving content from one section or page to another		9
Wording	Comments about odd language use and confusing passages		7
		Subtotal	**41**
Comments about Information			
More information	Comments about topics that needed more information or emphasis; suggestions for additional information		53
Less information	Comments about topics that should be reduced, eliminated, or deemphasized; suggestions for eliminations		16
Contradictions	Comments about information in one part of the *Guide* that seemed to contradict information in another part, or that contradicts information from another common source or from experience		16
Elaboration	Comments in which participants elaborate on a topic or provide an opinion that is not meant as a direct suggestion for changes or improvement to the *Guide* but rather to inform the interviewer		35
		Subtotal	**120**
		Total	**180**

redundant, excess, and ambiguous information. Factors related to context and readers who interfere with salience include expectancy bias, whether the information is solicited or unsolicited, whether it is framed negatively or positively, and to what extent it accords with audiences' prior knowledge. According to Civan, Doctor, and Wolf (2005), "[S]alient information is more likely to draw attention, be given more consideration, and have a stronger effect on risk-related behavior than less salient information" (p. 927).

We next discuss participants' problem comments that highlight issues related to salience. Additional comments about the *Guide* and other topics available from transcripts of the sessions—, positive, negative, and neutral—, add to the picture of residents' experiences and contextual knowledge.

Comments about Use

After reading the *Guide*, most of the participants expressed appreciation for the content overall, which was generally judged to be useful and informative. Many participants commented that the *Guide* was a good reminder about how to prepare for hurricanes and what to do in the event of a storm. Some indicated that they learned something they didn't know before reading it. Of the *Guide* overall, people said, for example:

> Very good information about how to prepare, what to expect … before and after the hurricane is there.
> It's a good informational source—an excellent information source … especially for visitors, I believe.
> I think it's well organized. I think it's well written, and I think it's full of information. And uh, there are a few things that don't apply to me directly, like boating, but I'm sure that's necessary for other people.
> I think it's really informative, and it is true to things that you need to know. Especially if you're a visitor and you don't, you're not used to hurricanes. Has a lot of information that would be useful.
> I mean it's informative but I think a lot of people right here won't. … We never … we've only left once. … Like it's good to know all that and just remind yourself.

While overall reactions to the *Guide* were favorable, some of the comments raise questions about audience. Problem comments included, for example, "It seems like [the] evacuation order only addresses vacationers"; "[M]ore information for visitors—two brochures?" "[V]isitors will leave, residents will know"; and "[P]eople may not take the time to read this." These and similar comments suggest that participants thought the *Guide* would be more useful for visitors or new residents than for themselves. Participants' observations about audience reflect two influences—the content of the *Guide* itself, which includes information for both residents and tourists, and the residents' conviction that they already know what they need to do.

In terms of the overall salience of the document, residents who believe they are sufficiently informed, and who quickly pick up on the references to vacationers, may not take the time to read the *Guide*. The critical information for visitors—"ALL VACATIONERS MUST EVACUATE"—is one of the most prominent pieces of information on the two-page center spread that constitutes the bulk of the information about what to do in various conditions, such as a hurricane watch, hurricane warning, or evacuation order. For residents, messages aimed at tourists are irrelevant, a perception that may "cloud how a person analyzes information" (Albers, 2007, p. 82), thereby decreasing the salience of information that otherwise might be important for them. By the same token, the *Guide* may be too detailed for visitors, who would need to look for information that applies to them amidst information about emergency preparedness, reentry procedures, and poststorm conditions, all of which are irrelevant for most of them. The information meant for residents is excess information for vacationers, and the level of detail may even signal to vacationers that the document does not address them. Attempting to inform both the tourist and resident audiences for the *Guide* may result in neither audience reading it.

Residents' perceptions of their experiences and coping skills is another aspect of opinions that the *Guide* is not useful for them. Even though many commented positively about it, residents' ideas about what they need are bound up in how they identify themselves and others. Participants drew distinctions between "locals," the group to which they belong, and "tourists" or "visitors." Participants characterized locals as knowledgeable about the area, storm risks, and managing during emergencies. They viewed themselves as "seasoned" by years of living in the area to know what to do and where to go. For example, participants were able to talk specifically about local conditions ("[F]looding's really bad by the sound"), evacuation routes that drivers were likely to have problems with ("[O]n the beach, stay in the middle"), and how long evacuating from different points might take ("Hurricane Isabel, it took four hours to get from here to [another] City. That's a long time").

By contrast, participants characterized visitors as uninformed and ill-equipped to deal with storms. "Visitors don't understand the enormity of a storm and its effects," as one local put it. The guide is seen as more important for the uninitiated "because," another resident told us, "folks come here and haven't a clue, on vacation, what hurricanes are about, and I think this [the *Guide*] gives them an idea that … if we issue evacuation orders, why it's important that you leave and that there are other things to worry about other than just getting our visitors off the islands." Visitors are seen by some locals as a hindrance during an emergency. One resident said of tourists' presence:

> [Y]ou really don't need people that don't need to be here. And I think you probably could almost emphasize here that the priority four [people who are last to be allowed back to the area after an evacuation] would include the vacationers [laughs] so that they understand, you know, or on their separate page, that they will not be considered a priority. They will not.

A number of participants put plusses by the *Guide*'s missive, "All vacationers must evacuate," because tourists are "unknowledgeable about hurricanes and so, they need to be somewhat forced to leave." Residents, on the other hand, decide whether to leave or stay.

The *Guide* does contain important information for residents. But despite their information needs, residents may reject material that isn't salient for them if features of the text draw their attention to messages that aren't pertinent. When audiences don't perceive that their needs and interests are central, they may ignore information completely. The challenge is to create documents that target audiences will attend to, not "consider throwing in the junk mail," as one participant indicated he might if he received the version of the *Guide* we tested. When the needs, perspectives, and distribution channels for multiple audiences seem at odds, it might be worth preparing separate messages for each.

Comments about Design and Writing

Participants pointed out a number of problems with the design, organization, and wording in the *Guide* that interfered with information salience. Problem comments about the design and writing focused on the overall look, graphics, fonts, organization of information, and wording (e.g., confusing passages or word choice).

Design

The comments about design ranged from general impressions to comments about more specific features, such as the use of spot color and emphasis. Comments suggest salience problems related to the perceived importance of information. One participant summed up concerns:

> I think there was a little bit too much uniformity to it. For example, all of this [several sections across two pages] was given somewhat equal weight. It doesn't have equal weight. It should be, there should be, a bold statement at the top and supporting statements underneath so that there is a priority.

The overall visual impression and readability, which involves features such as chunking, font size, and white space, can also enhance or detract from information. Participants also commented on specific problems with design. Several of the most often noted problems include the following.

- *The front page of the* Guide *does not convey the importance of the message.* Several participants were critical of the front page, which they said would not grab their attention. Two specific remarks from participants were that the front page is too dark and dull and that the title doesn't stand out. The title and subtitle, which include the words *hurricane, survival,* and *threatening,* may be the most important content on the page, but they are the least visible.

- *Figure–ground contrast in several sections of the document is insufficient.* Shaded areas throughout the document highlight content, for example, important warnings. Participants pointed out that these areas didn't provide enough contrast for optimal readability or for gaining readers' attention. In one instance, the message was "DON"T WAIT TO EVACUATE! Routes become very congested." In another, the message in blue was "PREPARE TO SUSTAIN YOURSELF FOR AT LEAST 72 HOURS. There might not be power, water, rescue, or medical services available." These messages seem too important to risk obscuring with poor figure–ground contrast.

- *The font size of the body text throughout the brochure is too small.* Some sections included text with a font size as small as eight points, and the use of font sizes was inconsistent throughout the document, which includes several sections of fairly dense chunks of text. Several people said that the font sizes overall would be a challenge for older people or people with low vision. Others commented on the density of the text in particular sections, indicating the need for additional white space.

Low salience in design means that readers may fail to recognize the relative importance of pieces of information or the value of the overall message.

Organization

Similarly, how information is organized impacts salience. Placement and hierarchy of information help people decide what information is important. Readers also want to easily find information that they need. Comments indicated that some information audiences deemed important was different from the information that the document's organization suggested was important.

For example, people suggested the designers *"make a real* [hurricane evacuation] *checklist on a separate page."* The checklist in the *Guide* is a bulleted list of items that people should prepare in advance of any emergency. In the draft we tested, the checklist is in a small shaded area (less than a quarter page) bordered by dotted lines that suggest that people can cut it out and save it. But the participants, who have likely seen many such checklists, didn't perceive this one as usable in part because of the size. In addition, if they cut it out, they would cut in half the Saffir-Simpson Hurricane Scale provided on the reverse. Most participants commented positively about the inclusion of the Saffir-Simpson Scale, one of the features of the *Guide* that they most valued overall and that therefore had salience for them. In fact, another comment about organization that several people offered was to "move hurricane categories to the front page."

Another part of the document that participants suggested be placed more prominently was "Reentry Procedures" about which "people may need to be reminded." They also pointed out that the font size used in part of the section was too small, making an important piece of information easy to miss or hard to read. Reentry takes place after people have been evacuated for a storm; the storm has passed, and the emergency management officials believe that most of the area is safe enough for people to return. In the County, reentry prioritizes different types of residents who need permits for access. The *Guide* informs people about the reentry process and how to get permits. According to residents, the reentry permitting process, and reentry itself, has created problems for them in the past. Participants told us that they had problems getting back into the area because they didn't know about the permits, that they couldn't reach anyone using the phone number provided for reentry information, and that people who didn't need to be in the area reentered too early.

The fact that reentry seems to be a hot-button issue for residents could explain why many people commented on that section. Information about reentry is salient for residents because of their experience, not because of the way it is presented in the *Guide*. Designers can better meet audience needs by learning what information is important to them and organizing information in ways that highlight those issues.

Wording

Wording, word choices, and readability received the least attention. However, one issue that received several comments and a mix of plusses and minuses exemplifies challenges with redundancy as a factor in salience. Two lists of words were associated with the phrases "What you should do when a hurricane is threatening" and "What you should do when an evacuation is ordered." These phrases are prominently displayed in headings at the top of the inside pages of the *Guide* and serve to separate the related information into two sections. The words associated with the first phrase are "monitor, listen, prepare, inspect, evaluate, notify, evacuate." Those associated with the second are "listen, prepare, check, pack, notify, leave early." As one participant put it, "[T]he list of actions is great but hard to memorize and is too vague." The number of words may pose a problem as well as the inconsistent overlaps and differences between the two lists. In addition, the terms are not tied directly to any other information; for example, the words don't appear as first words in items on a list anywhere. Some participants put plusses by the words lists— they liked the idea, but didn't really know what some of the words were meant to remind them about (e.g. "notify" whom, and when?). One participant suggested choosing words that start with letters that form "sayable" acronyms.

The intent behind including the word lists might be to build in the type of redundancy that can add to information salience through repetition of concepts or information (Albers, 2007, p. 81), but in this case, the words are not meaningful, nor were they defined for the context or connected to other content. In that sense, the word lists may simply add irrelevant or excess information that may detract from the message.

Document evaluations can reveal problems with information presentation that interfere with salience and thus usefulness and usability. If text is difficult to read because of design elements such as font size, shading, and text density, messages may go unread or unheeded. The overall appeal of the document is also a factor in whether the audience will even take the time to pick it up. But comments about design, organization, and writing also provide insight into what topics the audience for the document deems most important to them and how audience experience serves as a factor in salience.

Comments about Information

The majority of participants' problem comments focused on the need for more or less information about topics, indicating that participants may view information on some topics as incomplete, excessive, or redundant.

More and Less Information

By far, the most commonly recorded problem comments (fifty-three) focused on adding or emphasizing information that participants thought was missing from the *Guide*. These types of comments were relatively straightforward and most often came in the form of suggestions for adding or emphasizing information in several sections. For example:

- *Before the hurricane season*: Emphasize planning, people forget to plan or wait until the last moment; emphasize taking photos for insurance if that's important.

- *What to expect following a direct hit of a hurricane*: Add downed power lines, mosquitoes, more information about supplies and services that will not be available; indicate that it may take 2–3 weeks to get back to normal ("People may lose their patience"); emphasize staying out of the water because of germs, diseases, and pollution ("Think septic tank").

- *If you EVACUATE*: Add information about hydroplaning; clarify the one-way evacuation pattern on the map; emphasize leaving (and "Discourage staying more"); emphasize that the County has only two evacuation routes; emphasize leaving early, evacuation traffic problems, and how long an evacuation takes; and emphasize the importance of evacuation planning and having a destination.

- *If you stay*: Add that people stay at their own risk; add that people need to have means of escaping from homes in case of flooding; add "Buy a generator."

- *Hurricane evacuation checklist*: Add "Bring documents, drinking water, and extra batteries"; and keep a full tank of gas.

Participants also had suggestions for eliminating information they viewed as unnecessary. The item most cited for deletion was a highlighted box that listed the names for potential 2006 hurricanes. This information was outdated at the time of the study, conducted in 2007. It was also unnecessary for the audiences' safety and took up space on the page that the participants thought could be better used. A number of residents also suggested deleting information listed under "Boating Precautions" for similar reasons and because it didn't apply to them.

Messages that contain irrelevant and excess information can add unnecessarily to readers' cognitive loads, causing them to feel overwhelmed by information and unsure about which information is most important. When information is missing, the omission may seem like a mistake or lack of knowledge on the part of the information provider that can erode trust in the source.

Contradictions

Participants identified information problems in the *Guide* that we labeled *contradictions*, which generally fell into one of four types: conflicts with experience, information insufficiency, internal inconsistency, and conflicts of knowledge.

Conflicts with experience. Some contradictions were caused when information didn't conform to participants' previous experience or mental models. For example, the *Guide* instructs people preparing for hurricane season to "have enough food and water for 3 days per person," but several people commented that three days of supplies was not enough. The information conflicted with participants' experiential knowledge.

Information insufficiency. Contradictions also occurred when information lacked sufficient clarity, support, or detail for residents to understand or accept it. For example, several people commented on a statement that there are no Red Cross–approved shelters in the County. They asked whether that meant there were no shelters or just none approved by the Red Cross. Participants needed clarification to form an accurate understanding about the information pertaining to shelters, an important topic to people in hurricane-prone areas particularly since sheltering has been an issue in recent crises such as Hurricane Katrina.

Internal inconsistency. Other contradictions arose when information in one part or passage of the *Guide* conflicted with information in other parts or passages. These contradictions create ambiguity that requires readers to reconcile conflicting information to determine appropriate meaning and subsequent action. Disambiguation requires understanding "the interrelations of information within a situation's context" and the goals of those creating and using the information, which are not always clear (Albers, 2007, p. 82).

One important example that many participants mentioned is that the *Guide* seems to present evacuating or staying after a mandatory evacuation order is issued as two equal options between which residents may choose. By contrast, information directed to the visitor audience, which is to "leave immediately," unambiguously presents evacuation as the only option. In addition, evacuation is cast in more negative language than the information about staying. For example, the subheading under "If You EVACUATE" reads "DON'T WAIT TO EVACUATE! Routes become very congested." The subheading and several items in the bulleted list that follows begin with negative words, including "DON'T," "Avoid," and "Never." On the other hand, the subheading under "If You Stay" reads "PREPARE TO SUSTAIN YOURSELF FOR AT LEAST 72 HOURS. There might not be power, water, rescue, or medical services available." Neither the subheading nor any of the bullet points that follow begin with negative words.

The choice of positive or negative words is a form of framing that may have consequences for information salience. Albers (2007) notes,

> Information salience influences how people perceive risk. Presenting the same information but expressed in positive or negative terms effects how a person responds to it. A strong framing effect exists. Framing effects occur when the same information is evaluated differently when it is presented in gain or loss "frames." (p. 83)

Albers refers specifically to the ways that people interpret probabilistic information: "for example, a medical treatment can be described as '60% effective' (gain frame) or as '40% ineffective' (loss frame)" (p. 83), with peoples' preference being the gain frame. We suggest that positive and negative language, such as that used in the *Guide*'s discussion of evacuation and staying, may have a similar effect. Given that the residents often indicated a preference for staying based on a variety of factors—evacuation is "a pain," it's time consuming, it's expensive, people drive for hours or get evacuated into worse weather, and people want to stay to protect their property—the more positive language with which staying is framed may provide additional encouragement for that choice. The framing here, together with the sense that evacuating and staying are offered as equal options, appeared contradictory to participants. If the official policy is to get people out of the area in times of hurricanes, it would seem that the evacuation option should be presented as preferable.

However, other factors in the context that influence how this information is presented in the *Guide* are not readily apparent. Mandatory evacuation doesn't mean that officials will or can remove people from their homes. People are not forced to leave. Mandatory evacuation does mean that people who stay are, literally, on their own—they cannot expect rescue services during or immediately following a hurricane. This consequence is not stated directly. Instead, the *Guide* lists conditions that people who stay will likely face, including flooding, loss of power, and other problems. In addition, evacuation is not completely without risk, a fact that officials and many residents understand.

Decisions by officials to order an evacuation and of residents about whether to comply are based on complex sets of judgments and calculation of risks. Emergency management weighs many factors in determining whether and when to issue a mandatory evacuation. Nevertheless, though participants expressed their dislike of evacuating, many stressed that people should be encouraged in every way possible to leave and provided suggestions for including more information about the realities of staying during a severe event. The apparent ambiguity in the *Guide* may not be the presentation of evacuation as a choice, but the way that choice is framed.

Conflicts of Knowledge

Contradictions also arise when people misinterpret information. The causes may be the audiences' levels of previous knowledge or differences among mental models among the public, the professionals who prepare information for them, and/or the experts who provide information. These contradictions may be discovered fairly easily when, for example, people say they don't understand information. However, people may believe they understand information that they don't and make decisions on the basis of incorrect assumptions. To ferret out these types of problems, information designers need to listen to audiences' interpretations. An example from our study centers on participants' interpretation of the Saffir-Simpson Hurricane Scale.

The Saffir-Simpson Scale, which is the only information in the *Guide* that might be considered scientific or technical, rates the severity of hurricanes in categories from 1 to 5. The categories are defined by winds speed, potential storm surge, and descriptions of the kinds of damage that might be expected. For example, the description of a Category 3 storm is as follows:

> Winds 111–130 mph. Storm surge 9–12 ft. Foliage torn from trees; large trees blown down. Practically all poorly constructed signs blown down. Some damage to roofing materials of buildings; some wind and door damage. Some structural damage to small buildings. Mobile homes destroyed. Serious flooding at coast and many smaller structures near coast destroyed; larger structures near coast damaged by battering waves and floating debris. Low-lying escape routes inland cut by rising water 3 to 5 hours before hurricane center arrives.

The study participants were generally positive about the inclusion of the Saffir-Simpson Scale, though participants were less familiar with the storm categories than might be expected for residents of a coastal area that observes hurricane season for six months of every year. Residents said, for example:

> But you know what, it shows the surge with each category so a lot of people might not know that; it would be good to know what to expect.
> I thought that was good. Describes each category and mph of winds, and storm surge ... because it might say on the radio "Category 1," but not many people understand that means 74 mph, that's pretty high.

> A lot of people don't know. You hear on the news Category 3 storm, this is a good like reference to say OK Category 3 [pause]. Yeah, just so they know the severity I guess. Some people may hear Category 5 and may not know how severe. Now, I think all we have to say is Katrina.
>
> I think it's important to explain what people can expect for each category of a hurricane. I wouldn't have thought of something like that, but that's definitely important because [pause] you know, if they're calling for a Category 1 or 2, people might think it's really not that bad, but once you read the paragraph, you can see that, you know, something's gonna happen outside. It's not, like, 100 percent safe.

In the last comment, the speaker's realization about the severity of the Category 1 and 2 storms is important because many residents we spoke with during these and other interviews mention Category 3 as a sort of trigger for evacuating. For example, several said the following:

> Unless it's bigger than a 4, most of the locals won't evacuate; they'll stay here. It's a lot of hassle to evacuate.
>
> I guess in my experience ... you have to go through the wind speed to understand just how strong it is. And that I would not stay for ... you know, I used to say, "Oh, I'll stay no matter what." But, you know, I don't want to be here for a Category 3.
>
> [About evacuating] [T]hey'll look and say, "I might as well just stay here if it's only going to be a Category 2 or 3 hurricane." So, they weigh the risk of actually leaving to actually getting trapped on the highway.

Hurricane Isabel, the last severe hurricane to impact the North Carolina coast, hit the Outer Banks as a Category 2 storm, cutting a 2,000-foot inlet across Hatteras Island. The storm and related flooding were ultimately responsible for thirty-five deaths and over $3 billion in damage in North Carolina alone. A mandatory evacuation was initiated on the coast for Isabel, but about half of the people we interviewed had stayed. Isabel was a serious and dangerous storm at a Category 2. Consequently, that many people consider a Category 3 as the signal to leave may be of concern. Hurricane Katrina made landfall as a Category 3 storm.

Though residents were positive about the inclusion of the Saffir-Simpson Scale in the *Guide* and it seemed to have salience for them, many cannot interpret the measurements of wind speed and storm surge in a meaningful way. Nor do all residents have the experiential basis for fully comprehending the descriptive elements meant to illustrate potential effects, although that information is helpful. Thus, the information may not convey the seriousness of storms that the document developers intend by including it. Another problem is that the categories, which are general guidelines, may crowd out other, more context-specific considerations. A storm's effect on an area has as much to do with features of an area and the amount of rain and flooding—which in fact causes most storm-related deaths—as it does with the initial winds and surge of landfall. In fact, the most serious problems arising from

Katrina were not during the storm, but in the aftermath. If people focus too much on storm categories that they don't actually understand, they may fail to apprehend other cues about conditions and appropriate action.

Problems we labeled as contradictions may be caused by information that fails to accord with audiences' experiences, lacks clarity or support, is ambiguous, or doesn't present knowledge in a way that audiences can use or interpret. Contradiction may lead audiences of risk and emergency information to distrust the message and/or the source, in which case they may reject subsequent information and/or ignore instructions. Audiences may also fail to understand information, even if they don't realize it.

Elaborations

Some comments from participants constituted elaborations on information primarily for the benefit of the interviewers. Participants elaborated when they thought interviewers needed contextual information to understand their points. These comments provide insight into the audience's prior knowledge, which "influences how they interpret information," and related expectancy biases, which lead people to interpret information in particular ways when "a person's prior knowledge about a subject affects how they evaluate the information credibility" (Albers, 2007, p. 82).

Conclusion: Document Usability
Evaluation as Audience Analysis

Using the plus-minus method helped isolate problems with the *Hurricane Survival Guide* that can be corrected to improve the document. The *Guide* was minimally revised on the basis of information provided to County officials. More importantly, the usability evaluation yielded a number of insights about attitudes toward, experiences with, and perceptions about risk from severe weather as it affects this particular community that may be helpful in the development of additional preparedness information beyond the *Guide*.

These types of insights are helpful in assisting information designers to better identify and address gaps in audience knowledge and increase information salience, which is determined not only by how people manage features and content of a text but also by the experiences and perceptions that audiences bring to their engagement with the text. As Albers (2007) notes, what counts as salience depends on the reader's mental models and information needs, and the "subconscious quality value the reader has assigned to it" (p. 84). To learn about audiences' mental models and value assignments requires that information designers communicate with representative members of the audience.

But we can take the engagement a step further. Seeger (2006) has noted that best practices in crisis, risk, and emergency communication call for listening to the public's concerns and partnering with the public. "Ideally, the public can serve as a resource, rather than a burden, in risk and crisis management. Thus, crisis communication best practices would emphasize a dialogic approach" (Seeger, p. 238). But that level of participation with the public happens all too infrequently (see Sandman, 2006). It's possible that a practice as low-tech as engaging the public in evaluating the information meant for them will result in both better information and better lines of communication between residents and information providers who often serve as the face and voice of emergency management.

In addition, taking the time to listen to and learn about the audience can help communicators develop an empathetic approach to audiences' needs. People who provide information about risks and emergencies "should demonstrate appropriate levels of compassion, concern, and empathy" to "enhance the credibility of the message and enhance the perceived legitimacy of the messenger both before and after an event" (Seeger, 2006, p. 241). Accordingly, knowledge of audiences' experiences, attitudes, and informational needs should be an important component of the CERC model. Information providers need additional testing methods for the variety of sources available, but the usability evaluation serves not only to gather information about information quality. Usability evaluation may also provide a deeper understanding of audience that can be incorporated into future development of preparedness information.

Reader Take-Aways

Technical communicators are increasingly involved in developing and providing risk and emergency information. Situations in which people rely on these types of information are complex and rapidly changing. In general, the knowledge of experts and the purposes of emergency management professionals drive considerations about the content and form of risk and emergency messages. When the end user is the public, usability evaluation of texts—whether print or electronic—can serve as a useful form of audience analysis to ensure that the needs, attitudes, and perceptions of the audience are addressed.

- To facilitate the kinds of actions they want people to take, communicators should conduct audience analyses to develop a sense of what information their audiences will seek and use.
- Audience analysis can be accomplished using the plus-minus method of document evaluation. The plus-minus method allows communicators to not only isolate problems in tests, but also gain insight into audiences' concerns and motivations.

- By attending to the participants' comments about the text, and by understanding the contextual basis for those comments, information developers may be able to take a more empathetic approach to meeting audiences' needs.

- Technical and professional communicators who are involved in risk and emergency communication can become a resource to others in emergency management and policy making because of the insights they can offer about audiences' behaviors and uses of information.

References

Albers, M. (2007, October). Information salience and interpreting information. In *Proceedings of SIGDOC'07*, El Paso, TX (pp. 80–86). New York: Association for Computing Machinery.

Baron, J., Hershey, J. C., & Kunreuther, H. (2000). Determinants of priority for risk reductions: The role of worry. *Risk Analysis, 20*(4), 413–427.

Brown, K. M., Lindenberger, J. H., & Bryant, C. A. (2008). Using pretesting to ensure your messages and materials are on strategy. *Health Promotion Practice, 9*, 116.

Centers for Disease Control, Crisis, and Emergency Risk Communication (CDC). (2002). *CERC training materials*. Retrieved from http://www.bt.cdc.gov/cerc/pdf/CERC-SEPT02.pdf

Civan, A., Doctor, J. N., & Wolf, F. M. (2005). What makes a good format: Frameworks for evaluating the effect of graphic risk formats on consumers' risk-related behavior. In C. P. Friedman, J. Ash, & P Tarczy-Hornoch (Eds.), *AMIA Annual Symposium Proceedings 2005* (p. 927). Bethesda, MD: American Medical Informatics Association.

Cole, T. W., & Fellows, K. L. (2008). Risk communication failure: A case study of New Orleans and Hurricane Katrina. *Southern Communication Journal, 73*(3), 211–228.

Couzijn, M. (1999). Learning to write by observation of writing and reading processes: Effects on learning and transfer. *Learning and Instruction, 9*, 109–142.

De Jong, M. (1998). *Reader feedback in text design: Validity of the plus-minus method for the pretesting of public information brochures*. Atlanta, GA: Rodopi.

De Jong, M., & Lentz, L. (2007, October). Professional writers and empathy: Exploring the barriers to anticipating reader problems. In *Proceedings IPCC 2007 Conference, IEEE Professional Communication Society*, Seattle, WA. Piscataway, NJ: IEEE.

De Jong, M., & Rijnks, D. (2006). Dynamics of iterative reader feedback: An analysis of two successive plus-minus evaluation studies. *Journal of Business and Technical Communication, 20*(2), 159–176.

De Jong, M., & Schellens, P. J. (1997). Reader-focused text evaluation: An overview of goals and methods. *Journal of Business and Technical Communication, 11*, 402–432.

De Jong, M., & Schellens, P. J. (2000). Toward a document evaluation methodology: What does research tell us about the validity and reliability of evaluation methods? *IEEE Transactions on Professional Communication, 43*, 242–260.

De Jong, M., & Schellens, P. J. (2001). Optimizing public information brochures: Formative evaluation in document design processes. In D. Janssen & R. Neutelings (Eds.), *Reading and writing public documents* (pp. 59–83). Amsterdam: John Benjamins.

De Jong, M. D. T., & Lentz, L. R. (1996). Expert judgments versus reader feedback: A comparison of text evaluation techniques. *Journal of Technical Writing and Communication, 26*, 507–519.

Delort, J.-Y. (2009). Automatically characterizing linguistic salience using readers' feedback. *Journal of Digital Information, 10*(1), 22.

Gigerenzer, G., Hertwig, R., van den Broek, E., Fasolo, B., & Katsikopoulos, K. V. (2005). A 30% Chance of Rain Tomorrow: How does the public understand probabilistic weather forecasts? *Risk Analysis, 25*(3), 623–629.

Grabill, J. T., & Simmons, W. M. (1998). Toward a critical rhetoric of risk communication: Producing citizens and the role of technical communicators. *Technical Communication Quarterly, 7*(4), 415–441.

Haddow, K. S., & Haddow, G. (2008). *Disaster communications in a changing media world.* San Francisco: Butterworth-Heinemann.

Handmer, J., & Proudley, B. (2007). Communicating uncertainty via probabilities: The case of weather forecasts. *Environmental Hazards, 7*, 79–87.

Keller, C., Siegrist, M., & Gutscher, H. (2006). The role of the affect and availability heuristics in risk communication. *Risk Analysis, 26*(3), 631–639

Lazo, J. K., Kinnell, J. C., & Fisher, A. (2000). Expert and layperson perceptions of ecosystem risk. *Risk Analysis, 20*(2), 179–193.

Lentz, L., & de Jong, M. (1997). The evaluation of text quality: Expert-focused and reader-focused methods compared. *IEEE Transactions on Professional Communication, 40*, 224–234.

Luhmann, N. (1993). *Risk: A sociological theory.* New York: Walter De Gruyter.

Lundgren, R., & McMakin, A. (2004). *Risk communications: A handbook for communicating environmental, safety, and health risks* (3rd ed.). Columbus, OH: Battelle.

Masuda, J. R., & Garvin, T. (2006). Place, culture, and the social amplification of risk. *Risk Analysis, 26*(2), 437–454.

Morgan, M. G., Fischhoff, B., Bostrom, A., & Atman, C. J. (2002). *Risk communication: A mental models approach.* Cambridge: Cambridge University Press.

Paton, D. (2007). Preparing for natural hazards: The role of community trust. *Disaster Prevention and Management, 16*(3), 370–379.

Renn, O. (2008). *Risk governance: Coping with uncertainty in a complex world.* London: Earthscan.

Riechard, D. E., & Peterson, S. J. (1998). Perception of environmental risk related to gender, community socioeconomic setting, and age. *Journal of Environmental Education, 30*(1), 11–29.

Ropeik, D., & Slovic, P. (2003). Risk communication: A neglected tool in protecting public health. *Risk in Perspective, 11*(2), 1–4.

Rosati, S., & Saba, A. (2004). The perception of risks associated with food-related hazards and the perceived reliability of sources of information. *International Journal of Food Science and Technology, 39*, 491–500.

Sandman, P. M. (2006). Crisis communication best practices: Some quibbles and additions. *Journal of Applied Communication Research, 34*(3), 257–262.

Sapp, S. G. (2003). A comparison of alternative theoretical explanations of consumer food safety assessments. *International Journal of Consumer Studies, 27*(1), 34–39.

Schriver, K. A. (1992). Teaching writers to anticipate readers' needs: A classroom-evaluated pedagogy. *Written Communication, 9,* 179–208.

Schriver, K. A. (1997). *Dynamics in document design: Creating text for readers.* New York: John Wiley.

Seeger, M. W. (2006). Best practices in crisis communication: An expert panel process. *Journal of Applied Communication Research, 34*(3), 232–244.

Sjöberg, L. (2000). Factors in risk perception. *Risk analysis, 20*(1), 1–22.

Swaney, J. H., Janik, C., Bond, S. J., & Hayes, J. R. (1999). Editing for comprehension: Improving the process through reading protocols. In E. R. Steinberg (Ed.), *Plain language: Principles and practice* (pp. 173–203). Detroit, MI: Wayne State University Press.

Taylor-Goolby, P., & Zinn, J. (2006). *Risk in social science.* New York: Oxford University Press.

Ward, H., Smith, C., Kain, D., Crawford, T., & Howard, J. (2007, July 22–26). Emergency communications and risk perceptions in North Carolina's Coastal Zone. In *Conference Proceeding. Coastal Zone 07, 2007,* Portland, OR. Charleston, SC: NOAA Coastal Services Center.

Weiss, E. H. (1988). Usability: Stereotypes and traps. In E. Barrett (Ed.), *Text, context and hypertext* (pp. 175–185). Cambridge, MA:

15

Usability Testing, User Goals, and Engagement in Educational Games

Jason Cootey

Utah State University

CONTENTS

Introduction 334
Games Studied... 337
    Aristotle's Assassins 337
    Avalanche Software 338
Usability of Education Games.. 338
    Good and Bad Educational Games 338
    Gameplay and Engagement... 340
    Usability and Game Testing 341
        Usability Testing Defined 342
        Playtesting Defined 343
        Quality Assurance Defined 345
Methodology.. 346
    Observation Methods... 347
    Survey Methods 348
Findings.. 350
    Observation Findings... 350
    Survey Findings 351
Results... 354
Discussion 358
Conclusion 359
Reader Take-Aways 359
Acknowledgments... 360
References... 360

Abstract

In the spring of 2007, I designed a usability test at utah state University. In the process, the technical communicators–turned–usability testers discovered that the test conditions demanded a redefinition of *usability testing*. The game

333

was designed to be a fun, Greek adventure game with learning objectives embedded in the gameplay. These learning objectives demand the evaluation of a more complex system than usable interfaces; an interactive test scenario called "Otis in Stageira" was fashioned to measure whether players could learn rhetorical strategies of persuasion during gameplay. After the test, the development team was pleased to find players overcame the persuasion-oriented gameplay challenges. However, the test shared qualities with playtesting and quality assurance testing; consequently, I was less certain that *usability test* was an appropriate label. As an intersection of testing methods, Otis in Stageira challenges traditional definitions of users' goals in complex systems.

Introduction

In the spring of 2007, I implemented a usability test to answer a single question: If the game Aristotle's Assassins is designed to teach persuasion strategies, can the user use the game to learn persuasion or does the user simply play the game without learning anything? I discovered a complex system like a computer game, several significant factors challenge user goals: (1) Programmers predetermine user goals for every aspect of the game, (2) the computer game is supposed to be challenging and difficult to learn, and (3) there is no product of users' agency because the game is its own end. Consequently, computer games push at the traditional definitions of *user goals*; usability is centered on users' goals, an object's ease of use, and interfaces that serve users' agency. Most notably, testing Aristotle's Assassins blurred the boundaries of three kinds of tests relevant to game testing: usability testing, playtesting, and quality assurance testing. However, rather than dismantle each test type and create a veritable quagmire that serves no purpose at all, this chapter seeks to demonstrate that user goals can be the same as the system goals in complex systems, without being system centered (Johnson, 1998).

Laitinen (2006) identifies some exigency for my own argument; he describes a usability finding from his own game testing: "No feedback is given if the player cannot pick [up] an item" (p. 69). Apparently, the test participant discovered when the inventory is already full, the game does not signal the failure to store more items. Because this seems like a quality assurance issue, rather than usability testing, Laitinen concedes that an "interesting topic for the future studies would be to compare the usability methods to the traditional quality assurance methods used in game development" (p. 74). Consequently, the interactive test scenario called Otis in Stageira is a good candidate for the further study Laitinen suggests. Laitinen describes how the absence of inventory feedback impeded players in performing a task meant to reach their own goal, even if inventory management is a system goal. In the same way, the absence of educational cues in Aristotle's Assassins can impede players in learning the persuasion strategy of identification, even if I coded the educational cues.

One important distinction to make at the beginning of this chapter is the usage of users and players. Usability-testing research references users. Computer game research references players. Rather than attempt to use both words when I write about usability and games, respectively, I plan to only use the word *user*. Consequently, the user plays Aristotle's Assassins and is a usability participant in a test version of Aristotle's Assassins.

Aristotle's Assassins is an educational computer game that immerses users in a Greek environment in which users can learn about Greek history, mythology, music, rhetoric, and philosophy (see Figure 15.1). At the same time, users work through a fictional narrative about the assassination of Aristotle. In the spring of 2007, I questioned whether the game's interface helped users reach the educational goals or if the learning objectives were too buried in the gameplay. I fashioned Otis in Stageira to measure whether users could learn rhetorical strategies of persuasion during gameplay; the test was merely one learning objective in one scene already built into the story of Aristotle's Assassins. Otis in Stageira demonstrates that users could overcome the persuasion-oriented gameplay challenges. Yet at the time, I was not certain Otis in Stageira was a usability test.

Another important distinction is the development of Aristotle's Assassins itself. Aristotle's Assassins is an unpublished development project. The Learning Games Initiative (LGI) is an interdisciplinary and interinstitutional research collective. Aristotle's Assassins is a LGI development project. Dr. Ryan Moeller funded and directed the project at Utah State University. Under Dr. Moeller's direction, I took on various team roles throughout the project, depending on what was needed at any particular phase. Rather than confuse this chapter with the details of the multiple student teams and various LGI contributors, I simply refer to a generic Aristotle's Assassins development team.

FIGURE 15.1
Greek polis. Location for local politics and discourse in ancient Greece.

Whether educational goals are buried or not, the users' educational goals are generated by the programmer. Users might have agency while performing tasks in the game, but they cannot set their own goals. By engaging with the game, the user meets educational goals in a precoded system. Therefore, the users' goal is engagement with an educational game. Yet that educational goal is unattainable without the precoded goal of the system. This is why games are a complex system. Grice and Ridgway (1993) formulate usability in terms that can apply to game testing: "[R]eally successful usability testing is tied to task performance, whether it be task walkthroughs or laboratory testing with video cameras and stopwatches" (p. 430). A computer game is really a long string of tasks in which performance is obstructed at increasing degrees until the goal of the game is attained. A designer's challenge is to make those tasks worth engaging.

Testing engagement is the complex issue that made me question usability testing. Even though engagement happens while using some object, engagement isn't the task; engagement is how a user feels about the task. While playtesting might assess how users feel about a task, playtesting is about the preprogrammed system goals and not goals outside the system. While the educational take-aways are outside the system, the game was coded to engage users with an educational experience. Engagement might be quality assurance testing because there is no educational experience if Otis in Stageira breaks under the rigor of human agency. Yet, the test's objective was not to break the game. However, that doesn't simply leave usability testing as the only option left; after all, testing engagement is complex.

Aristotle's Assassins is an interactive game that is an increasingly complex system. Based upon the work of Albers (2003) and Mirel (2003, 2004), Janice Redish (2007) defines *increasingly complex systems* as "work that domain experts do when solving open-ended, unstructured, complex problems involving extensive and recursive decision-making" (p. 102). The recursive decisions for complex problems flag computer games as complex systems. In addition, Redish identifies ways in which complex systems differ from typical usability testing (p. 103). For instance, Redish mentions systems in which "there may be no way to know at the time of analysis if the result one gets is right or wrong" (p. 103); computer games give immediate feedback, but that feedback may or may not help users complete the goal of the game.

To get to where a challenging engagement can shape user goals, as I suggest, this chapter needs to identify the significant difference that engagement makes in good or bad educational games. With a clear sense of engagement, it is worth using traditional examples to show where usability, playtesting, and quality assurance differ before I blur the boundaries. Finally, in a complex system like Otis in Stageira, the user goals are the system goals.

Games Studied

There are two development teams that feature in the case examples I use to discretely distinguish usability testing, playtesting, and quality assurance testing. The first is the Utah State University Aristotle's Assassins development team. The second is a professional developer—Avalanche Software. Avalanche Software test practices are a worthwhile contrast to the test practices of Utah State's student development team.

Aristotle's Assassins

I started managing development of Aristotle's Assassins for Dr. Ryan Moeller at the Utah State University Department of English in the spring of 2006. I have worked with various student developers during ongoing stages of development. The project ended in the spring of 2008. Transient student development teams are not like professional development studios; consequently, the completion of a beta version demo game is a proud achievement.

Aristotle's Assassins is actually a modification—a *mod*—of the popular computer game Neverwinter Nights (2002). The mod turns the fantasy, medieval world of Neverwinter Nights into a thoroughly Greek environment—buildings, music, and togas (see Figure 15.2). The High Concept Document (2006) for Aristotle's Assassins states, "*Aristotle's Assassins* is a role-playing adventure game set amid the political turmoil of Ancient Greece. Players uncover a conspiracy to assassinate Aristotle, but the game really gets hot after players realize *they* are the assassins." While the game does feature a

FIGURE 15.2
Neverwinter nights: The game can be 'modded' by the gaming community.

fictional storyline about a secret cult of assassins seeking to topple Greece, users also interact with the history, mythology, politics, and philosophy of ancient Greece.

Avalanche Software

Avalanche Software started in 1995 with a handful of programmers and the development of Ultimate Mortal Kombat 3 (1996) on the Sega Genesis. Since then, Avalanche Software is responsible for the Tak computer game series (2003–2005) and 25 to Life (2006). In fact, *Tak* is a popular Nickelodeon TV series. In April 2005, Disney purchased Avalanche Software; since that time, Avalanche has done games based on the films *Chicken Little* (2006), *Meet the Robinsons* (2007), and other upcoming Disney/Pixar movies. Avalanche Software employs a development team of approximately two hundred game designers, artists, and programmers.

Project manager Troy Leavitt heads development at Avalanche Software and carefully filters proprietary information as he graciously works with me. I use examples that Leavitt (personal interview, October 19, 2007) gave me to identify test strategies that industry developers, like Avalanche Software, use in general practice.

Usability of Education Games

There are three foundational concepts that need to be clear before I can support an argument that suggests users' goals can be the goals of the system in usability testing. First, I want to be clear about what makes a good educational game. Second, I want to identify the importance of engagement in good educational games. Finally, I want to clearly identify usability testing, playtesting, and quality assurance testing before I blur the boundaries.

Good and Bad Educational Games

The engagement that makes games fun and motivating is also the reason why computer games are promising for education. However, the long history of educational computer games does not meet that promise. The concerns that led to the Otis in Stageira usability test stemmed from the difference between good and bad educational computer games.

Gee (2003) raises the stakes for good and bad educational games when he compares learning in school classrooms to the learning that users experience in computer games: "[I]t turns out that the theory of learning in good video games is close to what I believe are the best theories of learning" (p. 7). Gee is amazed at the effort users put into learning a game despite how increasingly difficult a game becomes. In fact, Johnson (2005) makes a similar reflection about learning activities in even monotonous game activities: "If you practically have to lock kids in their room to get them to do their math homework, and threaten to ground them to get them to take out the trash, then why are they willing to spend six months smithing in *Ultima*?" (p. 28). Computer games take 50–100 hours to complete, and a game's relentless learning curve does not abate until the game ends. Gee wonders why students are not willing to make a similar effort to learn increasingly difficult subjects at school.

Gee's (2003) book draws thirty-six learning principles from computer games that help the pedagogy of classroom educators. In fact, Gee goes so far as to advocate the belief that computer games can potentially do a better job of meeting the educational needs of students. He sets the stage for an educational game that plays like the gameplay-driven games similar to Tomb Raider (1996) and Grand Theft Auto (1997), yet also educates like the content-driven educational games similar to Jumpstart-Preschool (2003) and Oregon Trail (1985). However, if games are such effective teachers, then there should be engaging educational games instead of books promising what classrooms can learn from video games.

Content versus gameplay is not a semantic quibble; there are significant pedagogical consequences for a failure to utilize games according to their educational potential. After all, students don't need the cost of computer labs with licensed copies of Jumpstart so that kids can count the fish on the screen; they can count fish in a coloring book for thousands of dollars cheaper. Adriana de Souza e Silva and Girlie C. Delacruz (2006) attempt to diagnose the failure of the educational video game:

> Congruent with traditional didactic instruction that tends to promote rote memorization and regurgitation of basic facts and skills, video games such as Math Blaster™ (2005) were used for drill and practice to help master basic arithmetic and spelling skills. (p. 240)

The authors claim that educational games are often no better than memorization and regurgitation in classrooms. Therefore, the problem isn't so much that kids can count the fish in a coloring book as easily as on a computer; the problem is that educators want computer games to be like a coloring book that manages the memorization and regurgitation.

Unfortunately, most educational games favor interactive content over good gameplay. However, the playing is what makes computer games meet the promises made by researchers. To describe educational games, Prensky (2001) writes extensively about the importance of what he calls *digital game-based learning*: "[C]omputer games are such a powerful motivator for kids that we are crazy not to be using them in schools" (p. 187). The book details what digital game-based learning means, what kinds of learning it involves, and how to implement digital game-based learning. The challenge then is to make an exciting, engaging, educational game.

Prensky (2001) is clear about what the digital game-based learning game looks like:

> [I]t should feel just like a ... computer game, all the way through. But the content and the context will have been cleverly designed to put you in a learning situation about some particular area or subject matter. (p. 146)

Prensky suggests content and context are separate from the learning situation. He believes that content and context obstruct the feeling that the educational game is a computer game all the way through. Rather, a gameplay-driven learning situation includes the content of a subject matter and the context in which the user encounters that content; the priority of developers should be the design of gameplay that engages and educates users.

Gameplay and Engagement

Gee (2003) admits three realizations he had for why games possess engaging qualities:

> [1. The computer game] requires the user to learn and think in ways in which I am not adept ... [2.] learning is or should be both frustrating and life enhancing ... [3.] game designers keep making the games longer and more challenging (and introduce new things in new ones), and still manage to get them learned. (p. 6)

These three realizations are key to the thirty-six principles Gee extracts from computer games; however, not all the principles are necessary for the intersection of usability, playtesting, and quality assurance testing. In fact, I chose only six principles that highlight the meaning-making process users undergo to make sense of a game's semiotic language, to understand the learning relationship the user has with the game itself, and to discover what needs to be accomplished in the computer game. Consequently, I was interested in whether the Otis in Stageira test was as engaging as Gee or Prensky expect. After all, students need to experience the learning, handle the learning, and immerse themselves with learning; students need engagement.

Engagement constitutes the critical difference between content-driven and gameplay-driven games. Engagement is immersion plus a challenging learning situation. The content of that challenging learning situation might be strictly educational; however, game designer Raph Koster (2005) suggests that all computer game challenges must be challenging learning situations in order to be fun. Of course, both content and gameplay engage users, but the question is how they engage users. Engagement is a form of "participation" and requires people to join in, rather than be passive spectators. Unlike the users of instructions and Internet browsers, the computer game user is an active agent.

As an active agent, the user is a significant part of the game experience; the user can control the game tasks and determine the meaning of the game goals. Ken McAllister (2004) defines the *agent* as follows: "[S]omeone—or some collective—is always behind the management of meaning" (p. 45). For McAllister, meaning in video games is a cultural matrix in which dialectic conflicts not only define what the games mean but also what the agents mean to each other and by themselves; in other words, the agent has the choice to build autonomous meaning. Consequently, an educational computer game is more than a bunch of educational content; an educational game is engaging because the user-agent controls the learning situation, as well as the meaning of the content.

Aristotle's Assassins is an attempt to engage users in a learning situation that teaches, even while it focuses on the gameplay available to users through the onscreen avatar. Students who play the test version Otis in Stageira to learn persuasive strategies could potentially end up reading lots of instructional dialogue, drilling comprehension of the five canons of rhetoric, or resolving quiz-like situations to make progress in the game. However, there is no essential difference between such a game and the games that add no qualitative value to educational games. Prensky (2001) makes an important observation about the need to overcome this obstacle in the development of educational games: "[T]he issue is not to pull our kids away from the computer … [but to make] it worth their being there" (p. 199). Consequently, the Otis in Stageira test demonstrates that Aristotle's Assassins is a game students can play that requires persuasion as the gameplay.

Usability and Game Testing

The difference between user goals and tasks may be a foundational concept in usability; however, the concept is worth revisiting before I shift the definition. Carol Barnum (2002) argues a simple way to distinguish between goals and tasks: "Tasks are things that users do: steps they must take, processes they must complete, acts they must perform—to achieve a goal or objective. Developers sometimes focus on the tasks and lose sight of the user's goals" (pp. 91–92). Goals and tasks are important to usability testing because they are not only what users do, but also what users will use the product to do. Developers who simply create a sequence or selection of tasks forget that the object being designed is not an end—but a tool.

My problem is that games are a tool, the user doesn't have a goal independent of the game, and the game is the end. Therefore, with computer games, the task is the end. This is why computer games are a complex system and the user goals are the system goals.

When Barnum (2002) states that usability testing is "the process that involves live feedback from actual users performing real tasks" (p. 9), and when Dumas and Redish (1994) argue, "Usability means that the people who use the product can do so quickly and easily to accomplish their own tasks"

(p. 4), then a usability test on a computer game like Aristotle's Assassins should focus on users using the game as a tool to perform their own set of tasks, in order to reach their own goals. However, there are three ways in which a computer game does not seem to meet the criteria that Barnum and Dumas and Redish suggest:

1. Computer games are programmed systems in which the user cannot perform real tasks; users can only seek solutions to fictional gameplay challenges.
2. The solutions to gameplay challenges are not quick and easy to accomplish; they are meant to be challenges that require a great deal of work.
3. There is no product of users' agency because the game is its own end.

Barnum (2002) and Dumas and Redish (1994) make the definition of user tasks and user goals seem to be straightforward. However, the Otis in Stageira test is an intersection of multiple test methods. In other words, when the user goal is no longer outside the system and not necessarily the users' own goal, then engagement with the preprogrammed educational goals of an interactive system pushes at the definition of user goals. Usability testing, playtesting, and quality assurance testing are discrete methodologies that do not typically overlap. Therefore, a careful definition of each test method simply highlights the location of the intersection on which Otis in Stageira rests. In addition to a careful definition of each test method, an example from both the development of Aristotle's Assassins and development at Avalanche Software will further illuminate the test methods. Table 15.1 is meant to group the test methods, project names, and example references that are key to the test definitions.

Usability Testing Defined

Barnum (2002) states a simple litmus test for usability testing: "[T]he focus is on the user, not the product" (p. 6). User tasks and user goals are both important to usability testing because they are issues unique to a user, rather than either the developers or the product itself. Barnum explains, "Usefulness is

TABLE 15.1

Overview: Organizes Test Methods, Project Names, and Example References

	Aristotle's Assassins	Avalanche Software
Usability Testing	Installation instructions	Wii remote
Playtesting	Playgroups	Pizza parties
Quality Assurance (QA) Testing	Internal bug testing	External QA

defined in terms of the user's need for the product in context of the user's goals" (p. 6). A user's tasks and goals are not bound or preset by the design of the product or the goals of the developer.

Dumas and Redish (1994) define *usability testing* with four parameters (p. 4):

1. Usability means focusing on users.
2. People use products to be productive.
3. Users are busy people trying to accomplish tasks.
4. Users decide when a product is easy to use.

The priority of these four points is to highlight the value of the user's motives, user's priorities, user's needs, user's tasks, and user's goals. This is all completely at the expense of the product's functions, even the designers' vision of the product's purpose.

An example of clear usability testing was a test my team conducted on the Aristotle's Assassins installation instructions. Unlike a computer game, instructions map out a procedural sequence from which a user cannot deviate. In fact, deviations are signals of an instruction manual's failure. The Installation Instructions test tracked how users navigated the document in order to successfully reach the goal—install Aristotle's Assassins.

Avalanche Software project manager Troy Leavitt (personal interview, October 2007) was able to share an example of usability testing he conducted with the Wii Remote. The Nintendo Wii remote control is a kinetic interface, rather than the traditional joystick buttons. For instance, Avalanche Software is responsible for Hannah Montana: World Spotlight Tour (2007) game. Unlike other dance-styled computer games, for Hannah Montana the Wii remote uses the functionality of hand motions instead of a dance pad. Consequently, Avalanche Software had to conduct usability testing to identify how users moved the Wii remote. The problem was the precision with which the computer code must register motion signals. The dexterous capacity of different users meant that the remote would always be tilted at imprecise angles. Consequently, Leavitt (personal interview, October 2007) said the central question was "What is the user trying to do?" After observations, the Avalanche development team was able to program a general range of angles and directions that met the needs of a wider range of users.

Playtesting Defined

In *Game Testing All in One*, Schultz, Bryant, and Langdell (2005) argue that playtesting focuses on a different question than more numeric, quantifiable tests; the playtest answers the question "[D]oes the game work well?" rather than simply "[D]oes the game work?" (p. 296). The authors suggest that playtesting answers a host of questions, rather than seeking binary data:

Is the game too easy?

Is the game too hard?

Is the game easy to learn?

Are the controls intuitive?

Is the interface clear and easy to navigate?

Is the game fun? (Schultz et al., 2005, p. 296)

In order to answer these questions, playtesting observes what users did in the game, how well the game supported their agency, and how users chose to solve gameplay challenges. There are similarities with usability testing, but playtesting fails Barnum's litmus test: The purpose of the test is about the product.

There were several cases of playtesting with the development of Aristotle's Assassins. The objective of the playgroup tests was to let users into the game and observe them working though the gameplay challenges. The typical findings from playgroup tests was the various program flaws that interfered with what the users sought to do in the game. For instance, one playgroup was unanimous about the high complexity of the central challenge (see Figure 15.3) preventing users from making progress in the game.

The Avalanche Software development team frequently hosts pizza parties in their design studios. Kids play the current game under development, eat pizza, and showcase what they either like or don't like for the designers crowding the walls of the room. Troy Leavitt (personal interview, October 2007) states that the designers often ask questions such as "Why are you doing this?" or "How did you do that?" They hope to discover what is fun or frustrating so they can improve the game. However, Leavitt states there are challenges with children and playtesting. For instance, children need

FIGURE 15.3
Complex plot: Too many clues and too dense.

to articulate their feedback in helpful ways. Leavitt described one boy who stated that the test game "sucked worse" than another game he had played. The challenge was to get the boy to explain why the test game sucked worse and what exactly *sucked worse* really means.

Quality Assurance Defined

Quality assurance (or QA) is a very formal process of testing in computer game development. The objective is to list the bugs and to rank them according to level of importance. There are internal QA and external QA. In the first case, the development team identifies and fixes the bugs. In the second case, a contracted third party identifies and fixes the bugs. This is often procedural, methodical work. Schultz et al. (2005) argue that the rigorous requirements of quality assurance testing are the most difficult part of the process:

> You are given assignments that you're expected to complete thoroughly and on time—even if it means you have to stay late to get them done.... Everything you do when you play is for a purpose—either to explore some area of the game, check that a specific rule is being enforced, or look for a particular kind of problem. (pp. 19–20)

Often in game development, quality assurance is considered user testing. For instance, in the "User Testing and Tuning" section of their textbook, Adams and Rollings (2007) describe user testing by explaining that "the quality assurance department will create a test plan for the level and begin formal testing, known as alpha testing ... when QA considers the level to be thoroughly tested, they may make it available for beta testing (testing by end-users)" (p. 424). Because alpha user testing precedes the inclusion of users, the user testing Adams and Rollings describe is system centered, rather than user centered. Dumas and Redish (1994) critique this perspective of user testing: "Other developers only represent the users if the product is a tool for developers" (p. 5). In other words, unless professional QA testers are the target market for the game, the QA testers do not constitute the end user. Users discover the bug while playing a game according to their own agency and user goals, while a developer finds a bug by procedure.

The Aristotle's Assassins development team used internal bug-testing tests to perform quality assurance. Informal internal bug-testing tests were an ongoing cycle of design, check, and fix during any point in development. All the internal bug-testing tests combined involve nearly 150 bugs in the Aristotle's Assassins demo. Table 15.2 gives three bugs as an example of the kinds of findings from the quality assurance testing. For instance, Bug 4 of Table 15.2 details a condition in which the game's quest journal does not work.

Avalanche Software also performs internal testing, but their major quality assurance testing is all performed by a third-party service. During personal correspondence in November 2006, Leavitt provided a quality assurance walkthrough document for the game Chicken Little 2. Each of the items

TABLE 15.2

Bug Test Examples; Three of the 131 Bugs Found in Aristotle's Assassins

Bug Number	Bug Rank	Bug Description
4	High	Error in journal (blank entry) while talking with Juliun (second time journal is opened).
33	High	The prisoner doesn't know that I failed Mesaulius's questions when I'm sent to the prison cell.
99	Medium	Medium Attis servant should emphasize returning the amulet.

assumes that the third-party QA team is not familiar with the game; therefore, the walkthrough's 749 items identify all the steps for game setup and gameplay features. Consequently, the QA service is able to rigorously test each item from as many angles as possible. Table 15.3 provides three examples of QA testing at Avalanche Software.

Methodology

In the spring of 2007, the Aristotle's Assassins development team wanted to test whether the educational content written into the gameplay actually managed to teach users about ancient Greece. The team chose one scenario from the game—Otis in Stageira—to specifically test whether users learned persuasive strategies while completing the side quest. The objective of the game scenario was for participants to persuade Otis not to become a miner (see Figure 15.4).

The test version only involves a dozen characters and takes less than fifty minutes to play. Users began the scenario in the polis, or political arena, of Stageira. The user had to first discover what was happening in the polis.

TABLE 15.3

QA Walkthrough Examples: Three Examples from Avalanche Software

Deliverable	Description	Test Criteria
Hogknocker	Grenade-like exploder with area of effect. This is the base Secondary for Runt.	Secondary weapons are fired with the L2 button.
Destroy one of three turrets on third of three Shield Generators	Each shield generator has three turrets. Once all of the turrets are destroyed, the shield generator will fall out of the sky.	Can the user destroy turret? Does the turret fire at the user?
Enter the Ruin Sewers	Destroy the enemy hover tanks and turrets to open the next gate.	Gate opens when enemies in the area are destroyed.

FIGURE 15.4
The polis: Otis speaks with a citizen on the polis in the Otis in Stageira test.

Once the user was given the task to dissuade Otis from the life of a miner, the user had to seek out Otis. However, Otis was not willing to listen to the advice of a stranger. The user had to then wander the polis to discover more about Otis in order to succeed. There are other ways to persuade Otis. The user can try bribery, force, and threats. Fifty minutes might not seem like a lot of time for students to learn this much gameplay, as well as persuade Otis; however, authors like Stephen Kline, Nick Dyer-Witheford, and Greig De Peuter (2003) suggest that "a digitally empowered generation emerging from the global multimedia matrix" (p. 16) is primed to engage in a Greek, 3D environment for a full hour of gameplay.

Otis in Stageira is situated on the Greek polis of the city. Users need to circle the polis and converse with all the citizens there. After Otis swears to stay away from the mines, the user can access several areas promised as incentives: the fairy forest, the Minotaur's cave, and the marketplace. The reward system was identified at the beginning; I designed an array of rewards to appeal to as many students as possible. The user can access a market that is designed for users who want to amass a fortune trading minerals. The user can go to the chambers above the trading market to find a slave girl on sale for a million dollars. There is also the mine owner who offers the slave as a reward for killing the Minotaur. There is a fairy forest to explore with a fairy queen. Users can try to fight fairies, but the fairies are quite hostile.

Observation Methods

Students from two Intermediate Composition courses at Utah State University came to a designated computer lab in which Aristotle's Assassins had been previously installed. There are a total of forty registered students for both classes, an even distribution of males and females (although sex was not a controlled variable in the test), and thirty-five students present for the day of the test. Students each chose a computer and followed instructor directions to load the game. In order to see what work the students would do to learn persuasion, the instructor did not provide instructions on what to do when the game began.

I preloaded the software and game mod onto Department of English lab computers. I also tested the workstations to ensure that the computer game was working. I had previously instructed my classes to meet at the lab, rather than the regular classroom. I obtained the necessary release forms and walked the students through the steps necessary to load the game. I circled the room, taking notes about what the students were doing. I was particularly interested in what preprogrammed alternative persuasion strategies they would uncover (See Figure 15.5.). I elaborated my notes after the test according to recommendations from Emerson, Fretz, and Shaw (1995): "A word or two written at the moment or soon afterwards will jog the memory later in the day and enable the fieldworker to catch significant actions and to construct evocative descriptions of the scene" (p. 20). Finally, I was careful to watch for students who completed the game.

Survey Methods

Among Gee's (2003) thirty-six learning principles, there are six that seem most relevant to a game that seeks to let users play the gameplay, rather than the content, and still be educational. The six learning principles are the

FIGURE 15.5
Otis the miner: Dialogue interface to persuade Otis in Otis in Stageira.

active, critical learning principal; psychosocial moratorium principle; committed learning principle; identity principle; self-knowledge principle; and probing principle. Table 15.4 represents the six learning principles, along with Gee's definition.

There are three terms that Gee (2003) uses to frame the six principles I have selected: semiotic domain, learning and identity, and situated meaning and learning.

- Gee redefines traditional communicational systems as the *semiotic domain*: "Today images, symbols, graphics, diagrams, artifacts, and many other visual symbols are particularly significant" (p. 13); in other words, the meaning making of the agent occurs at many cognitive levels, rather than only linguistically or visually.

TABLE 15.4

Learning Principles: Gee's (2003) Descriptions for the Six Selected Learning Principles

Learning Principle	James Paul Gee's Definitions
Active, critical learning principle	All aspects of the learning environment (including the ways in which the semiotic domain is designed and presented) are set up to encourage active and critical, not passive, learning. (p. 49)
Psychosocial moratorium principle	Learners can take risks in a space where real-world consequences are lowered. (p. 67)
Committed learning principle	Learners participate in an extended engagement (lots of effort and practice) as extensions of their real-world identities in relation to a virtual identity to which they feel some commitment and a virtual world that they find compelling. (p. 67)
Identity principle	Learning involves taking on and playing with identities in such a way that the learner has real choices (in developing the virtual identity) and ample opportunity to meditate on the relationship between new identities and old ones. There is a tripartite play of identities as learners relate, and reflect on, their multiple real-world identities, a virtual identity, and a projective identity. (p. 67)
Self-knowledge principle	The virtual world is constructed in such a way that learners learn not only about the domain but also about themselves and their current and potential capacities. (p. 67)
The probing principle	Learning is a cycle of probing the world (doing something); reflecting in and on this action—and, on this basis, forming a hypothesis; reprobing the world to test this hypothesis; and then accepting or rethinking the hypothesis. (p. 107)

- By *learning and identity*, Gee means learners are "willing to commit themselves fully to the learning in terms of time, effort, and active engagement.... [They must be willing to] learn, use, and value the new semiotic domain" (p. 59). The concept requires an engaging game that demands users learn the game and learn the content.

- *Situated meaning and learning* suggest that "humans learn, think, and solve problems by reflecting [and connecting] their previous embodied experiences in the world" (p. 74); situated meaning is how students bridge game system goals and educational goals to make meaning in their own learning situation.

Findings

Both classes completed the test on time. They completed a survey in the last ten minutes of their respective test sessions. With the exception of one student, all students fully participated during the complete session.

Observation Findings

I observed that students implemented hostile persuasion techniques in the beginning; however, hostile actions were only one conversation option. The students quickly discovered that all citizens on the polis became hostile if one citizen was attacked; students would run from all eleven citizens in angry pursuit. Toward the end of the test, students were persuading Otis.

Students were discouraged to discover that the test had to be restarted when key plot citizens were dead. In fact, there was one girl who found a high-level Paladin Knight available in the premade character selection; she butchered the entire polis (see Figure 15.6) in a matter of seconds. Everyone in the town of Stageira was dead. I watched her run that Paladin Knight around the test environment until she turned to me and asked what she was supposed to do next.

I eventually found students wandering the fairy forests (see Figure 15.7), bragging that they found the cave of the Minotaur, and browsing the shops of the marketplace. While I designed each of the reward systems with depth, students were content to wander the areas. For instance, no one raised the million dollars necessary to buy the slave, even if they found the market.

The presence of students in the forests and mines was clear evidence that they convinced Otis to stay away from the mines. In addition, students were sufficiently engaged to continue playing after the completion of the game. This latter point is significant because the Otis in Stageira test is a good educational computer game. The survey findings are more specific about whether students were sufficiently engaged to learn persuasive strategies.

FIGURE 15.6
Polis massacre: Paladin leaves bodies on the polis in Otis in Stageira.

FIGURE 15.7
Fairy forest: The user's avatar accesses the reward system in Otis in Stageira.

Survey Findings

There are two kinds of questions in the survey—short answer and rating.

The students scored themselves in response to the survey's three numerical questions. I wanted numbers that could show a difference between someone who hates gaming but had fun with my game. I also wanted numbers to show how motivated students were to persuade Otis. The rating questions are as follows:

- Rate yourself as a video game user on a scale of one to ten.
- Rate how fun playing Otis in Stageira was.
- How important was it for you to help Otis?

I split the rating results into three different figures based upon answers to the first question. Consequently, the ratings are clustered by how much a student enjoys playing computer games.

Of course, Figure 15.8 showcases the ratings of the students who like computer games the least. This group of students is the overwhelming majority of the class and therefore the results in this figure are the most exciting. For instance, Student 2 scored a 1 on everything. Student 8 didn't care at all about Otis—and scored a 0 for investment. However, Student 4 scored a 1 for the fun value of the game and a 10 for investment in Otis's predicament.

The mean scores are the most interesting. The mean enjoyment of computer games was 2.14 for students who rated 1 to 4. However, the mean for enjoyment of the test game and investment in Otis's story was 4.19 and 4.42, respectively—more than game enjoyment.

Figure 15.9 represents the ratings for casual gamers. The casual gamers have the most erratic scores. As a group, they disliked the game test more than the nongamers, even if they like games more and were more invested in Otis.

The students who enjoy computer games the most are displayed in Figure 15.10. These students scored the highest in all three questions. Student 1 did not like the fun of the game test, even if the student was really invested in Otis. However, Student 6 gave perfect 10s to love of games and investment in Otis, and scored the fun of my game test at an 11—higher than that student's love for games.

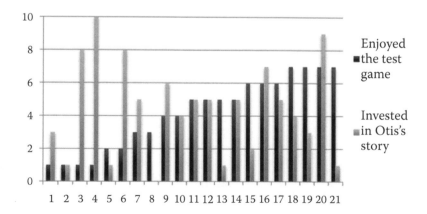

FIGURE 15.8
Survey numbers: Students who enjoy gaming 1–4 on a scale of 10.

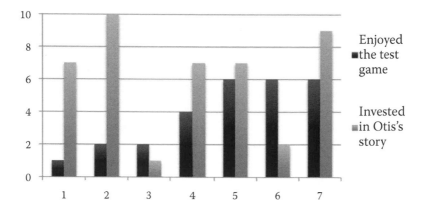

FIGURE 15.9
Survey numbers: Students who enjoy gaming 5–7 on a scale of 10.

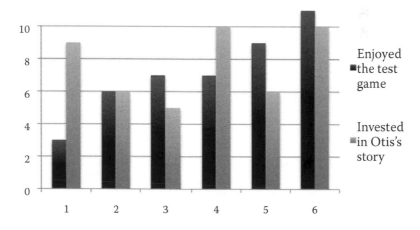

FIGURE 15.10
Survey numbers: Students who enjoy gaming 8–10 on a scale of 10.

In addition to the numerical survey questions, there are also short-answer questions. The list of survey questions and responses in Table 15.5 showcases the variety of responses from students. I was pleased to see that so many students were able to understand persuasion and articulated how they used persuasion.

A few of these answers are particularly compelling, insofar as I want evidence that students are learning strategies of persuasion. For instance, one student understood how the persuasion happened and indicated that this is either new knowledge or a new strategy: "I got to know more about him—I don't typically take time to get to know my audience." Some students were not very helpful in their responses ("persuade Otis by doing whatever Otis wanted me to do"); however, other students were really insightful: "it is ok to assert yourself a bit more" or "gained his trust and explained a better alternative." The students were able to articulate the persuasive strategies they were using as they used the game during the test.

Results

There is something to be learned about user goals in regard to the six learning principles (Table 15.4). The learning principles highlight how users' goals are engagement and that is where the boundaries of usability testing, playtesting, and quality assurance blur. However, the findings of Otis in Stageira do form a picture of how exactly the students did in fact "use" the game after all. Gee's six learning principles work to present those findings.

> *Active, critical learning principle.* Students were motivated. I had built a fairy forest and a Minotaur cave to reward students who persuaded Otis. In most cases, students found these bonus areas exciting destinations. In some cases, students indicated that they played the test game simply to access the fairy forest. Observation indicates that students were very motivated. They were laughing, watching each other's progress, helping one another, and persistently attempting to master the game. The students remained active while they pushed and prodded the interface until they reached their user goal. Students 8, 19, and 21 (see Figure 15.8) dislike games, dislike Otis, and rated playing Otis in Stageira significantly higher than their other ratings. Even Students 8, 19, and 21 played until they learned persuasion strategies. Consequently, students were motivated by the engaging, usable system.

> *The psychosocial moratorium principle.* Students did not take many risks; consequently, they were not as proactive in their learning as Gee or Prensky might desire. However, I coded risks that students could take. There was an option to convince Otis to actually become a miner; the alternative was placed to provide a measurable way to assess exploration. However, there were less than a handful of students who discovered this option. In fact, students did not seem too interested in the alternative conversational options at all. There was one student who wrote about quicker, more hostile options: "Sometimes I wanted to use force hoping it would be more productive and quick." Otherwise, students were too keen on the fairy forest and Minotaur cave to worry much about conversational options. The usability test reveals that optional strategies didn't interfere with the user goal but didn't help students learn about persuasion either.

Committed learning principle. Twenty-four out of thirty-five students (Table 15.5, Question 11) were invested enough in the game to comment on how they felt about Otis's mining ambition and desire for true freedom: "it was kind of weak," "I thought he was an idiot," "he was being irresponsible and not listening," "I thought he could have used my advice," "Why would a kid want to be a miner?" and "Unrealistic, is there really true freedom in mining?" The survey ratings suggest commitment and engagement, too. Twenty-two out of thirty-five students rated their computer game interest lower than or equal to their interest in helping Otis. For instance, while Student 4 (Figure 15.8) rated computer games at 1 and rated Otis in Stageira at 1, Student 4 also felt really engaged in the plight of young Otis—giving this a rating of 10. There are twenty-one other examples. The conclusion is that students were engaged and motivated by the Otis narrative, rather than just the fairy forest. That is an important distinction. After all, educational goals aren't very engaging if everyone only wants to find the fairy forest. Consequently, the user goal was

TABLE 15.5

Survey Results: Students Answered Questions about Their Test Experience

1. What feature of the game made you want to play more?
 - I just wanted to find out what the fairy queen did.

2. What did you do to learn to control the game?
 - Listen to who Otis wanted me to talk to.

3. What was the hardest part of persuading Otis?
 - Having to talk to others.
 - Going around to different people trying to figure out what to say.
 - Finding everyone to talk to.
 - We tried each different response the first time and nothing helped! Then when we came back, it changed.
 - Persuading Otis!
 - He was stubborn, it took several times to persuade him.
 - The dang kid wouldn't listen.

4. You likely have persuaded people to do stuff before; did you learn anything about your own persuasive skills?
 - People can't be told what to do.
 - Just do what the person wants.
 - Use the advice of others.
 - It can be ok to assert yourself a bit more.
 - To be stern, positive, friendly.
 - Not to be rude.
 - A little bit.
 - Tell the truth.
 - I don't like to make people angry, I like to have more info than I need going in.

5. Was there any time while playing that you forgot you were playing a game? When was it?
 - Yes, entering the forest.
 - Oh yea! When I was talking to people.

6. Did you do something to persuade Otis that you don't typically do when you try to convince people?
 - I got to know more about him—I don't typically take time to get to know my audience.

7. There were many conversational options for you to choose. Did you experiment with the less productive options? Were there enough less productive options for you?
 - Yes, there was enough to make it difficult.

8. How did you feel about selecting statements that would not help you achieve your goals?
 - Sometimes I wanted to use force hoping it would be more productive and quick.

TABLE 15.5 *(continued)*

Survey Results: Students Answered Questions about Their Test Experience

9. How is it that you finally managed to persuade Otis? What did you have to do?
- Talk to others to learn info.
- Become his friend, Talk to other people and tell him to buy the mine instead.
- Go to different people.
- I talked to a million people and made him realized he was being tricked.
- Get him to trust me.
- Talk to a series of people.
- Be nice and listen to his need.
- Get to know more about him by talking to the politician.
- Get to know him by talking to this other guy, I didn't know he was rich.
- Have him trust me and kindly persuade him.
- Show him that I knew him well enough.
- Gained his trust, learned about him.
- Gave him bigger perspective from someone he trusts.

10. What were the persuasive steps you took in persuading Otis?
- Gathering Intelligence.
- Got him to see an option.
- Went and spoke with others to find out how Otis was.
- You had to talk to a bunch of other people to know how to persuade him.
- We had to get to know him better and gain his trust.
- Learning about him. Talking to him, talking to others.
- Learned about his background.
- Whatever Otis wanted me to do.
- Finding out about him, and telling him his options.
- Talked to many people and told him that he should ask his dad to buy the mine.
- Found info about him.
- Talking to everyone, listen to Otis and talk to his dad.
- Gained his trust and explained a better alternative.
- Talked to other people, got to know about Otis more.
- Found info and relayed it to him.

11. Describe how you felt about Otis's conflict over a mining career.
- He was a feisty little guy.
- He needed to learn more.
- It was kind of weak.
- It didn't make sense.
- I didn't understand why a little boy wanted to be a miner so bad.
- I thought he could have used my advice.
- At first I felt bad trying to convince him to not follow his dream.
- He shouldn't do it. Listen to your dad.
- He was a baby.
- His own decision, but he should weigh all his options.
- I think he was just young and didn't know what to do.
- Well, what else did he want besides freedom?
- I thought he was an idiot.

11. *(continued)*
- Better to own than work blue collar.
- Unrealistic, is there really freedom in mining?
- It was a no brainer, his dad is a very rich man, why be a miner.
- He was being irresponsible and not listening.
- He just wants freedom and to be convinced properly.
- He was too young to decide anything like that.
- He was a kid and he thought like a kid.
- He thought the grass was greener on the other side.
- It was pathetic.
- He needs to get all the facts.
- Why would a kid want to be a miner? Was my thought.

as much persuading Otis as it was getting to the forest or Minotaur cave. So users' own goal was engagement with the game beyond the preprogrammed goal; however, users' own goal was also the preprogrammed educational goal of the game.

Identity principle. While there were a majority of students who chose not to answer survey questions about their user-agent identity, there was a consensus among the students who did answer the question. Students seemed to forget they were playing a game while engaging characters in conversation. The compelling thing about these responses is that the conversation interface was both the gameplay and the actual learning situation in which students learned persuasion strategies. Students were learning persuasion strategies partly because of their engagement and partly because of the interface. Consequently, the Otis in Stageira usability test cannot test the usability of the interface without also testing the engagement.

Self-knowledge principle. There were a significant number of students who took the self-knowledge questions seriously. For instance, I hoped to discover whether students would be able to recognize and articulate their growth from using this game. While other students didn't answer or wrote something sarcastic, many students articulated how they persuaded Otis: "gathering intelligence," "talk to others to learn info," "get him to trust me," and "learned about his background." Those students clearly learned persuasion and clearly attributed that learning to the game strategies they used to persuade Otis in the first place. However, while usability testing determines whether a user can use some object to obtain insightful self-knowledge, that particular knowledge set is not the goal of the system. Otis in Stageira seeks to test whether the user used the game to learn the knowledge I programmed into the game. However, if Gee's self-knowledge principle means anything, it means that once users invest identity into the semiotic domain of a game, the knowledge that results is users' own goal.

The probing principle. While there were a few students who were able to itemize a list of the various motives they were able to attribute to Otis, the vast majority of the class either left the question blank or wrote something that did not answer the question. Obviously, the question (Table 5, Question 2) based on this principle was inefficient. Students simply did not have enough information to know what the question really asked. What I hoped to discover was what thinking the students did as they probed the game. If students had answered questions about probing the controls and probing the game content, then I could compare the usability of obvious user tools, like game controls, to less obvious user tools, like probing persuasion strategies.

Discussion

The problem posed in this chapter is whether testing an educational game system like Otis in Stageira, in which the users' goal is both tied to the system and preprogrammed by the programmer, can be usability testing. The answer is yes; however, that is only the case if users' goals are engagement with the educational goals of the system itself. Yet, engagement is where usability testing transgresses boundaries with both playtesting and quality assurance testing. Each one of the test methods alone, with discrete examples, seems clear and distinct. However, the Otis in Stageira test pushes at the boundaries.

A playtest is really about engagement; without immersion or challenge, a game isn't fun—isn't engaging. The survey results highlight students' engagement; in fact, the numerical ratings show even students who hate computer games had fun. Yet, the Otis in Stageira test actually was not playtesting. A playtest simply seeks to identify with what users engage while they play, rather than whether they learned persuasive strategies while playing.

At the same time, quality assurance is engagement too because bugs easily obstruct immersion and challenge. Bugs are like the fourth wall of fiction; they break the suspension of disbelief and they significantly interfere with the challenge (because either the game simply doesn't work or the user can exploit too many cheats). For instance, a student picked the lock to the fairy forest gate and bypassed the Otis in Stageira scenario. However, the Otis in Stageira test was not quality assurance because quality assurance simply seeks to flag features that do not work.

However, Otis in Stageira passes Barnum's litmus test; the test is about the user and not the product. My purpose was to verify what users did to reach their goals, even if I programmed those goals myself. In other words, Otis in Stageira tested users' engagement, rather than the product.

The key to engagement in a complex system like computer games is McAllister's "agent." The agent is an active influence on the game system. A computer game is nothing but lines of code and visual representation until the human agent makes meaning of the system. Gee offers three terms that can be somewhat of a grammar that describes the meaning making of McAllister's agent: semiotic domain, learning and identity, and situated meaning and learning. These three concepts are the grounds for the learning principles in Table 15.4 and serve to highlight why an agent's goals are also the educational goals of Otis in Stageira.

For instance, the semiotic domain suggests that the agent cannot even function in the game world without ascribing meaning to the visuals, audio, controller inputs, and gameplay feedback. However, in ascribing meaning to the semiotic domain, the human agent must necessarily situate the personal experience that supports that meaning. Consequently, the users in the Otis in Stageira test were situating their identity in the polis of the computer game.

Insofar as the educational goals are a critical part of the semiotic domain, users' goals include making meaning of those goals.

The findings confirm that students were meaning-making agents in Otis in Stageira. The survey demonstrates that students learned strategies to persuade Otis because they were engaged in the system. They were occupied defining the semiotic domain, identifying themselves as an agent in the game system, and situating their own experience in that semiotic domain. The survey also demonstrates that Otis in Stageira is a good educational game because content was embedded in engaging gameplay and an engaging learning situation. Users' engagement is the reason why they used the game to learn persuasion; their engagement is why users' goals can be the same as the system goals in a complex system.

Conclusion

Traditional usability testing would reject that the users' goal is both tied to the system and preprogrammed by the programmer. However, complex systems are different; for instance, a computer game has no goals outside the system. Consequently, engagement within the system is the users' goal. Therefore, usability testing transgresses boundaries with other kinds of system testing—playtesting and quality assurance testing. I conducted my own usability testing with a test scenario I called Otis in Stageira. The Otis in Stageira test pushes at the boundaries of all three test methods. Yet, the Otis in Stageira test is neither playtesting nor quality assurance testing. However, Otis in Stageira passes Barnum's litmus test; the test is about the user and not the product. The findings confirm that users' engagement is the reason why Otis in Stageira qualifies as usability testing. Researchers need to expand the definition of users' goals to include the system goals in a complex system.

Reader Take-Aways

The Otis in Stageira test could have been a usability test, a playtest, or a quality assurance test. Yet, by first understanding the concept of engagement in complex, closed systems like educational computer games, the definition of user goals shifts to include engagement with the system itself.

- Even while usability testing is concerned with users' own goals and determined by the system, the engagement of a user in a complex system can preclude objectives outside the system and align with the preprogrammed goals of the system itself.

- The boundaries between playtesting, usability testing, and QA testing are indistinct, and usability professionals should apply that ambiguity creatively (and, at the same time, judiciously) to improve the product.
- It's possible (and often desirable, if not essential) to test usability in complex systems by having the user achieve a higher-level goal and letting the user choose his or her own path rather than focusing only on the narrower, task-focused goals of traditional usability testing.

Acknowledgments

The Learning Games Initiative is an interdisciplinary and interinstitutional research collective co-directed at the University of Arizona Department of English by Dr. Ken McAllister and at Arizona State University Department of Communications Studies by Dr. Judd Ruggill. Dr. Ryan Moeller directs the Learning Games Initiative chapter at the Utah State University Department of English. Dr. Moeller is principal investigator on two Utah State University research grants that funded Aristotle's Assassins. Credit also goes to multiple undergraduate and graduate students at the University of Arizona, Arizona State University, and Utah State University for work on the concept team, several development teams, usability test team, and the graphic team.

References

Adams, E., & Rollings, A. (2007). *Fundamentals of game design*. Upper Saddle River, NJ: Pearson Prentice Hall.

Albers, M. J. (2003). Complex problem solving and content analysis. In M. J. Albers & B. Mazur (Eds.), *Content and complexity: Information design in technical communication* (pp. 263–284), Mahwah, NJ: Lawrence Erlbaum.

Barnum, C. (2002). *Usability testing and research*. New York: Longman.

de Souza e Silva, A., & Delacruz, G. C. (2006). Hybrid reality games reframed. *Games and Culture, 1*(3), 231–251.

Dumas, J. S., & Redish, J. C. (1994). *A practical guide to usability testing*. Portland, OR: Intellect.

Emerson, R. M., Fretz, R. I., & Shaw, L. L. (1995). *Writing ethnographic fieldnotes*. Chicago: University of Chicago Press.

Gee, J. P. (2003). *What video games have to teach us about learning and literacy*. New York: Palgrave MacMillan.

Grice, R. A., & Ridgway, L. S. (1993). Usability and hypermedia: Toward a set of usability criteria and measures. *Technical Communication, 40*(3), 429–437.

Johnson, R. R. (1998). *User-centered technology: A rhetorical theory for computers and other mundane artifacts*. Albany: State University of New York.

Johnson, S. (2005). *Everything bad is good for you*. New York: Riverhead.

Kline, S., Dyer-Witheford, N., & De Peuter, G. (2003). *Digital play: The interaction of technology, culture, and marketing*. Montreal: McGill-Queen's University Press.

Koster, R. (2005). *A theory of fun*. Scottsdale, AZ: Paraglyph.

Laitinen, S. (2006, February). Do usability expert evaluation and testing provide novel and useful data for game development? *Journal of Usability Studies*, 2(1), 64–75.

Learning Games Initiatives. (2006). High Concept Document. Aristotle's Assassins. http://lgi.usu.edu/projects/aristotle.htm

McAllister, K. (2004). *Game work: Language, power and computer game culture*. Tuscaloosa: University of Alabama Press.

Mirel, B. (2003). Dynamic usability: Designing usefulness into systems for complex tasks, in M. J. Albers and B. Mazur (Eds.)., *Content and complexity: Information design in technical communication*. Mahwah, NJ; Lawrence Erlbaum Associates, pp. 233–262.

Mirel, B. (2004). *Interaction design for complex problem solving: Developing useful and usable software*. San Francisco: Morgan Kaufmann.

Prensky, M. (2001). *Digital game-based learning*. St. Paul, MN: Paragon.

Redish, J. (2007). Expanding usability testing to evaluate complex systems. *Journal of Usability Studies*, 2(3), 102–111.

Schultz, C. P., Bryant, R., & Langdell, T. (2005). *Game testing all in one*. Boston: Thompson Course Technology. Retrieved from http://proquest.safaribooksonline.com/1592003737

Index

A

Abstraction, 50, 52, 238, 296–297
Activity theory, 181–206
  cultural-historical activity theory, 187–193
  operationalizing context, 185–193
  time frame, 201–203
  USDA food guide pyramids, 193–203
Analyzing user experience, 89–108
Anesthetists, 140, 142
API design process, 223–250
  abstraction level, 238
  application use, 228–230
  class, 227
  cognitive dimension, 238–240
  consistency, 239
  definition, 226–230
  design, 240–245
  development, 240–245
  difficulties with API use, 235–240
  domain correspondence, 240
  elements, 227
  examples, 228
  function, 227
  header file, 227
  heuristic evaluations, 242
  interface, 227
  learning style, 238
  measuring usability, 234–235
  parameter, 227
  peer reviews, usability, 242–243
  penetrability, 239
  premature commitment, 239
  progressive evaluation, 239
  role expressiveness, 239
  significance, 230–233
  structure, 227
  symbolic constant, 227
  task focused, 239
  technical writers, 241–245
  terms, 227
  usability, 240–245
  viscosity, 239
  working framework, 238
Application programming interface. *See* API
Ash, Tim, 32
Audiences
  complex information systems, 30–32
  identifying, 34
  understanding, 34

B

Background of participants, electronic commerce websites, 168–169
Background of products, educational publishing, 68–70

C

Call centers, 274
Category, industrial, 147
Cause-and-effect relationship, 126, 160
Characteristics of group, 142
Cognition, 38, 41, 169, 238–240, 294
Collaborative work environments, 286–289
Comfort zone, 122
  information, 122
Committed learning principle, 349, 356
Communication of risk, 305–332
  audiences, 310–311
  conflicts of knowledge, 326–328
  crisis, 308–310
  design, 320–321
  document analysis, 316–328
  document usability evaluation protocol, 315–316
  documents, 311–313
  elaborations, 328
  emergency risk communication, 308–310

feedback-driven audience analysis,
      312–313
Hurricane Survival Guide, testing
      with, 313–328
   organization, 321–322
   pilot study participants, 314–315
   wording, 322–323
Complexity of language, 47–66
   composition instruction, 61–64
      course redesign, 62–63
      usability, 63–64
   dynamic information language, 50
   formal written English, 57–61
      cognitive load learning theory,
         60–61
      comprehensible input, 59–60
      language acquisition, 58–59
   historical language, 50
   multidimensional language, 50
   nonlinear response language, 50
   open-ended language, 50
   test environment, 52–53
Comprehension of complexity, 1–86
   audiences
      identifying, 34
      understanding, 34
   composition instruction
      complex system, 63–64
      course redesign, 62–63
      usability, 63–64
   educational publishing, 67–86
      collaboration, 79–82
      failure, opportunity in, 70–73
      innovative product
         development, 73
      product background, 68–70
      product development, 82–84
      rhetoric, 79–82
      unexpected complexity, 73–79
   exigency, complex information
      systems, 30–32
   formal written English
      cognitive load learning theory,
         60–61
      comprehensible input, 59–60
      language acquisition, 58–61
   genre theory
      content evaluation, 25–26
      extended, 26

heuristic application
   cognitive walkthrough, 40
   NASA, 42–43
   post-test interviews, 40–41
   practical application, 41–43
language complexity, 47–66
   composition instruction, 61–64
   dynamic information language, 50
   formal written English, 57–61
   historical language, 50
   multidimensional language, 50
   nonlinear response language, 50
   open-ended language, 50
   test environment, 52–53
   test procedure, 52–53
purpose, complex information
   systems, 30–32
rhetorical theory, web-based texts,
   17–45
   Ash, Tim, 32
   audience, complex information
      systems, 30–32
   audience analysis complexity, 33–34
   audience need, expectation, 32–35
   conceptual structures, complex
      information systems, 37–38
   demands on author, 38–39
   describing genre, 29–30
   design discussion, 23
   exigency, 28–29
   expectations, author, 38–39
   filters in websites, 25
   genre-based heuristic, 26–27
   genres in websites, 25
   heuristic application, 39–43
   identifying audiences, 34
   incomplete rhetorical filters, 24–25
   individual pages, 31–32
   Landing Page Optimization, 32
   Letting Go of the Words, 34
   metrics, 19–20
   navigation discussion, 23
   physical structure, 38
   purpose, 28–29
   Redish, Janice, 34–35
   rhetorical expectations, 32
   rhetorical need, 32–33
   schema, combining components
      into, 27–39

social demands of author, 38–39
software evolution, 19
text, conceptual structure, 35–38
web design, conceptual
structures, 36–37
writers' ability to evaluate texts, 23
writing, usability, relationship,
22–25
writing quality, 23–24
selected results, 76–78
test methodology, 74–76
unexpected complexity,
implications, 78
usability issues, 3–16
conceptual layer, 13–14
definition of usability, 8–11
lexiconal layer, 11–12
pragmatic layer, 11–12
semantic layer, 12–13
syntactic layer, 11–12
writing quality, discussing, 24
Conceptual usability layer, 13–14
Consistency, 160, 238–240, 242, 257
Content relationships, 109–131
Contextual awareness usability, 109–131
contextual awareness, 112–117
data analysis, 126
data collection, 124–126
quality issues, 125–126
test plan, 120–124
contextual awareness, 124
information relationships, 123–124
information salience, 123
mental models, 122–123
usability tests, 117–120
knowledge-based measures, 118
performance-based measures, 118
Contextual complexity, 283–286
Continuous usability evaluation, 133–155
Correspondence, domain, 237, 240
Critical learning, 348–349, 354
Cultural-historical activity theory, 187–193

D

Data analysis, 126
Data collection, 124–126
Decision paths, 137
Definition of usability, 8–11

Demands on authors, 38–39
Designing for complexity, 179–250
Development
innovative product, 73
product, 73
Discount testing, 89, 103–104
vs. ecological approach, 103–105
Domain, semiotic, 349, 357–358
Domain correspondence, 237, 240
Domain service, process evaluation, 136
Dynamic information, 19, 49–50

E

Ecological analysis, 89–108
discount testing, *vs.* ecological
approach, 103–105
evaluation of complexity, 100–101
hypotheticals, 97–100
other methods, 103–105
small-scale usability test, 101–103
user web, 93–95
Educational games, 333–361
committed learning principle, 349,
356
education game usability, 338–346
engagement, 340–341
game testing, 341–346
playtesting, 343–345
quality assurance, 345–346
games studied, 337–338
Aristotle's assassins, 337–338
avalanche software, 338
identity principle, 349, 356
methodology, 346–350
observation findings, 350–351
observation methods, 347–348
probing principle, 349, 356
psychosocial moratorium principle,
349, 356
self-knowledge principle, 349, 356
situated meaning, 350
survey findings, 351–353
survey methods, 348–350
Educational publishing, 67–86
collaboration, 79–82
failure, opportunity in, 70–73
innovative product development, 73
product background, 68–70

product development, 67–86
  collaboration, 79–82
  failure, opportunity in, 70–73
  innovative product
      development, 73
  product background, 68–70
  product development, 82–84
  rhetoric, 79–82
  unexpected complexity, 73–79
rhetoric, 79–82
unexpected complexity, 73–79
  implications, 78
  selected results, 76–78
  test methodology, 74–76
Elaboration, API, 239
Electronic commerce websites, 157–177
  design of usability tests, 162–172
      conducting tests, 171–172
      research instrument, 171
      scenario and/or task, 164–165
      setting, 171–172
      website selection, 162–164
  participants
      age, 169–170
      cultural background, 168–169
      gender, 167–168
      identifying, 165–170
      language proficiency, 168–169
      number of, 165–166
      recruiting, 165–170
      selection, 166–170
  reliability, 160
  research approach, 160–161
  validity, 160
  website usability testing, 159–160
Elements of information, 14, 87, 111–113,
      118, 121, 123, 310
Expressiveness, 237, 239
  role, 237, 239

F

Failure, opportunity in, 70–73
Feedback-driven audience analysis,
      312–313
Food guide pyramids, 193–203
Framework, working, 238
Functionality layer, 12–13
Future behavior, 125

G

Games, educational, 333–361
  committed learning principle, 349, 356
  education game usability, 338–346
      engagement, 340–341
      game testing, 341–346
      playtesting, 343–345
      quality assurance, 345–346
  games studied, 337–338
      Aristotle's assassins, 337–338
      avalanche software, 338
  identity principle, 349, 356
  methodology, 346–350
  observation findings, 350–351
  observation methods, 347–348
  probing principle, 349, 356
  psychosocial moratorium principle,
      349, 356
  self-knowledge principle, 349, 356
  situated meaning, 350
  survey findings, 351–353
  survey methods, 348–350
Genomics research, 207–222
  collaborative, complex work, 215
  customization, 217–218
  developmental methodology, 213–215
      contextual inquiries, 214–215
      design requirements, 214
      field studies, 214–215
  domain expertise, 213
  field studies, 213
  genome project, 210–220
  methodologies, 218–220
      contextual analysis, 219–220
      field observation, 218
      user involvement, 218–219
  paradox of structure, 216–217
  single user interface, 211
  unified user interface, 211
  unifying structure, 216–217
  usability challenges, 215–218
Group characteristics, 142

H

Health care usability evaluation,
      138–142
  ASUR model, 140

design of system, 138–140
overview of evaluation results,
140–142
usability evaluation techniques, 138
Heuristic evaluations, API design
process, 242
Historical viewpoints, 50, 187, 199–201
Hotel group reservation study, 270–274
Hurricane Survival Guide, testing with,
313–328
Hypotheticals, 97–100

I

Identity, 348–349, 356–358
Increasingly complex systems, 133–155
Industrial category, 147
Industrial products, 147
Information
    comfort zone, 122
    dynamic, 19, 49–50
    elements of, 14, 87, 111–113
    integrated complex, 113
    interaction with, 113
    levels of, 122
    relevant, 112, 114, 122–123, 126–127
Information elements, 14, 87, 111–113,
    118, 121, 123, 310
Information needs, 3–4, 116–117,
    122–128, 306, 320
Information relationships, 109–131
    contextual awareness, 112–117
    data analysis, 126
    data collection, 124–126
    quality issues, 125–126
    test plan, 120–124
        contextual awareness, 124
        information relationships, 123–124
        information salience, 123
        mental models, 122–123
    test plans, 123–124
    usability tests, 117–120
        knowledge-based measures, 118
        performance-based measures, 118
Information salience, test plans, 123
Innovative product development, 73
Integrated complex information, 113
Interaction styles, 281–303
Interaction with information, 113

K

Knowledge-based measures, 118–120

L

Landing Page Optimization, 32
Language complexity, 47–66
    composition instruction, 61–64
        complex system, 63–64
        course redesign, 62–63
        usability, 63–64
    dynamic information language, 50
    formal written English, 57–61
        cognitive load learning theory,
        60–61
        comprehensible input, 59–60
        language acquisition, 58–61
    historical language, 50
    multidimensional language, 50
    nonlinear response language, 50
    open-ended language, 50
    test environment, 52–53
    test procedure, 52–53
Learning, 348–349, 354, 356
    committed, 349, 356
    critical, 348–349, 354
    styles of, 63, 238
Learning styles, 63, 238
Levels of abstraction, 238
Levels of relevant information, 122
Lexiconal layer, 11–12
Loyalty program enrollment, hotel
    group study, 267–268

M

Management, 142–149, 152–158, 329
Mapping usability, 89–108
    discount testing, *vs.* ecological
        approach, 103–105
    evaluation of complexity, 100–101
    hypotheticals, 97–100
    other methods, 103–105
    small-scale usability test, 101–103
    user web, 93–95
Meaning, situated, 349–350, 358
Measures
    knowledge-based, 118–120

performance-based, 118–120
Mental models, test plans, 122–123
Moratorium principle, 348–349, 356
    psychosocial, 348–349, 356
Multiple viewpoints, 113
MyPyramid. *See* USDA food guide
    pyramids
MyPyramid graphic, 183

N

Network monitoring, management,
    269–270
Nonlinear response, 49–50
Nonroutine situation, 115–116
Nurses, 6–7, 10, 139–142

O

Operationalizing context, 185–193

P

Participants, usability tests
    age, 169–170
    cultural background, 168–169
    gender, 167–168
    identifying, 165–170
    language proficiency, 168–169
    number of, 165–166
    recruiting, 165–170
    selection, 166–170
Penetrability, 239
Performance-based measures, 118–120
Post-test questionnaires, 256–257
PPM. *See* Project portfolio management
Pragmatic layer, 11–12
Premature commitment, 239
    API design process, 239
Principles
    committed learning, 349, 356
    identity, 348–349, 356
    learning, 348–349, 354, 356
    moratorium, 348–349, 356
    probing, 348–349, 357
    psychosocial moratorium, 348–349,
        356
    self-knowledge, 348–349, 357
Probing principle, 348–349, 356–357
Product background, 68–70

Product development, 73, 82–84
    educational publishing, 67–86
        collaboration, 79–82
        failure, opportunity in, 70–73
        innovative product
            development, 73
        product background, 68–70
        product development, 82–84
        rhetoric, 79–82
        unexpected complexity, 73–79
    unexpected complexity
        implications, 78
        selected results, 76–78
        test methodology, 74–76
Product reaction cards, 257–265
    learning to use, 264–265
    methods to show results, 265
Products, industrial, 147
Progressive evaluation, 239
Project portfolio management, 142–153
    assumptions, 150–151
    continuous evaluation, 146–149
    domain services, 151
    experimental usability evaluation,
        149–150
    increasingly complex systems for, 145
    service coverage, 151–153
    tasks, 144–145
Psychosocial moratorium principle,
    348–349, 356
Publishing, educational, 67–86
    collaboration, 79–82
    failure, opportunity in, 70–73
    innovative product development, 73
    product background, 68–70
    product development, 67–86
        collaboration, 79–82
        failure, opportunity in, 70–73
        innovative product
            development, 73
        product background, 68–70
        product development, 82–84
        rhetoric, 79–82
        unexpected complexity, 73–79
    rhetoric, 79–82
    unexpected complexity, 73–79
        implications, 78
        selected results, 76–78
        test methodology, 74–76

Purpose, complex information systems, 30–32
Pyramid, food guide, 193–203

Q

Quality issues, 125–126

R

Relationships, 5, 68, 87, 109–129, 283, 349
    cause-and-effect, 126, 160
    informational, 109–131
        contextual awareness, 112–117
        data analysis, 126
        data collection, 124–126
        quality issues, 125–126
        test plan, 120–124
        usability tests, 117–120
    test information, 125
    test plan
        contextual awareness, 124
        information relationships, 123–124
        information salience, 123
        mental models, 122–123
    usability tests
        knowledge-based measures, 118
        performance-based measures, 118
Relevant information, 112, 114, 126–127
Research approach, 160–161
Response, nonlinear, 49–50
Rhetoric, 79–82
Rhetorical theory, 17–45
    Ash, Tim, 32
    audience, complex information
        systems, 30–32
    audience analysis complexity, 33–34
    audience need, expectation, 32–35
    audiences
        identifying, 34
        understanding, 34
    conceptual structures, complex
        information systems, 37–38
    demands on author, 38–39
    describing genre, 29–30
    design discussion, 23
    exigency, 28–29
        complex information systems,
            30–32

expectations, author, 38–39
filters in websites, 25
genre-based heuristic, 26–27
genre theory
    content evaluation, 25–26
    extended, 26
genres in websites, 25
heuristic application, 39–43
    cognitive walkthrough, 40
    NASA, 42–43
    post-test interviews, 40–41
    practical application, 41–43
identifying audiences, 34
incomplete rhetorical filters, 24–25
individual pages, 31–32
Landing Page Optimization, 32
Letting Go of the Words, 34
metrics, 19–20
navigation discussion, 23
physical structure, 38
purpose, 28–29
    complex information systems,
        30–32
Redish, Janice, 34–35
rhetorical expectations, 32
rhetorical need, 32–33
schema, combining components into,
    27–39
social demands of author, 38–39
software evolution, 19
text, conceptual structure, 35–38
web design, conceptual structures,
    36–37
writers' ability to evaluate texts, 23
writing, usability, relationship, 22–25
writing quality, 23–24
    discussing, 24
Risk communication, 305–332
    audiences, 310–311
    conflicts of knowledge, 326–328
    crisis, 308–310
    design, 320–321
    document analysis, 316–328
    document usability evaluation
        protocol, 315–316
    documents, 311–313
    elaborations, 328
    emergency risk communication,
        308–310

feedback-driven audience analysis,
312–313
Hurricane Survival Guide, testing
with, 313–328
organization, 321–322
pilot study participants, 314–315
wording, 322–323
Role expressiveness, 237, 239

S

Salience, test information, 125
Satisfaction ratings, self-rated
questionnaires, 255–256
Self-knowledge principle, 348–349,
356–357
Self-rated questionnaires, satisfaction
ratings, 255–256
Semantic layer, 12–13
Semantic usability layer, 12–13
Semiotic domain, 349, 357–358
Service coverage, project portfolio
management, 151–153
Situated meaning, 349–350, 358
Small-scale usability test, 101–103
Sociotechnical systems, 294–297
Styles of learning, 63, 238
Surgeons, 142
Symbolic constant, API design
process, 227
Syntactic layer, 11–12
Syntactic usability layer, 11–12

T

Task focused, 239
Teacher professional development
website, 268–269
Technical complexity, 283
Technical writers, API design process,
241–245
Technology, 282–289, 291–302
Test information relationships, 125
Test information salience, 125
Test plans, 120–124
contextual awareness, 124
information relationships, 123–124
information salience, 123
mental models, 122–123

Testing, 109–131
discount, 89, 103–104
knowledge-based measures, 118
performance-based measures, 118
typical discount, 104
Theorizing complexity, 87–177
analyzing user experience, 89–108
content relationships, 109–131
contextual awareness, 109–131
continuous usability evaluation,
133–155
data collection, 124–126
design for usability testing, 157–177
designing usability test, 162–172
discount testing, *vs.* ecological
approach, 103–105
ecological framework, 89–108
ecology, usability and, 93–103
electronic commerce websites,
157–177
health care, 138–142
increasingly complex systems, 133–155
mapping usability, 89–108
methodology, usability evaluation,
135–137
project portfolio management, 142–153
research approach, 160–161
test plan, 120–124
testing complex information, 109–131
usability, information relationships,
109–131
validity, usability testing research, 160
Theory of usability, rhetorical
theory with
Ash, Tim, 32
audience, complex information
systems, 30–32
audiences
identifying, 34
understanding, 34
demands on author, 38–39
filters in websites, 25
Landing Page Optimization, 32
purpose, 28–29
complex information systems,
30–32
Transportation, 147, 197, 200
Typical discount testing, 104

U

Unexpected complexity, 73–79
    implications, 78
    selected results, 76–78
    test methodology, 74–76
Unified user interface, 211
Usability, information relationships,
        109–131
USDA food guide pyramids, 193–203

V

Validity, usability testing research, 160
Viewpoints, 113, 116
    historical, 50, 187, 199–201
    multiple, 113
Virtual environments, 289–294
Viscosity, 239
    API, 239
    API design process, 239

W

Web-based texts, 17–45
    Ash, Tim, 32
    audience analysis complexity, 33–34
    audience need, expectation, 32–35
    conceptual structures, complex
        information systems, 37–38
    describing genre, 29–30
    design discussion, 23
    exigency, 28–29
        complex information systems,
            30–32
    expectations, author, 38–39
    genre-based heuristic, 26–27
    genre theory
        content evaluation, 25–26
        extended, 26
    genres in websites, 25
    heuristic application, 39–43
        cognitive walkthrough, 40
        NASA, 42–43
        post-test interviews, 40–41
        practical application, 41–43
    identifying audiences, 34

    incomplete rhetorical filters, 24–25
    individual pages, 31–32
    Letting Go of the Words, 34
    metrics, 19–20
    navigation discussion, 23
    physical structure, 38
    Redish, Janice, 34–35
    rhetorical expectations, 32
    rhetorical need, 32–33
    schema, combining components into,
        27–39
    social demands of author, 38–39
    software evolution, 19
    text, conceptual structure, 35–38
    web design, conceptual structures,
        36–37
    writers' ability to evaluate texts, 23
    writing, usability, relationship, 22–25
    writing quality, 23–24
        discussing, 24
Websites, electronic commerce, 157–177
    design of usability tests, 162–172
        conducting tests, 171–172
        research instrument, 171
        scenario and/or task, 164–165
        setting, 171–172
        website selection, 162–164
    E-commerce websites, 158–160
        website usability testing, 159–160
    participants
        age, 169–170
        cultural background, 168–169
        gender, 167–168
        identifying, 165–170
        language proficiency, 168–169
        number of, 165–166
        recruiting, 165–170
        selection, 166–170
    reliability, 160
    research approach, 160–161
    validity, 160
Working frameworks, 238

Z

Zone of comfort, 122

Printed and bound by CPI Group (UK) Ltd, Croydon, CR0 4YY

23/10/2024

01777671-0015